Our Living Legacy

Proceedings of a Symposium
on
Biological Diversity

OUR LIVING LEGACY

Proceedings of a Symposium on Biological Diversity

Edited by

M.A. Fenger, E.H. Miller, J.A. Johnson and E.J.R. Williams

ROYAL
BRITISH
COLUMBIA
MUSEUM

Published by the Royal British Columbia Museum, 675 Belleville Street, Victoria, British Columbia, V8V 1X4, Canada, in association with the Ministry of Environment, Lands and Parks, through funding provided by the Corporate Resources Inventory Initiative.

Printed in Canada.

The cover illustrations are of British Columbian deep-sea fishes, highlighting our extremely poor knowledge of the taxonomy and distribution of many life forms in the province. The two specimens on which the illustrations were based were caught in 1972 and still represent, respectively, only the first and second records for the province: *Sternoptyx pseudobscura* (front cover) and *Chirolophis tarsodes* (back cover).

Cover illustrations by Patricia Drukker-Brammall, courtesy of Alex Peden [see *Syesis* 7:47-62 (1974)].
Cover design by Chris Tyrrell, RBCM.
Typeset in Palatino 10/12.
Typesetting and page composition by Gerry Truscott, RBCM.

Canadian Cataloguing in Publication Data
Main entry under title:
Our living legacy

    ISBN 0-7718-9355-8

    1. Biological diversity conservation –
Congresses. I. Fenger, M.A., 1949 -   . II. Royal
British Columbia Museum.

QH75.A1097 1993   333.95   C93-092162-3

Table of Contents

Diversity at Risk

Strategies

What Can We Do?

Public Expectations

Our Living Legacy

Proceedings of a Symposium
on
Biological Diversity

Foreword

The Need

"The diversity of life forms, so numerous that we have yet to identify most of them, is the greatest wonder of this planet" (Wilson 1988: v). This wonder, coupled with concern, provided the impetus for this symposium, held in the spring of 1991. The need for a broadly based symposium was first identified during provincial workshops on developing a strategy for conserving old-growth forests in British Columbia. In these workshops, it was emphasized that old-growth forests carry only a portion of B.C.'s biological diversity. Clearly there was a need to address the conservation of biological diversity more completely. There was also a need to summarize current understanding of provincial biodiversity, to communicate provincial responsibility and action in a more public forum and to place provincial issues in their broad ecological and geographic contexts.

An Overview of the Symposium

David Suzuki, the keynote speaker, opened the symposium with a global overview stressing the interactions among biological, social and economic systems. Wade Davis provided a spellbinding account of many cultural and ethnobotanical aspects of biodiversity. Jack Ward Thomas discussed experiences in the United States, with particular reference to endangered species legislation. And Bristol Foster placed British Columbia in a global context.

In the section on "Principles", Chris Pielou outlined some complexities in measuring biodiversity and suggested some straightforward ways to do it. She was followed by Larry Harris, who emphasized the importance of ecosystem scale and migration routes, using examples of endangered insects and mammals. A chilling comparison of geologic extinctions to current global extinction rates was given by Geoff Scudder. And Tom Ledig treated genetic variation, species losses and the unmourned "secret extinctions" of genetic material lost with the disappearance of local populations. A paper was later added by Ted Miller on the critical roles played by systematics and taxonomy in biodiversity initiatives.

In the section on "Diversity in British Columbia", Jim Pojar summarized the diverse and unique terrestrial ecosystems of B.C. and compared our flora and fauna to those of other jurisdictions. Similar treatment was given for marine ecosystems by Verena Tunnicliffe, and for freshwater and estuarine diversity by Don McPhail.

The section on "Diversity at Risk" began with Bill Munro describing B.C.'s and Canada's systems for classifying species and subspecies at risk. We have appended a current list of Red- and Blue-listed vertebrate species and subspecies considered to be threatened and vulnerable, courtesy of Bill Harper. Chris Chanway and Val Marshall emphasized the ecological importance of biodiversity in soils and how little we know about it. Bill Harper and Debbi Hlady provided an example of a conservation strategy being implemented in the South Okanagan area of B.C.

Bill Wareham opened the "Strategies" section with a discussion of the need for reserves and the broad context for these within the landscape. Tom Gunton then addressed the reformation of B.C.'s planning systems. Colin Rankin reviewed types of legislation aimed at protection of biological diversity. And Jim Walker provided an overview of B.C.'s provincial conservation strategies.

In the section on "What Can We Do", Hal Salwasser focused on biodiversity conservation in the United States with emphasis on the United States Forest Service, Ken Lertzman discussed the roles of continuing and new research and Andrew Harcombe described the emerging role of B.C.'s Conservation Data Centre.

The symposium concluded with a Panel Discussion addressing public, government, industrial and native views on conservation, with Fred Bunnell providing the concluding remarks.

We recommend a book that is complementary to this one: *Biodiversity in British Columbia: Our Changing Environment*, edited by L.E. Harding and E.A. McCullum (Environment Canada, Pacific and Yukon Region, Vancouver, B.C., 1993).

Updates and Future Directions since the Symposium

In the two years that have elapsed since the symposium, significant advances for conserving biological diversity have been made globally, nationally and provincially. Some of these are:
- The Global Summit on the Environment and Development was held in Rio De Janeiro in 1992, with Canada as one of the first nations to sign the biodiversity convention.
- The "Protected Areas Strategy" has emerged from the "Parks

and Wilderness for the 90s Strategy" with an expanded goal to protect viable representative examples of natural diversity and special natural, cultural and recreational features of the province.

- The provincial "Old-growth Strategy" has been completed and the implementation of old-growth reserves has been transferred to the Protected Areas Strategy .
- Legislation has been developed on environmental impact assessment. A forest practices code and legislation on protection of species, subspecies and habitats, are under development.
- The Commission on Resources and Environment has been formed to make recommendations on broad land-use decisions, with an initial focus on regional planning and the implementation of Protected Areas.
- A renewed commitment has been made to improve inventory to support planning through an integrated approach under the provincial Corporate Resource Inventory Initiative.
- Guidelines have been developed for conserving biological diversity at the broad-planning landscape level and at the forest-stand level.
- Forest harvesting levels are under review. Significant reduction has been implemented in some areas in order to reach biologically sustainable harvest levels.
- Initiation of regular "State of the Environment" reporting has begun, with many biological elements used as indicators.

The term "biological diversity" is used more widely in B.C. than it was when this symposium was held two years ago. In these two years the level of debate has been raised and attitudes have been changed. The term is better understood, and its importance in maintaining functional ecosystems and all their parts is better accepted as a fundamental underpinning in filling stewardship obligations and in land-use decisions from planning to guidelines and implementation. There is a recognition and acceptance of the need to better integrate biological, social and economic information.

Despite changes in attitudes and many new initiatives, there are few on-the-ground examples of changes in management practices that adequately address conservation of biodiversity. The editors hope this publication will increase awareness and concern, and help to maintain the impetus of provincial and national initatives and to extend the new attitudes to routine land management.

Throughout the symposium there was a recognition that conservation of biological diversity can only be achieved if it is factored into decision-making in the political, social and economic arenas. In addition, basic biological information has to be improved to provide diagnostic criteria in integrated land-use and resource-development decisions. Continued support for research, inventory, assessment techniques, monitoring and communication is necessary to change the present practices that continue to threaten the elements and functioning of ecosystems. Sustaining biodiversity can be achieved only through knowledge, understanding and appreciation of what E.O. Wilson called "the greatest wonder of this planet".

Format of Presentations

The editors' task has been to make the presentations inviting, readable and consistent, to recreate the excitement generated throughout the symposium. Most of the speakers prepared written papers; transcripts were used for others and for the panel discussion. Abstracts were added when absent, and the editors take responsibility for any changes in emphasis that may have been introduced. To standardize the use of common and scientific names, we used the references listed below as much as possible, though we were liberal with authors' use of local or exotic names. Depending on how technical the context was, we used scientific or well-established common names or both. Common names of species are capitalized (e.g., Hairy Manzanita), but group names are not (e.g., manzanita), regardless of context.

References

Banks, R.C., R.W. McDiarmid and A.L. Gardner. 1987. Checklist of vertebrates of the United States, the U.S. Territories, and Canada. U.S. Fish and Wildl. Serv., Resource Publication 166, Washington, DC.

Cannings, R.A., and A.P. Harcombe (Editors). 1990. The vertebrates of British Columbia. Scientific and English names. Royal B.C. Museum. Herit. Rec. 20, Victoria, BC.

Meidinger, D. 1988. Recommended vernacular names for common plants of British Columbia. B.C. Ministries of Forests and Lands, Res. Rep. RR7002-HQ (rev.), Victoria, BC.

Nagorsen, D.W. 1990. The mammals of British Columbia: a taxonomic catalogue. Royal B.C. Museum, Memoir 4, Victoria, BC.

Straley, G.B., R.L. Taylor and G.W. Douglas. 1985. The rare vascular plants of British Columbia. Syllogeus (Nat. Mus. Can.), 59.

Acknowledgements

We are indebted to the many people who contributed to the success of the symposium and to the publication of its proceedings: Fred Bunnell, Rob Cannings, Adolf Ceska, Sharon Chow, Laura Darling, Don Eastman, Evelyn Hamilton, Andrew Harcombe, Bill Harper, Richard Hebda, Ken Lertzman, Andy MacKinnon, Del Meidinger, Brian Nyberg, Bob Ogilvie, Rick Page, Jim Pojar and Bill Pollard. Gerry Truscott did the final editing and layout, and oversaw countless details of publication.

The symposium was sponsored and organized by the Association of B.C. Professional Foresters, the Association of Professional Biologists of B.C., B.C. Ministry of Environment, B.C. Ministry of Forests, B.C. Ministry of Parks, MacMillan Bloedel, the Royal B.C. Museum, the Sierra Club and the University of British Columbia.

Mike Fenger
Ted Miller
Jackie Johnson
Liz Williams
February 1993

Welcoming Remarks

Bill Barkley
Royal British Columbia Museum

On behalf of the Minister responsible for the Museum, the Minister of Municipal Affairs, Recreation, and Culture, the Honourable Lyle Hanson, I'd like to welcome you to British Columbia's Provincial Museum, the Royal British Columbia Museum. It's a pleasure for my staff and me to see all of you here today talking about a topic that's very important to this Museum and to all of us who live on this planet.

It's timely that we should have these discussions right now. Museums play a fundamental role in studies of biological diversity. They are at once research centres, repositories of millions of plant and animal specimens, and a point of contact between the work of the scientist and the general public. Museum curators are experts in identifications, in names and in classification of plants and animals. They collect them, they skin them, they stuff them, they pickle them, they preserve them, and they share their findings with individuals and groups through publications and meetings of scientific colleagues. Examples of such sharing of information showing some of this museum's taxonomic publications and identification guides are on display in the foyer. We are very proud of our publication series, a series that has been produced here for many years.

Some museum publications are in scientific literature, describing new species and subspecies and analysing their evolutionary relationships. As an example of such basic work, a deepsea eel pout of the genus *Pachycara* was described as a new species by our Curator of Fishes, Alex Peden (see Fig. 9 in Miller, this volume - Eds). Reference specimens of new species such as *Pachycara* are deposited at major institutions throughout North America and in some cases throughout the world. Specimens of the new *Pachycara* species are now at the Smithsonian Institution and other major institutions.

Publications that build on this basic taxonomic research are exemplified by the museum publication, *The Mammals of British Columbia: A Taxonomic Catalogue.* This catalogue, compiled by David Nagorsen, our Curator of Mammals, is a history of names of B.C. mammals. It is an essential work for ecologists, foresters, environmental consultants, educators, wildlife managers and others. As a small example of the importance of such information, consider these two specimens from our museum's collection. One was collected in 1936 by a well-known scientist, Dr Ian McTaggart-Cowan. The other was collected in 1956 by someone whose name many will recognize, Charlie Guiguet. The genus *Peromyscus* is an ecologically important part of what some of us call the truffle web. These mice disperse the spores of mycorrhizal fungi that are so important to forest ecology and tree growth. Recent museum studies using chromosomes and bones indicate that there are two species of *Peromyscus* in southwestern British Columbia: the Deer Mouse *Peromyscus maniculatus* and the Columbian Mouse, *Peromyscus oreas*. What are the ecological roles of these two species in the truffle web of the coastal forests of British Columbia? How do they differ in function where the species overlap in the coastal forest area? Such questions can only be asked on a firm taxonomic base and can only be answered after accurately identifying field collections.

Over the next two days you will hear and participate in an array of talks whose topics will underscore how vital it is for all parties concerned with biodiversity to interact and collaborate. The Royal B.C. Museum is one of those parties. Use our collection, use our expertise, consult with our curators, use our publications, and above and beyond all else, leave your specimens at the museum so that we can preserve them and their data for future generations.

Randy Chipps
Becher Bay Indian Band

The principles of faith, harmony and the continuance of everyday philosophy within the world order that we live in are the reason we are here. We bring together thoughts and ideas of how the quality of life and health could be married in a fruitful blend that will become reality within our own flesh. The thoughts and the dreams of each individual can be welcomed to this meeting with the warmth and generosity that our grandfathers have allowed us to maintain deep within our hearts.

I wish to share many thoughts with you. I do that by allowing you to join me in my convention, my ceremony, as each day I walk, breathe and share my mind and my spirit with everyone I meet. If I appear angry or upset, let it be known that it is my difficulty. Remind yourselves of this difficulty we're sharing, that sometimes we do not understand what is being said. As we move through this day and tomorrow, we bring our fellowship of humanity into this world order that we share. We all walk as brothers and sisters, just as in this coastal area some ten generations ago lived, side by side, people speaking different languages, with different cultures from here to the California coast—sharing, caring, interacting when we needed to. With this present world order we need even more to share, because of our rapid communication and travel.

As I reflect on what needs to happen, I offer you a welcome song from my tradition to your tradition, with a reminder that all our traditions stem from the same fatherhood, the fatherhood of humanity. I thank my good wife for joining me here today as well as all of my brothers and sisters. I listen to the heart beat of humanity of Grandmother Earth and remind myself of Grandfather Sun and the four directions: east, where spirituality and softness take us; south, to the calm, warming south winds; west, where dreams and reality start to manifest; and north, where we find wisdom, understanding, clarity. If I look up, Grandfather Sun smiles at me. If I look across, I see Grandmother Earth, whose bosom I traverse as I travel back to my home.

> Grandfather, enter my heart, my soul.
> Remove the knife edge from my tongue.
> May I praise all that I meet today with a blessing of fellowship.
> May I share my tears and my laughter with others, without shame,
> With a strong hope that I will remember why I am here.
> As I travel homeward, oh Grandfather, spiritual creator,
> I await the blessing of my return back to my home land.
> Welcome to this very wonderful symposium.

Lee Doney
B.C. Round Table

I also welcome you to the conference on biodiversity, but as others have welcomed you so adequately, my welcome is a touch redundant. But I do welcome you and I will take this opportunity to advertise the Round Table and what we're doing.

The Round Table is supportive of and very interested in the concept of biodiversity. Those who have read our first report, "A Better Way", designed to stimulate discussion, will know that a statement of development is in our six principles. We recognize the concept of biodiversity and the importance of biodiversity to achieving sustainability. We too will discuss the issue of biodiversity in more detail in an upcoming report on land and water usage in British Columbia that is just now being produced. If I'm not overly optimistic, we will have this report out this spring. It will be a touchstone for round-table discussions on land and water use. The whole issue of biodiversity is important not only for aesthetic and ecological reasons, but also for environmental and economic reasons.

The other issue I want to bring to your attention is that the Round Table will be touring the province this spring (i.e., 1991) to discuss these and other concepts with you and to hear your points of view. I would like to encourage you to come forward and help the Round Table understand the importance of biological diversity to British Columbia. The Round Table will be an area and a focus where we can bring together and translate information into recommendations to government that will result in public policy action and decision making in the government. Likewise, you will see many members of the Round Table at this symposium during the next days, and if you get a chance to talk to them, I encourage you to do so.

So without further ado, I wish you the best in your deliberations. I look forward to what will come from this conference because it's important to discuss what biodiversity really is and what it means to British Columbia.

The Honourable Cliff Serwa
Minister of Environment

On behalf of the Premier, I am pleased to be able to open this symposium on biological diversity and welcome all of you, speakers and delegates, to Victoria.

The adjectives "beautiful" and "supernatural" have been applied to British Columbia for many years now. I don't need to tell you that it is the exceptional biological diversity of our province that makes it possible to describe it in such glowing terms. These natural systems— our forests, grasslands, marine and freshwater ecosystems—provide us with outdoor recreation, lifestyles, and to a large extent our present and future material wealth.

Biological diversity is a relatively new term. It is a broad concept that describes the variety of life, the processes that maintain life and the way each relates to the other. We are only just beginning to understand how little we know about these processes and relationships, how much there is to be learned in this area, and how important it is to gain that understanding. Clearly, a better appreciation of biological diversity is crucial to making wise decisions about land and resource use.

It follows that meetings like this one will help to equip us with the knowledge needed to plan for a sustainable future in British Columbia. It will help us to build on processes already underway that represent a positive approach to natural resource stewardship and the maintenance of diversity. For example, the Ecological Reserves Program, established 20 years ago, recognizes the scientific value and enables conservation of fragile, rare and representative biological systems. We need the kind of knowledge and expertise represented here today to help us set the right directions for this program.

We must also use this expertise to develop a provincial old-growth strategy to maintain representative old-growth ecosystems and their biological productivity. The task force developing this strategy has identified 121 areas that have special old-growth attributes. The process should result in a system of old-growth reserves representing the various ecosystems in the province and in greater attention to preserving old-growth values in the managed forest. A strength of the process is that the task force is composed of people representing a wide variety of interest groups, industry and government. Hopefully it will be a model for such co-operative decision-making in the future.

Parks Plan 90 and the Ministry of Forests' Wilderness Systems Plan are currently underway and are soliciting public comment. Such initiatives will also help monitor biological diversity through a system of reserves.

As a final example, in an area which, for me, is close to home, the South Okanagan Conservation Strategy has identified habitats and species that will need special conservation efforts if they are to be sustained.

Within the Ministry of Environment we are currently establishing a Conservation Data Centre, in co-operation with other agencies and The Nature Conservancy. This Data Centre will allow us better access to information on a variety of sensitive species, both for the agencies and for the public who will have access to the data.

Although reserves, parks and conservation areas are needed to maintain rare and fragile elements of biological diversity, the real opportunity for maintaining or enhancing biological diversity lies in understanding the limits, thresholds and functioning of natural ecosystems across the diverse landscapes of British Columbia. Most importantly, in the move to greater concern for biological diversity, I have noted over the years that agencies such as the Wildlife Branch, Fisheries and the Ministry of Forests are attempting to broaden their approach to the natural world to consider more than just the single resources that formed the focus of their responsibilities. I believe this new appreciation for all components of the natural world, by all agencies and by the private sector, is our greatest hope for the future.

Only by working with nature and with each other will we continue to have a beautiful, supernatural British Columbia.

I know you can look forward to a productive and rewarding series of discussions, and I'm confident the new insights gained from it will be invaluable in sustaining the biological resources of British Columbia.

Biodiversity for the Next Millennium

David Suzuki

Department of Zoology, University of British Columbia, Vancouver, British Columbia, Canada V6T 1Z4.

Abstract

Our continuing exploitation of the natural world with increasingly destructive technology constitutes a massive assault on the global life-support system of water, atmosphere, soil and biodiversity, and thus threatens our existence as a species. Eighty per cent of Canadians live in controlled urban environments with no connection to the natural world for water, food or garbage disposal. There is little understanding of the environmental carrying capacity as we create dams, roads and clearcuts. This planet has been recycling its materials through intricate webs and cycles for five billion years. The most important reason for maintaining biological diversity is to keep these webs functioning to maintain productive soil and clean water and air.

As a species we have a built-in need for other species. Yet we differ from other species with our human concepts of the future, loyalty, nations, justice, peace, economics and resource values. Culture and personal experience shape our beliefs and values, which may lead to conflicts in such places as the Stein Valley.

Global ecocrisis facts include: increasing human population (5.3 billion now, 10 billion in 40 years); declining food production; increasing toxic wastes; atmospheric changes (ozone depelotion, acid rain and greenhouse effect); destruction of tropical rainforests; and unprecedented species extinctions. Our inability to act, despite knowledge of the facts, is predicated on a distaste for bad news, an unwillingness to make sacrifices and massive denial. As a species we have evolved to respond to immediate physical threats rather than to the hazards listed above. We cling to sacred truths such as: humans exist apart from nature and create their own environment; science provides the knowledge and power to understand and control nature; a healthy economy must precede a healthy environment; and, steady economic growth is necessary for progress.

We need to act before more ecological thresholds are crossed and ecosystems

In *Our Living Legacy: Proceedings of a Symposium on Biological Diversity*, edited by M.A. Fenger, E.H. Miller, J.A. Johnson and E.J.R. Williams, pp. 7-20. Victoria, B.C.: Royal British Columbia Museum.

collapse. A long-term strategy to maximize the protection of biological diversity and allow us to maintain a healthy economy must be a high global priority. Achieving this goal requires massive restructuring of social priorities and actions of governments and industry. Actions required are reduction in population and consumption, creatively forgiving poor countries' debts and warning developing countries of our environmentally destructive path. Zero discharge limits for toxic wastes and green house gases should be enforced. Understanding the fallacies of the sacred truths requires a change in focus to the long-term from the present preoccupation with the moment.

As animals we need clean air, food, shelter, community, meaningful work and social justice. These entail different priorities than those of our highly mobile consumptive lifestyles, and we must look for common ground towards these shared goals.

Biodiversity for the Next Millennium

It is an honour to have been asked to participate in your symposium and I thank you for what I know was a rather courageous decision to include me in the program. Issues surrounding the conservation of biological diversity are among the most important and contentious of our time, and I hope that I can provide a perspective from my experience as a scientist who has entered the vulgar world of journalism and broadcasting.

I cannot begin a talk like this without first saying that no one I know in the environmental movement is against logging. The forest industry is a key element of this province's economy, history and culture and I hope that it will continue to play that role well into the future. I want every person who is a logger today to go on being a logger as long as he or she wants, and for logging to be a way to earn a living for future generations.

My perspective is that of a biologist aware of a massive assault on the planet that is causing terrifying changes in its life-support systems of water, atmosphere, soil and biological diversity that could very well render irrelevant much of the knowledge we have accumulated over time. I speak as a parent who remembers a radically different Canada when I was a child and who knows that we are not leaving anything like that richness, abundance and variety to future generations.

The important questions about wilderness and forests are not about whether or not to log or develop. I wish it were that simple. They have to do with how we see ourselves as biological beings living with tens of millions of other species on Earth. They have to do with our perspective: do we begin with the assumption that it is our right to exploit anything on Earth as we wish as quickly as we can? or do we start with humility, recognizing how little we know, how dependent we are on the natural world, yet how powerful and potentially destructive are our technologies? It is with the latter attitude that I would like to frame my thoughts and ideas.

Eighty per cent of Canadians live in cities and towns. We are an urban people. And in a human-created environment, it is easy to forget that we remain animals who are deeply rooted in the biological world. In towns, we have severed our obvious dependence on the natural world.

We live in temperature-controlled apartments and houses, so the seasonal variations have far less impact on us.

We buy food and goods at stores and supermarkets. Vegetables and fruit without blemishes and meat cleaned of blood, fur, scales and feathers have little to remind us of their biological origins. Many children today don't realize that meat is an animal's muscle or that vegetables grow in dirt.

We draw water by turning on a tap.

We eliminate body wastes by the flush of a toilet.

And we get rid of garbage by simply putting it out on the curb.

In boardrooms and government offices, business and political decisions are made to create dams, roads, massive clearcuts, pulp mills and developments that reduce our biological ties even more. As a consequence, human activity proceeds with little regard to such simple principles as "carrying capacity" of ecosystems or sustainable yields of renewable resources. The result is an impending ecological catastrophe that will force us back to reality. To appreciate where we are, we have to place the present in the proper perspective.

The history of this planet goes back about 5 billion years. Life arose perhaps 3.5 billion years ago. The planet is finite, and while there have been major changes in the chemical composition of the atmosphere (most recently in the accumulation of oxygen when plants evolved), basically the constituents of the air, water and soil have always existed. As they aren't replaced or made anew all the time, they have been endlessly recycled through all living forms since the beginning of time.

Until very recently, the pristine state of air, water and soil was maintained by the interaction of the global network of organisms that exploited them. One organism's wastes can be another organism's food in an intricate web of interconnections. Scientists do not and will never fully understand the components of ecosystems and their interactions that have kept this planet liveable. This is without question the most powerful argument for maintaining maximal biological diversity: we don't comprehend it, but we know that it is what keeps air, water and soil clean and productive.

Human beings arose as a species out of the web of life a mere 200,000 years ago or so. We are made up of the same basic atoms in the same proportion, the same macromolecules, and the same cellular plan as all other living things because we share common ancestors. The genetic blueprint is written in nucleic acids in a universal genetic code. According to scientists, our hominid ancestors inhabited the area along the Rift Valley of Africa. From the present populations of animals that occupy the Serengeti Plains, we can infer the kind of biological communities within which those first humans lived. They depended on other species not only for food, but also for their greater sensory acuity to warn of danger or as barometers of their surroundings. Our ancestors also depended on the animals and plants for *companionship.*

As a geneticist, I believe that we have a built-in need to be with other species. Any parent knows that young babies have an instant attraction to a flower or butterfly seen for the first time: there is no revulsion or fear. A child playing with a pet dog or cat gives evidence of a powerful need to be with another species, just as gardeners demonstrate a strong emotional satisfaction from working with plants. It is well documented that patients in hospitals, old age homes and mental institutions respond tangibly and positively to pet animals. Taken together, these observations make a case for our *biological need to be with other species.* Harvard biologist E. O. Wilson calls this *biophilia,* a love of other life. In discussing the "value" of biodiversity, we must not discount this requirement which, although of a spiritual nature, is nevertheless real.

The distinctive survival strategy evolved by our species was a highly developed brain. From that brain emerged self awareness, memory, curiosity, imagination, and creativity. The human brain had an unprecedented capacity for abstraction, to create ideas that existed only in a mind but became as real to that mind as a rock or a tree. The *future* is such a human invention. There is no such thing as a future except as an idea in our minds, yet by acting on this concept, we can make deliberate choices in the present and change that future.

It was this capacity for abstraction that enabled us to invent notions of tribes, loyalty, nations, justice, peace, war, love, equality, god, heaven and hell. And it was also responsible for creating economics, an idea that posits value in human activity and natural "resources" that we can exploit and which symbolize that value in currency. Money, only a physical representation of human ideas of value, has taken on an identity equivalent to the real thing. When biological diversity is depleted, all the money in the world won't substitute for it.

Nobel laureate Francois Jacob has said that the human mind has an innate need for *order,* a fear of chaos and the unexplained. Thus, we construct "world views" that are comprehensive explanations, however incredible, of the world around us. This confers a sense of understanding and control over the cosmic forces impinging on our lives. Culture arises within these world views and culture in turn shapes the minds of the people who live within them. Thus, a feedback mechanism exists whereby culture is created by human minds and then imposes filters that influence how those minds perceive the world.

Heredity, personal experience and culture all act to shape the perceptual lenses that select and edit impressions of our surroundings. This point cannot be overemphasized. What we see is conditioned by our own experience, beliefs and value systems. Ask a man and a woman to respond to words like "family" or "war", or query a company executive and a welfare recipient on the economy and the national debt. Their responses will be radically different.

The first time I saw the Stein Valley I was in a helicopter accompanied by a Native from Lytton. As we flew up that incredible ecosystem, my companion pointed to a place where a major battle had been fought between different Native groups long ago. He indicated sacred burial areas and showed me where the Grizzly Bears feed in the spring. The pilot chimed in, saying that the week before he had flown a group of foresters and industry executives along the same route and all they had talked about were wood volume, jobs and profit. Two groups of people looking at the same valley but seeing radically different things.

This is the challenge we face. In discussions of environmental issues, we each bring all of the baggage of our personal values and belief systems reinforced by our lifestyles and what we do for a living. Too often, debates rage over words that have totally different meanings to the people using them.

What are the global environmental facts, why can't we act on them and what can we do to start?

The Facts

Before we can begin a serious discussion about the biodiversity crisis we have to appreciate the dimensions of the global ecocrisis within which issues of biodiversity must be weighed. Human numbers and technological muscle power have increased to such a point that our species is changing the planet itself. These are a few of the issues.

Population

At the time of Jesus Christ's birth, about 250 million people were on Earth. It took almost 2,000 years to reach one billion in the last century. But in 150 years, we doubled twice. We're at 5.3 billion and will hit 10 billion in 40 years! Today, three human beings are added to the world's population every *second*, over a quarter of a million per day, and the rate is increasing. Every addition puts greater stress on the planet's biosphere.

Food Decline

World agricultural production peaked in 1984 and has been declining since. In large part, that is because we lose 25 billion tons of agricultural topsoil annually, the equivalent of Australia's entire wheat-growing capacity. The oceans, once thought to be limitless sources of protein, are being mined of their resources, the Atlantic cod being the latest casualty. With exploding populations and declining food, something has to give. At present, over 40,000 people in the Third World, three quarters of them under the age of five, die of starvation every day and that number will rise.

Toxic Wastes

Every five minutes around the clock a major shipment of toxic chemicals crosses a national border for disposal. Much of it is dumped in the Third World, and much is illegally poured into air, water and soil. In 1987 alone, American industries legally spread over one *billion* pounds of toxic chemicals into air, water and soil in the U.S. There is nowhere you can go on the entire planet to escape the toxic debris of human activity.

Atmospheric Change

Ozone is being depleted by human-created compounds like chlorofluorocarbons, acid rain is killing forests and sterilizing lakes in North America and Europe, and greenhouse gases that trap heat are accumulating in the upper atmosphere. Climate and weather are being affected and can no longer be depended upon, forests and agriculture are being negatively influenced, while oceans will rise as their increasingly warm water expands.

I have interviewed people from the Maldive and the Seychelle Islands who know that one consequence of global warming will be the loss of their islands from sea level rise. In November I talked to agronomists from Nigeria and Kenya who say that traditional farmers depend on seasonal cues to know when to plough, plant and harvest, but now they can no longer rely on these signals because the climate has changed. In a time of drastic atmospheric and climatic change, numerous plants and animals will not be able to adapt to the new conditions. Forests will not be able to move fast enough to keep ahead of the heat, and we cannot anticipate what particular species and genotypes will be able to survive after the changes. How can we practice forest management for a future when conditions under which trees will mature will be vastly different and constantly changing? The only hedge we have against the enormous uncertainty of the coming decades is maximal biological diversity.

Forest Destruction and Species Extinction

Only 30% of the planet's surface is land, and of the land only 5% is occupied by tropical rainforests. Yet they are the home to well over half of all species on Earth. They are being destroyed at the rate of an acre per *second* (a hectare every 2.5 seconds). The result is a change in weather and climate from the consequent desiccation, erosion and carbon dioxide release. As well, forest destruction is accompanied by an unprecedented extinction spasm that, depending on the estimated number of species there are, may be claiming 20,000 to 150,000 of them per year.

While tropical forests disappear, we in temperate regions are taking out old-growth forest at a rate that will see most of it gone within two decades. Yet our ignorance about forests is enormous. The canopy research platform built in the Carmanah Valley last year is, I am told, the world's first and only in temperate rainforests and the highest in the world. We are just beginning to look at the canopy ecology of tall forests.

These are facts, not speculation, and they should be the overriding reality within which we make our decisions and set society's priorities. The planet is under attack by the deadliest predator in the history of life on Earth and we cannot tell what the consequences of this attack will be in the future.

Why We Can't Act

In the face of the facts, why aren't we adopting the heroic measures so clearly needed to get off this destructive path? There are many reasons, not least of which is our distaste for bad news. No one wants to be told we're in big trouble, that we must change the way we live or, heaven forbid, that we must make sacrifices. So we experience massive denial, delay and blaming of the messengers. Suppose you bring your sick child to the doctor and are told she is dying but that her life might be saved by taking heroic action. Would you put off taking the news seriously rather than facing up to the facts? Of course not, yet when it comes to the ecosphere, we use every argument in the book to postpone a serious reckoning of the dimensions of the crisis.

Another problem is that while we have evolved superbly to respond to immediate physical threats that can be detected by our sensory organs, we can not sense the real hazards that face us today: global warming, ozone depletion, toxic chemicals in food and water, increased background radiation, acidity of rain, species extinction. So our senses seem to deny the reality of the hazards being discussed by scientists and environmental groups.

But the most difficult impediment to recognizing the problems and acting on them is the tenacity with which we cling to values and beliefs that blind us or even cause the crisis we face. I call them *sacred truths,* ironically, because they are neither sacred nor true. These are some of what I call sacred truths:

Because human beings are different from all other life forms, we lie outside of nature and create our own environment. But we remain biological beings who are just as dependent as any other species on clean air, water and food from the soil for our health and survival. By using the environment to dump our toxic wastes and by reducing the biodiversity on the planet, we compromise the very things that sustain us as animals.

Science provides the knowledge and power to understand nature and control it. The nature of science is that it does the opposite by necessarily fragmenting the world into isolated bits and pieces. The Newtonian concept of a mechanistic universe whose properties are the sum of its elementary parts has been discarded by physics through relativity and quantum mechanics. We now live in a probabilistic universe, not an absolute one, and it is clear that synergistic interactions between its components result in properties that cannot be predicted on the basis of studies of the individual components. Yet most scientists in medicine and biology, including forestry, continue to operate on the Newtonian ideas.

Even if it were possible to discover principles that govern higher levels of organization, how much do we know? I was in Australia last month and heard a report from the Daintree rainforest in Queensland. An entomologist sprayed an insecticide in a small area of the Daintree and was astounded to recover hundreds of insect species, virtually every one of which had never been seen by a human being before. He estimates there must be hundreds of thousands of yet undiscovered species in this small park, and to date only a few hundred are known. Terry Erwin at the Smithsonian Institute estimates that we have 30 million species on Earth, yet until now biologists have only identified 1.4 million!

And that only means that a dead specimen was given a name. It doesn't mean that we know anything about the basic biology of each species — its food needs, habitat, mode of reproduction, or interaction with other species. It means that all we have done is assign labels to perhaps 5% of the diversity of life on the planet. Consider this. I spent 25 years of my adult life studying one species of organism, an animal called *Drosophila melanogaster*, a fruit fly. This species has been a favourite of geneticists for over 70 years. I estimate that hundreds of billions of dollars have been spent studying this one species while tens of thousands of Ph.D.s have spent their entire lives studying it. Eight Nobel prizes have been won by people studying the fruit fly. The result of this massive investment is that we know a lot about this one kind of organism. Yet in spite of the money, person-years, and Nobel prizes, we still don't know how the fly survives the winter in Canada. We don't understand how it can distinguish a sibling species that only a handful of experts can identify. We don't know how a fly's egg is transformed into a larva, then a pupa, and finally an adult. And *Drosophila melanogaster* is one of tens of thousands of species of fruit flies. One thing some people learn after a lifetime of research is how little we know.

And so when foresters tell me, as many have, that we know enough to "manage" forests, I say that all of my years as a scientist have taught me that's just not true. Only time and nature have ever created forests. A forest to me is the sum total of the air, water, soil, micro-organisms, plants, animals, and of course, large trees. Taken together, they are a forest. A plantation or tree farm is a one-dimensional parody of what an ancient forest is.

We need a strong economy to afford a healthy environment. The economy is based on the physical world. Without a "clean" environment, there can be no economy because people won't survive. The economy is ecologically destructive when it is disconnected from nature. Air, water, soil and biological diversity are considered as "externalities" to current economic considerations. The role that ecosystems perform in maintaining the quality of the environment is not costed in our accounting systems, thereby literally driving us to destroy nature for economic returns.

For example, a standing old-growth forest performs countless "services" for the planet: exchanging oxygen for carbon dioxide; storing water and minimizing erosion, floods, storm damage and fires; modulating weather and climate; supporting wildlife. None of these ecological services has any value in conventional economics. So if an old-growth forest adds fibre at 3-4% a year, it makes *economic* sense to liquidate the forest (thereby losing the services performed by the standing forest for generations) and to bank or invest the money from the timber because you can earn 10% interest or more.

We must do everything we can to maintain steady economic growth.
Exponential growth of any system is a physical impossibility in a finite
world. We are currently mining the planet's resources because we have
the technological muscle to do it. We are far exceeding the carrying
capacity of the planet and are simply not practicing the kind of
conservation policies that ensure renewability of resources. As Stanford
ecologist Paul Ehrlich has said, "Steady mindless growth is the creed of
economists and cancer cells." When growth becomes synonymous with
progress, as it has in our society, and governments see growth as their
main preoccupation and measure of success, then there is no end. We
can never envision a time when we say, "Enough. We have more than
enough and will stop at this level."

We are mortgaging our children's futures in this rush to maximize
profit as quickly as possible. How can environmentalists be blamed for
lost jobs in the forest industry when the amount of cutting has doubled
over the past two decades while the number of jobs has declined by a
third? Continuing to log above the replacement level is absolute folly,
especially when the forestry practices are themselves destructive of
sustainable harvesting.

And what is the nature of this economic growth? Economists
provide us with economic indicators like GNP to measure the annual
increase in goods and services, and governments and businesses make
major decisions on the basis of the GNP. But the GNP is a perverse
notion that doesn't take into account the reason for the growth. Thus, as
a result of the Exxon Valdez oil spill, the U.S. GNP rose in 1989 partly
because Exxon spent a billion dollars cleaning it up.

Suppose a major accident occurred at the Darlington nuclear plant
next to Toronto and radioactivity spread across the city causing
thousands of people to get radiation sickness. To economists, that's
good because the GNP rises. You need more doctors, nurses, hospitals,
medicine and, no doubt, lawyers. And if dozens of people die, again the
GNP goes up — because of funeral parlours, gravediggers, flowers,
caskets. It is grotesque to have whole social and economic programs
based on the GNP.

*In a democracy, we elect people to represent us and lead us into the
future.* In a time when the major issues facing us have to do with
science, technology and the environment, most politicians come from
business and law backgrounds that do not prepare them for the serious
problems they will tackle. They will have to assess such issues as
sustainable yields, biodiversity, global warming, species extinction,
biotechnology, etc., but can't because they are scientifically illiterate. So
political priorities become skewed by economics and jurisdiction.

In our democratic system, politicians must set the time frame for their actions by the interval between elections. This is simply not long enough to deal with major environmental problems. So politicians prefer short-term megaprojects that give a temporary illusion of jobs and action.

Only people vote and only people hold political office. No one speaks for the other inhabitants of Earth. Who represents the birds, the whales, the trees, the air, the rivers, the forests? To the politician whose constituency is human, human priorities must be uppermost in his or her actions. Nature inevitably pays the price of human imperatives.

Finally, bureaucratic delineations of responsibility do not correspond to ecological realities, so we cannot deal with the "forest problem" or the "salmon issue" or "water" in an integrated way that makes biological sense. Consider the forests. Animals and plants don't pay attention to human-created borders but distribute themselves within ecological perimeters. It makes no biological sense to talk about biological diversity as if Canadian forests somehow end at the border and are independent of American forests. We have both federal and provincial forestry portfolios with different perspectives and priorities. But forests have broader implications that extend beyond a single province's Ministry of Forests. They impinge on Departments of Finance, Energy, Mines and Resources, Indian Affairs, Fisheries and Oceans, Tourism, Urban Development, Agriculture and Science, and Technology. By subdividing a biological issue, the forests, into a number of competing bureaucratic areas, we ensure that it will never be treated in a holistic, integrated way.

What Can We Do?

When the global dimensions of the ecocrisis are understood, it becomes crystal clear that we have little time to make decisions before the planet will have been so altered that nature will no longer have the resilience and capacity it has always had to renew itself. The changes mean that nature is becoming less predictable and dependable. As the temperature rises, as more ultraviolet radiation strikes the earth, as rainfall becomes acidic, as new toxic chemicals are bioamplified up the food chain, we simply cannot foresee long-term consequences. As when our distant ancestors became aware, unpredictability holds terrors for us once more. Biological diversity becomes the only hedge we have against our ignorance.

The recently noted mysterious decline of frogs around the world should be a stark warning of how little we know. Our hominid ancestors watched their non-human companions as indicators of the health of the environment just as coal miners observed canaries they kept in the mines. We should all pay attention to species that are global canaries. Paul Ehrlich has said that habitat destruction and species extinction are like rivets being randomly removed from an airplane. As more rivets are popped, we can't predict exactly what will happen but we can be certain the plane will eventually fall out of the skies.

And it's not just a matter of preserving representatives of each species. To return to my beloved fruit flies, in the 1960s population geneticists developed ways to look at genetic heterogeneity through the protein products of specific genes. To their surprise, samples from what they thought were homogeneous stocks revealed wide variation. Now, with more precise molecular techniques to look at DNA itself, it is clear that genetic variation is a characteristic of any wild species. Genetic diversity is a survival trait that confers the flexibility that enables populations of a species to adapt to environmental change. When samples of a single species are taken from different geographic locations, the characteristics of genetic variation in each also typically differ. Reducing an endangered species to a small number of protected survivors is like pickling them in a museum because, from the standpoint of genetic diversity, they are *artifacts*, not genuine wild populations. Similarly, pockets of wilderness do not preserve the diversity that in any way mimics the wild.

Genetic heterogeneity applies to whole ecosystems as well as to individual species. If each species has specific regional genotypes, then the sum total of genetic diversity represented in each watershed or forest is unique. The protection, then, of one or two watersheds on Vancouver Island while all of the others are logged means a severe reduction in the amount of genetic diversity that will never be recovered once lost. It is very important to link protected areas into a network via wilderness corridors (to allow a semblance of migration, for example) and to maximize both genetic exchange and diversity. A comprehensive biological strategy and vision that will both maximize protection of biodiversity *and* allow us to maintain a healthy economy must be a high priority for everyone who is concerned about the degradation of the global environment.

While it is also important to take individual responsibility to save the environment — walking more, driving less, recycling, composting, using cloth bags, installing fluorescent lights, etc. — these actions will *not* save the planet. If enough of us begin to tread more lightly on the earth, it may buy us a little more time to define the long-term solutions.

The goal that has to be achieved is easy to identify: a massive restructuring of social priorities and actions by government and industry. We have to help the poor countries cut their population growth while giving them access to more of the planet's resources. Through increasing foreign aid and by creatively forgiving debt, we must ensure that the developing world does not follow the environmentally destructive path we have been on. The industrialized countries have to cut back drastically on consumption, and on output of greenhouse gases and toxic wastes. We have to protect above all else air, water, soil and biodiversity by aiming at zero discharge. We have to change the very foundations of economics to include the real value of nature. Those are the goals. The hard part is getting there from where we are now.

To begin this reorientation of global priorities, we have to change our mind-sets and look at the world with new eyes. To do that, we should listen to our elders, those in our society who are in their seventies and eighties. Ask them what this province was like when they were five or six years old. They are a living record of the explosive changes that have taken place within the span of a single lifetime. They tell us that Canada has changed beyond recognition and that we can no longer blithely assume that such change is just "the price of progress" or that "there's plenty more where that came from".

We will not be able to act seriously until we understand the fallacies inherent in our "sacred truths". That means we have to examine some of our most deeply held beliefs and values and assess them against the ecological realities of the world that sustains us. Then we can start to bring our values and priorities into harmony with the productive capacity of Earth.

In the high-tech, frenetic world of modern industrial society, we have come to accept the necessity for instant gratification of our need for pleasure. In our hectic lives, we are preoccupied with the moment, with the next pay cheque, the coming dividends on investments. We have to think over a much longer time frame, reflecting on past generations and ahead to the kind of future we are leaving our children. Every extinction of a species or ecosystem, every compromise of ground water, forests, watersheds, prairie or arctic ecosystem, is a closing door for all future generations. If we do love our children as we say, we have to make our actions conform to that claim.

We must re-orient our priorities. First we have to ask what we absolutely need as animals — clean air, clean water, clean soil, biological diversity, social interaction, shelter, clothing. Since we are social animals, we have secondary requirements every bit as important for our well being — opportunity for love and companionship, family, sense of community, meaningful work, opportunity to experience wilderness, social justice, etc. You can see that in this way we can sort our priorities into different layers. In the process, highly mobile, consumptive and wasteful lifestyles are pushed far out of the essential layers of needs.

If the ecocrisis can be seen as an opportunity, it is to develop *spiritually* and to re-orient our personal and social priorities to a greater balance with the natural world that supports us.

I hope the fact that we have assembled for this symposium on biological diversity and the "biodiversity crisis" reflects a desire in all of us to find common ground, to establish a shared understanding of language, assumptions and goals. Then we can work together to find the solutions I believe are waiting to be discovered. Biological diversity keeps the planet going and offers the wonders that evoke our great love of nature. The writing is on the wall for many species, for many indigenous cultures and for many ways of making a living. We should be working to preserve them all.

Global Values of Biodiversity

The Global Importance of Biodiversity

Wade Davis

New York Botanical Garden, New York, New York, U.S.A. 10458, and 1073 Clyde Ave., Vancouver, British Columbia, Canada V7T 1E3.

Abstract
The 20th century will be remembered, not for its wars or technological advances, but rather as the era in which men and women stood by and either passively endorsed or supported the massive destruction of biological and cultural diversity on the planet. Nowhere is the current crisis more intense and the outcome more significant than in the tropical rainforests, home to the greatest concentration of biological wealth on the earth. Focusing on the Amazon, but referring to Sarawak and other regions of the paleotropics, this paper introduces the extraordinary diversity of the tropical rainforests, describes their current status, and discusses the implications of the development schemes that today ride roughshod over so many of these fragile forest ecosystems. Rates of deforestation and extinction are placed in historical context and the economic, ecological and spiritual consequences of the loss of biodiversity are detailed. Ethnobotany is introduced as a science that can assist biologists in identifying the economic potential of the living forest and thus can provide policy makers with a vital means for rationalizing conservation. The past and future contributions of indigenous peoples to medicine, agriculture and technology are heralded. It is suggested, however, that the ultimate gift of these traditional societies will not be natural products, but rather a new vision of life itself, a profoundly different way of living with the forest.

About two years ago I spoke at a symposium in Barbados and was fortunate to share the podium with two extraordinary scientists. The first to speak was Richard Leakey, the renowned anthropologist

In *Our Living Legacy: Proceedings of a Symposium on Biological Diversity*, edited by M.A. Fenger, E.H. Miller, J.A. Johnson and E.J.R. Williams, pp. 23-46. Victoria, B.C.: Royal British Columbia Museum.

who, with his father, drew from the dust and ashes of Africa the story of the birth of our species. The meeting concluded with astronaut Story Musgrave, the first physician to walk in space. It was an odd and moving juxtaposition of the end points of the human experience. Dr Musgrave recognized the irony and it saddened him. He told of what it had been like to know the beauty of the earth as seen from the heavens. There he was, suspended 320 km above the earth, travelling 29,000 km per hour with the golden visor of his helmet illuminated by a single sign, a small and fragile blue planet enveloped in a veil of clouds, floating, as he recalled, "in the velvet void of space". To have experienced that vision, he said, a sight made possible only by the brilliance of human technology, and to remember the blindness with which we as a species abuse our only home, was to know the purest sensation of horror.

Many believe that this image of the Earth, first brought home to us 30 years ago, will have a more profound impact on human thought than did the Copernican revolution of the 16th century, which transformed the philosophical foundations of the western world by revealing that the planet was not the centre of the universe.

From space we see not a limitless frontier nor the stunning products of humanity, but a single interactive sphere of life, a living organism composed of air, water and earth. It is this transcendent vision, more than any amount of scientific data, that teaches us that the Earth is a finite place that can endure our foolish ways for only so long.

In light of this new perspective, this new hope, the past and present deeds of human beings often appear inconceivably cruel and sordid. Just last week, while lecturing in the Midwest of the United States, I visited two places that in a different, more sensitive world, would surely be enshrined as memorials to the victims of the ecological catastrophes that occurred there. The first was the site of the last great nesting flock of Passenger Pigeons, a small stretch of woodland on the banks of the Green River near Mammoth Cave, Ohio. This story of extinction is well known. Yet until I stood in that cold, dark forest I, for one, had never sensed the full weight of the disaster, the scale and horror of it.

At one time Passenger Pigeons accounted for 40% of the entire bird population of North America. In 1870, at a time when their numbers were already greatly diminished, a single flock 1.6 km wide and 500 km long containing an estimated two billion birds passed over Cincinnati on the Ohio River.

Imagine such a sight. Assuming that each bird ate half a pint of seeds a day, a flock that size must have consumed over 17 million bushels of grain each day. Such sightings were not unusual. In 1813

James Audubon was travelling in a wagon from his home on the Ohio River to Louisville, some 90 km away, when a flock of Passenger Pigeons filled the sky so that the "light of noonday sun was obscured as by an eclipse". He reached Louisville at sunset and the birds still came. He estimated that the flock contained over one billion birds and it was but one of several columns of pigeons that blackened the sky that day.

Audubon visited roosting and nesting sites to find trees half a metre in diameter broken off at the ground by the weight of birds. He found dung so deep on the forest floor that he mistook it for snow. He once stood in the midst of a flock when the birds took flight and then landed. He compared the noise and confusion to that of a gale, the sound of their landing to thunder.

How did such an abundant species disappear? It is difficult now to imagine the ravages of man that over the course of half a century destroyed this species. Throughout the 19th century pigeon hunting was a full-time job for thousands of hunters. Pigeon meat was the mainstay for thousands of Americans and a single merchant in New York might sell as many as 18,000 birds a day. The more affluent classes slaughtered birds for recreation. It was not unusual for shooting clubs to go through 50,000 birds in a weekend competition; hundreds of thousands of live birds were catapulted to their death before the diminishing supply forced skeet shooters to turn to clay pigeons.

By 1896, a mere 50 years after the first serious impact of man, only 250,000 birds were left. In April of that year, the birds came together for one last nesting flock in the forest outside of Bowling Green, Ohio.

The telegraph wires hummed with the news and the hunters converged. In a final orgy of slaughter over 200,000 pigeons were killed and 40,000 mutilated; 100,000 chicks were destroyed. A mere 5,000 birds survived. The entire kill was to be shipped east but a derailment on the line left the dead birds rotting in their crates.

On 24 March 1900 the last Passenger Pigeon in the wild was shot by a young boy. On 1 September 1914, as the Battle of the Marne consumed the flower of European youth, the last Passenger Pigeon died in captivity.

When I left the scene of this final and impossible slaughter, I travelled west to Sioux City, Iowa, to speak at Buena Vista College. There I was fortunate to visit a remnant patch of tall grass prairie, a 75-hectare reserve that represents one of the largest remaining vestiges of an ecosystem that once carpeted North America from southern Canada to Texas. Again it was winter, and the cold wind blew through the coneflowers and the dozens of species of grass. The young biology student with me was familiar with every species on that extraordinary mosiac — they were like old friends to him. Yet as we walked through

25

that tired field my thoughts drifted from the plants to the horizon. I tried to imagine buffalo moving through the grass and the physics of waves as millions of animals crossed that prairie.

As late as 1871 buffalo outnumbered people in North America. In that year, one could stand on a bluff in the Dakotas and see nothing but buffalo in every direction for 50 km. Herds were so large that it took days for them to pass a single point. Wyatt Earp described one herd of a million animals stretched across a grazing area the size of Rhode Island. Within 9 years of that sighting, buffalo had vanished from the Plains.

The destruction of the buffalo resulted from a campaign of biological terrorism unparalleled in the history of the Americas. U.S. government policy was explicit. As General Philip Sheridan wrote at the time, "The buffalo hunters have done in the past two years more to settle the vexed Indian Question than the regular army has accomplished in the last 30 years. They are destroying the Indians' commissary. Send them powder and lead, and let them kill until they have exterminated the buffalo." Between 1850 and 1880 more than 75 million hides were sold to American dealers. No one knows how many more animals were slaughtered and left on the prairie. A decade after native resistance had collapsed, Sheridan advised Congress to mint a commemorative medal with a dead buffalo on one side, a dead Indian on the other.

I thought of this history as I stood in that tall grass prairie near Sioux City. What disturbed me most was to realize how effortlessly we have removed ourselves from this ecological tragedy. Today the people of Iowa, good and decent folk, live contentedly in a landscape of cornfields that are claustrophobic in their monotony. For them the time of the tall grass prairie, like the time of buffalo, has drifted from history into the realm of myth, as distant from their immediate lives as the fall of Rome or the battle of Troy. Yet the destruction occurred but a century ago, well within the lifetime of their grandfathers.

This capacity to forget, this fluidity of memory, is a frightening human trait. Several years ago I spent many months in Haiti, a country that as recently as the 1920s was 80% forested. Today less than 5% of the forest cover remains. I remember standing with a vodoun (voodoo) priest on a barren ridge, peering across a wasteland, a desolate valley of scrub and half-hearted trees. He waxed eloquent as if words alone might have squeezed beauty from that wretched sight. He could only think of angels, I of locusts. It was amazing. Though witness to an ecological holocaust that within this century had devastated his entire country, this man had managed to endure without losing his human dignity. Faced with nothing, he adorned his life with his imagination. This was inspiring but also terrifying.

People appear to be able to tolerate and adapt to almost any degree of environmental degradation.

As the people of Iowa today live without wild things, the people of Haiti scratch a living from soil that has never known the comfort of shade. In both instances we gain a frightening glimpse of what our species may do to the earth unless we change our ways.

If Haiti offers a disturbing image of what may happened to the earth, the tropical rainforests, imperiled as they are, represent the last best hope for the planet. Joseph Conrad wrote that the jungle was less a forest than a primeval mob, a remnant of an ancient era when vegetation rioted and consumed the earth. He referred to a time still known to our fathers, a time when the tropical rainforests of the earth stood immense, inviolable, a mantle of green stretching across entire continents. Today, sadly, that era is no more. In many parts of the tropics the clouds are of smoke, the scents are of grease and lube oil, and the sounds are of machinery — the buzz of chainsaws and the cacophony of enormous reptilian earth-movers hissing and moaning with exertion. It is a violent overture, like the opening notes of an opera about war, a war between humanity and the land, a wrenching terminal struggle to make the latter conform to the whims and designs of the former. The residue of war now colours the landscape of Borneo and Sumatra, Zaire and Madagascar, Costa Rica, Gabon, Indonesia, and a hundred other lands once covered in forest. Now the conflict has spread into the heart of the Amazon and it is there that the ultimate fate of the world's tropical rainforests will be decided.

Even for those of us from Canada, a country where landscape sweeps over the imagination and defines the essence of the national soul, it is difficult to grasp the size of the Amazon.

A marvelous tale is told of the travels of Francisco Orellana, the first European to traverse the length of the Amazon. In 1541, having crossed the Andes in search of the mythical land of El Dorado, Gonzalo Pizarro dispatched Orellana on a desperate search for food. Orellana sailed down the Rio Napo, a swift river in eastern Ecuador, and it is said that when he finally reached the confluence of the Rio Ucayali, as the upper Amazon is known in Peru, he went temporarily insane. Coming as he did from the parched landscape of Spain, he could not conceive that a river on God's earth could be so enormous. Little did he know what awaited him 3,200 km downstream where the river becomes a sea and the riverbanks, such as they are, lie nearly 200 km apart.

This story, apocryphal or not, tells of the central dilemma that confronts all travellers on their first visit to the Amazon. It is the issue of scale and the impossibility of imagining rainforests of such magnitude.

In the Amazon there are 5,000,000 km^2 of forested lands still wet with the innocence of birth, a vast expanse of biological wealth the size of the continental U.S., and somewhat larger than the face of a full moon.

The river itself is over 6,700 km long, just longer than the Nile and far more extensive, comprising as it does, over 80,000 km of navigable water spread across five Latin American nations.

Within the Amazon drainage are 20 rivers larger than the Rhine, and 11 of these flow more than 1,600 km without being disrupted by a single rapid. The river delta is enormous. If the mouth of the Amazon could be superimposed onto a map of Europe, the Eiffel Tower would sit on the south bank and the north bank would support the Tower of London.

Within the hundreds of islands that make up the delta, one named Marajo is by itself larger than Switzerland. Sedimentary deposits at the mouth are 3,700 metres deep and freshwater may be drunk from the sea 240 km beyond the shore. Tidal influences reach as far up the river as Obidos, a small city located just below the mouth of the Rio Trombetas, 400 km from the apex of the Amazon delta and a full 720 km from the sea.

The Amazon did not always flow into the Atlantic. Two hundred and fifty million years ago the South American continent was still attached to Africa, and the predecessor of the Amazon flowed east to west, draining an arc of massive highlands the remnants of which are now known as the Brazilian and Guiana shields. The river reached the Pacific Ocean somewhere along the shore of contemporary Ecuador. A hundred million years later the two southern continents split apart. Four million years ago the birth of the Andean Cordillera effectively dammed the river, creating a vast inland sea that covered much of what is now the Amazon basin. In time, these waters worked their way through the older formations to the east, and formed the modern channel of the Amazon.

The Rio Negro and the Rio Solimões, the two main branches that form the Amazon proper at Manaus, Brazil, are a legacy of these staggering geological events. The Rio Negro drains the northern half of the Amazon basin, rising in the ancient soils of the Guiana Shield, and its dark colour is due to a high concentration of humic matter, very little silt load and a tannin content equal to that of a well brewed cup of tea. The Solimões, and its affluents, by contrast, are born in ten thousand precipitous mountain valleys of the high Andes. Rich in sediments, these are the fabled milk rivers of Indian mythology, the source of rich nutrients that each year replenish the flood plain of the lower Amazon.

Rainfall in the Andes and water cut loose from the ice of thousands of Cordilleran glaciers drive the entire system.

One of the most remarkable features of the Amazon is that, while the river falls 4,300 metres in its first 1000 km, the elevation drop over the last 4,200 km is but 74 m, less than 2 cm per km. If the Washington monument stood at the mouth in Belem, its tip would be higher than any building in Iquitos, Peru, a sizeable city 3,200 km upriver. The Amazon doesn't flow to the sea, it is pushed by the annual runoff from the Andean Cordillera.

The Amazon contains 20% of the world's fresh water. In 24 hours the Amazon pumps as much fresh water into the Atlantic as the Thames does in an entire year. Two hundred million litres of water flow into the sea each second. That is enough water to provide 300 people with a bath each week for approximately 250 million years. If the U.S. Army Corps of Engineers could figure out a way to drain Lake Ontario and divert the channel of the Amazon, and no doubt they have such a plan, the lake could be refilled in three hours.

The seasonally replenished flood plain, which constitutes a mere 3% of the land, represents only one-half of the Amazonian reality. Beyond the borders of the flood plain, beyond the reach of the milk rivers, lies another world, *tierra firme*, the upland forests that cover so much of the basin. It is only here that one begins to sense the overwhelming grandeur, the power of the forest. It is a subtle thing — there are no cascades of orchids, no herds of ungulates as one might encounter on the Serengeti. Just a thousand shades of green, an infinitude of shape, form and texture that so clearly mock the terminology of temperate botany. If you close your eyes you can sense the constant hum of biological activity — evolution, if you will, working in overdrive.

The biological diversity of these tropical rainforests is staggering. Two square kilometres of Amazonian forest may be home to as many as 23,000 distinct forms of life. Brazil alone harbours more primate species and, in sheer numbers, more terrestrial vertebrate animals than any other nation. There are more species of fish in the Rio Negro than in all of Europe, more species of birds in Colombia than in any other country.

If all of New England has around 1,200 plant species, then the Amazon has more than 80,000. While half a hectare of woodland in British Columbia might have six species of trees, the same area in the Amazon could contain over 300 species. One researcher working in the tropical rain-forests of Borneo reported that in 10 one-hectare plots were over 700 species of trees, as many as are found in all of North America.

The insect fauna is especially rich. One researcher surveyed the canopy of 19 individuals of a certain tree species and found over 1,200 species of beetles. Based in part on this remarkable discovery, entomologists now believe that tropical rainforests harbour over 30 million species of insects. There are 10,000 species of ants alone, and at any one moment more than 1,000,000,000,000,000,000 ants are alive! In the

Amazon, ants make up over 30% of the total animal biosphere. Harvard entomologist E.O. Wilson found, in one tree stump in lowland Peru, more kinds of ants than occur in all the British Isles.

These figures, impressive as they are, give little indication of the biological drama constantly being played out in the tropical rainforests. Break open the trunk of a common Cecropia tree and find a colony of *Azteca* fire ants living inside the hollow internodes. The plant feeds the ants with tiny capsules of carbohydrate; the fire ants in exchange protect the tree from *Atta* ants, voracious leaf cutters capable of defoliating a tree in a matter of hours. Watch for these leaf cutters in the forest. Long trails of workers scurry to unknown destinations, each toting a cut section of a leaf like a sail on their back.

A closer look reveals the magnitude of their achievement. If you can imagine these creatures on a human scale, such that their one-centimetre length becomes two metres, you would note that each foraging ant runs along the trail for about 16 km at a speed of 25 km per hour. At the end of the trail, each ant picks up a leafy burden weighing some 350 kg and runs back at a speed of 24 km an hour. This marathon is then repeated, without pause for rest, dozens of times during the course of a day and night.

The ants do not eat the leaf fragments. They turn them into mulch which they use to grow the mushrooms that form the basis of their diet. One struggles to imagine a colony of three to four million ants, dwelling underground in thousands of chambers, cultivating in the darkness a mushroom found nowhere else in nature.

Consider the extraordinary pollination mechanism of the giant lily, *Victoria amazonica*. This famous plant, with its enormous leaves capable of supporting the weight of a small child, grows inside channels and standing bodies of water throughout much of the Amazonian flood plain. The simultaneous opening at dusk of its massive white blossoms is one of the most inspiring scenes in the Amazon. The exterior of the flower has four large sepals covered by sharp spines. Within are numerous petals arranged in a spiral with the smaller ones on the inside. Inside the petals is a whorl of thicker structures called staminodes. Next are the 300 stamens that carry the pollen. Inside the stamens is yet another whorl of floral parts that together with the other structures form what amounts to a tunnel leading to a large cavity at the base of which is the carpel, the female part of the flower. Lining the carpel is a ring of appendages that are full of starch and sugar.

When the flower buds are ready to open, they rise above the surface of the water and, precisely at sunset, triggered by the falling light, the flower opens with a speed that can be seen with the naked eye. The brilliant white petals stand erect and the flower's fragrance, which has been growing in strength since early afternoon, reaches its peak of intensity. At the same time, the metabolic processes that generate the odour raise the temperature of the central cavity of the blossom by precisely 11 °C above whatever the outside temperature happens to be. The combination of colour, scent and heat attracts a swarm of beetles that converge on the centre of the flower. As night falls and temperatures cool, the flower begins to close, trapping the beetles with a single night's supply of food in the starchy appendages of the carpel. By two o'clock in the morning the flower temperature has dropped and the petals begin to turn pink. By dawn the flower is completely closed and remains so for most of the day. In early afternoon the outer sepals and petals alone open. By now a deep shade of reddish purple, they warn other beetles to stay away. Last night's beetles, meanwhile, remain trapped in the inner cavity of the blossom. Then, just before dusk, the male anthers of the flower release pollen and the beetles, sticky with the juice of the flower and once again hungry, are finally allowed to go. In their haste to find yet another opening bloom with its generous offering of food, the beetles dash over the anthers, becoming covered with pollen that they carry to the stigma of another flower, thus pollinating the ovaries.

This sophisticated pollination mechanism is, in its complexity, not unusual for the plants of the Amazon. Indeed, a botanist would be hard pressed to invent a strategy of pollination or seed dispersal that some species had not already come up with. There are fruit-eating fish, seeds that float in the wind, succulent fruits designed for birds and primates, tough woody fruits for the massive rodents, fruits that explode, fruits that are carried by bats, seeds that swim and even seeds small enough to be dispersed by ants.

Perhaps the best symbol of the Amazon rainforest is the Three-toed Sloth, a gentle herbivore that dwells exclusively in the canopy of the forest. It moves literally at a snail's pace and this, together with its cryptic colouration, protects it from its only major predator, the Harpy Eagle. Viewed up close, the sloth appears as a hallucination, an ecosystem unto itself that softly vibrates with hundreds of ectoparasites. The sloth's mottled appearance is caused in part by a blue-green alga that lives symbiotically within its hollow hairs. A dozen varieties of arthropods burrow beneath its fur; a single sloth weighing a mere 5 kg may be home to over a thousand beetles.

The life cycles of these insects are completely tied to the daily round of the sloth. With its excruciatingly slow metabolism, the sloth defecates only once a week. (I had professors at Harvard who probably did the same, but it was not by choice and it was not an evolutionary advantage.) When it comes time for the sloth to defecate, the animal climbs down from the canopy, excavates a small depression at the foot of the tree, voids its faeces and then climbs back up. Mites, beetles and even a species of moth leap off the sloth, deposit an egg in the dung, and climb back on their host for a ride back up the tree. The eggs germinate and, in one way or another, the young insects find another sloth to call home.

You have to ask yourself a question. Why would this animal go down to the base of the tree, exposing itself to all forms of terrestrial predation, when it could just as easily defecate from the treetops? The answer provides an important clue to the immense complexity and subtlety of this ecosystem. Biologists have suggested that, in depositing the faeces at the base, the sloth enhances the nutrient regime of the host tree. That such a small amount of nitrogenous material might actually make a difference suggests that this cornucopia of life is far more fragile than it appears. In fact many ecologists have called the tropical forest a counterfeit paradise. The problem is soil. In many areas, there is essentially none.

Forests have two major strategies for preserving the nutrient load of the ecosystem. In the temperate zone, with the periodicity of the seasons and the resultant accumulation of rich organic debris, the biological wealth is in the soil itself. The tropical ecosystem is completely different. With constant high humidity, and annual temperatures hovering around 28°C, bacteria and micro-organisms break down plant matter virtually as soon as the leaves hit the forest floor.

Ninety per cent of the root tips in a tropical forest may be found in the top 10 cm of earth. Vital nutrients are immediately recycled into the vegetation. The biological wealth of this ecosystem is the living forest itself, an exceedingly complex mosaic of thousands of interacting and interdependent living organisms. It is a castle of immense biological sophistication built quite literally on a foundation of sand.

Removing this canopy sets in motion a chain reaction of biological destruction of cataclysmic consequences. Temperatures increase dramatically, relative humidity falls, rates of evapotranspiration drop precipitously, and the mycorrhizal mats that interlace the roots of forest trees enhancing their ability to absorb nutrients dry up and die. With the cushion of vegetation gone, torrential rains create erosion that leads to further loss of nutrients and chemical changes in the soil itself. In certain deforested areas of the Amazon, the precipitation of iron oxides in leached exposed soils has resulted in the deposition of miles upon miles of lateritic clays, a rock hard pavement of red earth from which not a weed will grow.

What percentage of the Amazon has suffered deforestation is a matter of debate. Estimates ranges from as low as 2% to a shocking 25%.

Experts agree, however, that the primary concern today is not the absolute amount of land that has been cleared but rather the rate at which deforestation is proceeding. Every minute 20 ha are cut. Each day 31,000 ha disappear. Each year 76,000 km^2 — an area of virgin rainforest three times the size of Belgium — is destroyed.

The effects of this deforestation will be felt continentally and globally. Since fully half of the precipitation in the Amazon is generated from evapotranspiration, we can expect rainfall in the basin to be reduced by as much as 50%. Worldwide clearing operations that burn the remnants of tropical forests put 52 trillion kg of carbon dioxide into the air each year, an amount roughly equal to 40% of all industrial emissions. The result is the Greenhouse Effect, a warming of the earth's atmosphere which, even by the most conservative estimates, promises unprecedented climate change and the increase of sea levels by as much as two metres, a rise that will inundate entire countries.

Perhaps most tragically, the destruction of the earth's tropical rainforests is resulting in massive loss of biological diversity. Although extinction is a global problem, tropical rainforests are particularly susceptible because species tend to occur in low densities with restricted ranges.

The impact in certain regions of the earth has already been devastating. In Madagascar, for example, 90% of the species are endemic yet only 7% of the forest remains undisturbed. The Atlantic forests of Brazil, another centre of high endemism, have been reduced to less than 2% of their original extent. Human activity is not only affecting individual species but also changing the actual conditions of life itself. Acid rain, global warming, the depletion of the ozone, the accumulation of synthetic compounds into the environment — these are conditions that represent changes in the actual chemistry of the biosphere.

The elimination of life, of course, is nothing new in the history of the earth. Mass extinctions marked the end of the Permian, Triassic and Cretaceous periods and other crises occurred in the Late Devonian and at the end of the Eocene. Shortly after the arrival of humans in South America 15,000 years ago, 45 of 120 genera of mammals became extinct. In general, however, over the last 600 million years speciation has outpaced extinction and the diversity of life has steadily increased.

What has changed in a disturbing way in the last 50 years is the rate of disappearance. Species extinction when compensated by speciation is a normal phenomenon. Massive abrupt species extinctions and the consequent biological impoverishment are not. The current wave of extinction is unprecedented in the last 60 million years, both in abruptness and in the total number of species that will be lost.

During the extinction of the dinosaurs, for example, an extinction occurred roughly every 1,000 years. Between 1600 and 1900 perhaps 75 species were driven to extinction because of human activities. Since 1960, within our lifetimes, extinction has claimed, at a conservative estimate, upwards of 1,000 species per year. E.O. Wilson believes that within the last 25 years of this century, one million species may disappear. That figure represents a loss of a species every 13 minutes, 110 each day, 40,000 over the course of a single year.

Does this loss of biodiversity matter? Biologists may scoff at this question, but providing an answer that makes sense to both the public and policy makers is, in fact, one of our most critical challenges. It is and always will be difficult to argue that the value of a single species is worth more than a particular development. Stanford biologist Paul Ehrlich explains the ecological significance of species diversity with a metaphor. Imagine, he writes, that as you are entering an airplane you notice a workman popping out rivets. The workman explains that the rivets can be sold for two dollars each and thus subsidize cheaper air fares. When questioned about the wisdom of the procedure, he responds that it has to be safe, as no wings have fallen off despite many rounds of de-riveting. This, in effect, is what we are doing to the biosphere through the erosion of biodiversity.

In a moment I will present an economic argument for diversity based on the insights gained through ethnobotany. But there is an important point to be made first. The value of a species, as Tom Lovejoy of the Smithsonian Institution has pointed out, is not simply that it may one day yield a pharmaceutical drug — though this may occur. The real issue is that there is not a single species that we can claim to understand.

Our knowledge is embarrassingly rudimentary. A species that has no use today may, as our own knowledge increases, yield astonishing insights. Who, in the 19th century, for example, would have realized the value of the humble *Penicillium* mould?

Consider the potential of every form of life. A single bacterium, E.O. Wilson reminds us, possesses about ten million bits of genetic information, a fungus one billion, an insect from one to ten billion depending on the species.

If the information in just one ant were translated into a code of English letters and printed in letters of standard size, the string of letters would stretch 1,600 km. One handful of earth contains information that would nearly fill all 15 editions of the Encyclopedia Britannica.

This is the true resonance of nature. Each incident of extinction represents far more than the disappearance of a form of life: it is the wanton loss of an evolutionary possibility and its irrevocable severance from the stream of divine desire.

Most of the species doomed for extinction have yet to be described by science. Estimates of the total number of species range from 3 to 30 million. Incredibly, despite 200 years of scientific research, we do not know the true number of species on the earth, even to the nearest order of magnitude. Though approximately 1.5 million species have been taxonomically classified, most forms of life do not even have a scientific name. Virtually every time a botanist or an entomologist goes to the tropics he or she brings back species new to science.

What do we gain from the massive destruction of the tropical rainforests? In 1974 Volkswagen acquired 10,000 km² of Brazilian rainforest, applied agent orange from the air and then torched the land, creating the largest man-made fire in recorded history. The most optimistic estimates of agronomists working at the sight suggested that cattle might be able to graze on the land for 12-14 years. Each animal cut loose upon the land requires a hectare of converted forest to yield a few dozen kg of meat. A single Brazil nut tree left standing in the forest produces 1,000 kg of protein-rich seeds each year. Incredibly, in no place today in the Amazon where forest land was converted to pasture before 1980 are cattle supported.

Growing cattle is one way of using the land, but when one considers the true potential of the rainforest, it appears to be rather like using a Van Gogh to kindle a campfire. It gets the job done but at such a high cost.

The result of this wanton destruction is extinction, not only of plants and animals but of human societies that have developed, over the course of thousands of years, an intimate knowledge of the forest and the natural products it contains. Largely responsible for the deforestation are development programs initiated by governmental and international agencies struggling to deal with problems of massive debt and chronic population growth, poverty and unemployment. Worldwide, 35,000 people starve to death each day. Forty per cent of human beings live in absolute poverty. The global population has doubled since 1950 and will double again by the middle of the next century. Each year there are 90 million new human mouths to feed, many of them in the developing countries that contain the remaining tracts of tropical rainforest.

Clearly the need for conservation, particularly of the tropical rainforests, must be balanced with the social and economic requirements of people. No protected area will endure if it conflicts with a people's fundamental struggle to survive. Inevitably, economic arguments will continue to dominate the public policy debate.

Conservationists must seek an environmental policy consistent with these economic realities by showing that the long-term income-generating potential of the standing forest equals or exceeds the short-term gain resulting from its destruction.

Economic botany and ethnobotany can contribute to this strategy in several ways. First, ethnoecological studies of Indian adaptation may provide models for profitable and environmentally sound multiple land-use management programs. Among the Mebêngôkre-Kayapô of central Brazil, for example, ethnobotanists have documented an indigenous system of integrated land management of remarkable sophistication. The biological use of insects, the manipulation of semi-domesticated plants, and the deliberate encouragement or transplanting of wild trees and medicinal plants along trailsides and in fields, are all elements of a complex sustainable agro-forestry system that stands in marked contrast to the crude and destructive patterns of modern land use in the tropics.

Second, ethnobotanists can invoke the considerable economic potential of as yet undiscovered or undeveloped natural products. This is by no means merely an academic pursuit. In the field of medicine alone about 25 - 50% of modern drugs are derived from plants, and most of these were first used as medicines or poisons in a folk context.

Plants are useful as poisons or medicines because they have evolved complex secondary compounds and alkaloids as chemical defenses against insect predation. These defensive chemicals, that in certain plants may constitute 10% of dry weight, interact harmfully with the biochemical apparatus of the insects. The same properties, however, can be exploited by the modern chemist for therapeutic purposes.

Once extracted and identified, a chemical compound derived from a medicinal plant can be exploited in various ways. Plant extracts may be used directly as pharmaceuticals. Alternatively, they may serve as templates for the chemical synthesis of related medicinal compounds. Finally, in many cases the natural product may be used as a tool in the process of drug development and testing. Thus, in seeking new drugs, researchers attempt first to identify any compound that is pharmacologically active. In many instances the difference between a medicine, a poison and a narcotic is dosage.

According to the World Health Organization, approximately 88% of the people in developing countries rely chiefly on traditional medicine for their primary health care needs. This high degree of dependence, together with many thousands of years of experimentation, have yielded numerous plants of true pharmacological worth. Worldwide there are about 121 clinically useful prescription drugs derived from 95 species of higher plants, 47 of which are native to the tropical rainforest. Roughly 40% of these drugs are used in North America. Plant ingredients valued in excess of U.S. $8 billion are included in a quarter of all prescriptions dispensed by community pharmacies in the U.S. and Canada. Of the 3,500 new chemical structures discovered in 1985, 2,619 were isolated from higher plants.

To date, only 5,000 out of about 250,000 - 300,000 species of higher plants have been studied in detail for their possible medical applications.

Knowledge of tropical rainforest plants is especially inadequate. Though 70% of all plants know to have antitumour properties have been found in the tropics, over 90% of the Neotropical flora has yet to be subjected to even a superficial chemical screen. At least 85% of the world flora of higher plants have yet to be examined for anticancer activity. In the Amazon, of 80,000 species of higher plants, a mere 470 have been investigated in detail.

Any practical strategy for expanding knowledge of this living pharmaceutical factory must include ethnobotanical work among the indigenous societies who best understand the forest. To attempt to assay the entire flora without the consultation of indigenous people would be logistically impossible, intellectually short-sighted and historically uninformed. Seventy-four per cent of the 121 biologically active plant-derived compounds currently in use worldwide were discovered in a folk context, the gifts to the modern world of the shaman and the witch, the healer and the herbalist, the magician and the priest.

Identifying both psychologically and cosmologically with the rainforest, and depending on that environment for virtually all their material needs, it is not surprising that indigenous peoples are exceptionally skilled naturalists. The extent of their knowledge, however, and the sophistication of their interpretations of biological relationships, can be astounding. The Waorani of the Ecuadorian Amazon, for example, not only recognize such conceptually complex phenomena as pollination and fruit dispersal, they understand and accurately predict animal behaviour. They anticipate flowering and fruiting cycles of all edible forest plants, know the preferred foods of most forest animals, and may even explain where any particular animal prefers to pass the night. Waorani hunters can detect the scent of animal urine at forty paces in the forest and accurately identify the species of animal from which it came.

This perspicacious knowledge of the forest, combined with an active process of experimentation, has led to the discovery of remarkable chemical properties of plants. Indigenous peoples of the Amazon, for example, employ dozens of different plant species as ichthyotoxins, or biodegradable fish poisons, some of which contain up to 20% rotenone in their roots. Placed into slow moving bodies of water, these poisons interfere with respiration in the gills of the fish. The fish float to the surface and are readily harvested. The peoples of Southeast Asia employ as a dart poison the latex of a tall forest tree, *Antiaris toxicaria*. The active ingredient is an extremely toxic cardiac glycoside that, once released into the bloodstream, precipitates lethal arrhythmias of the heart.

Other plants have yielded medicines or narcotics. The infamous Trees of the Evil Eagle, a species of the genus *Brugmansia* in the Solanaceae, contains tropane alkaloids that are useful in low dosage in the treatment of asthma. In higher dose, however, these drugs induce a frightening state of psychotic delirium marked by burning thirst and visions of hell fire, and ultimately stupor and death. All species of *Brugmansia* are psychotropic and may be smoked, ingested as decoctions or infusions, absorbed as enemas, or administered topically as salves and poultices. Sorcerers among the Yaqui of northern Mexico anoint their genitals, legs and feet with a salve based on crushed datura leaves and thus experience the sensation of flight. Many believe that the Yaqui acquired this practice from the Spaniards for, throughout Medieval Europe, witches commonly rubbed their bodies with hallucinogenic ointments made from belladonna, mandrake, henbane and datura. In fact, much of the behaviour associated with witches is as readily attributable to these drugs as to any spiritual communion with demons. A particularly efficient means of self-administering the drug for women is through the moist tissue of the vagina; the witch's broomstick or staff was considered a most effective applicator.

The common image of a haggard woman on a broomstick comes from the medieval belief that the witches rode their staffs each midnight to the sabbat, the orgiastic assembly of demons and sorcerers. In fact, it now appears that their journey was not through space, but across the hallucinatory landscape of their own minds.

Over 90 different species of plants in the Amazon yield curare, the flying death employed throughout the basin as an arrow and dart poison. The active principle is a muscle relaxant, d-tubocurare, which became an invaluable drug in modern surgery after it was isolated in the early 1940s. What is fascinating about these curare preparations, from an epistemological point of view, is that their elaboration involves a number of procedures that are either exceedingly complex or that yield a product whose use would not have been inherently obvious to the inventor. Curare is principally derived from a number of species in several genera of lianas (e.g., *Chondrodendron*, *Abuta*, *Curarea*). Generally the bark is rasped and placed in a funnel-shaped leaf compress suspended between two hunting spears. Cold water is then percolated through the the drippings collected in a ceramic pot. This dark-coloured liquid is slowly heated over a fire and brought to a frothy boil numerous times until the fluid thickens. It is then cooled and later reheated until a thick layer of viscous scum gradually forms on the surface. This scum is removed, the dart tips are spun in the viscid fluid and the darts are then carefully dried by the fire. The procedure itself is mundane. But the realization that this orally inactive substance, derived from a handful of the hundreds of forest lianas, could kill when administered intramuscularly is profound.

In the case of the ritual hallucinogen *ayahuasca*, the most important intoxicant of the Northwest Amazon, it is the sophistication of the actual preparation that is impressive. *Ayahuasca* is most frequently prepared from two species of large forest lianas of the genus *Banisteriopsis* (*B. caapi*, *B. inebrians*), though as many as seven species may be used. The drug may be made in various ways, but generally the fresh bark is scraped from the stem then boiled for several hours until a thick, bitter liquid is produced. The active compounds are the ß-carbolines, harmine and harmaline, whose subjective effects are suggested by the fact that when they were isolated first, they were known as telepathine.

Significantly, the psychoactive effects of *ayahuasca* are enhanced dramatically by adding a number of subsidiary plants. This important feature of many folk preparations is attributed in part to the fact that different chemical compounds in relatively small concentrations may effectively potentiate each other. In the case of *ayahuasca*, some 21 admixtures in morphologically dissimilar genera in distinct families have been identified to date. Three of these, the shrubs *Psychotria viridis* and *P. carthaginensis* and the liana *Diplopterys cabrerana*, are of particular interest. All three of these plants contain tryptamines, powerful psychoactive compounds that are orally inactive because of the activity of an enzyme, monoamine oxidase (MAO), found in the human gut. Tryptamines can be taken orally only if combined with an MAO inhibitor. The ß-carbolines found in *Banisteriopsis caapi* are inhibitors of precisely the sort required to potentiate tryptamines. Thus when these three admixture plants are combined with the *ayahuasca* liana, the result is a powerful synergistic effect, a biochemical version of the whole being greater than the sum of the parts.

Just how, in a flora of thousands of species, the indigenous peoples managed to identify and combine in this sophisticated manner morphologically dissimilar plants with these unique and complementary chemical properties is one of the most important questions in ethnobotany.

The traditional explanation has been "trial and error", a notion that upon closer examination appears to be little more than a euphemistic veil obscuring our inability to provide an adequate answer. The indigenous peoples, by contrast, have rich cosmological explanations that are, from an emic perspective, inherently logical.

How the shaman classifies different varieties of *ayahuasca* is yet another botanical enigma. The Tukano of the Colombia Vaupés, for example, consistently identify and name six kinds of yagé, which they distinguish by specific characteristics associated with the nature and intensity of the visions the plants induce. The Ingano of the upper Putumayo recognize seven varieties that they identify by the tone and key of the sacred incantations that the plants sign on the night of a full moon.

To the Western taxonomist, all of these indigenous varieties represent the same species and are morphologically indistinguishable.

Like most ritual hallucinogens, *ayahuasca* is a sacred medicine and a vital component of the shaman's repertoire, enabling him to diagnose illness, ward off evil, prophesy the future. But for the peoples of the northwest Amazon, it is far more. *Ayahuasca* is the visionary medium through which humanity orients itself in the cosmos. Under the cloak of the visions, the user of *ayahuasca* encounters the gods, the primordial beings and the first humans, even as he embraces for good and for bad the wild creatures of the forest and the powers of the night. Lifted out of the body, the shaman enters a distant realm, soaring as a bird beyond the Milky Way, or descending the sacred rivers in supernatural canoes run by demons to reach distant lands where they may reconquer lost or stolen souls and work their deeds of mystical rescue.

Several species of trees in the nutmeg family are the source of yet another important hallucinogenic snuff used in the western Amazon and adjacent parts of the Orinoco basin of South America. This sacred drug is known as epená, the semen of the sun. Taken as a snuff, this tryptamine-based hallucinogen induces not merely the suspension of reality, but the complete dissolution of the material world as we know it. The subjective effects of inhaling epená may be likened to being shot out of a rifle barrel lined in baroque paintings only to land on a sea of electricity.

The source of this most remarkable hallucinogen is the blood-red resin of the trees *Virola calophylla*, *V. calophylloidea*, *V. elongata* and *V. theiodora*, the last being by far the most commonly used. Preparations once again vary. The nomadic Makú ingest the resin directly, while other tribes, notably the Witoto and Bora, swallow pellets made from a paste of the resin. The drug is taken as a snuff by the Barasana, Makuna, Tukano, Kabuyaré, Kuripako and Puinave of eastern Colombia and various groups of the Yanomama in the upper Orinoco. To prepare the snuff, the bark is removed from the trees in early morning and the soft inner layers are scraped. The shavings are kneaded in cold water that is subsequently filtered and boiled down to a thick syrup which, when dried, is pulverized and mixed with the ashes of the bark of wild cacao.

As in the case of *ayahuasca*, several admixture plants, including *Justicia pectoralis* var. *stenophylla*, may be added to potentiate the preparation.

Perhaps the most significant gift of the rainforest to the modern pharmacopoeia is coca, a sacred plant of the Inca civilization that has played a critical role in the biological, cultural, and spiritual adaptation of the South American Indians for over 4,000 years. Originally native to the moist montane forests of the eastern flank of the Andes in Bolivia, cultivated coca had already become, in pre-Columbian times, differentiated by artificial selection into the four varieties employed by indigenous peoples today.

Erythroxylum coca var. *coca*, found today growing at elevations between 500 and 2,000 metres on the eastern slopes of the Andes in Bolivia and Peru, is the major commercial source of coca leaves and of most of the world's cocaine supply. Amazonian coca (*E. coca* var. *ipadu*), a recently differentiated close relative, is commonly grown today on a small scale from cuttings by numerous indigenous groups of the northwest Amazon. Curiously, though the two varieties are interfertile and contain many similar chemical constituents, Amazonian coca has a consistently lower cocaine content, generally half that found in other cultivated varieties. Lowland peoples, particularly the Desana groups, and the Witoto and Bora, have adjusted to this discrepancy by developing a unique means of maximizing absorption of the nutritional and stimulant properties of the leaf. Wherever coca is chewed, native peoples have discovered means of changing the acidity of the mouth such that the cocaine content of the leaves may be absorbed. In most regions this entails adding some alkaline material (burnt seashells, limestone, various forms of ash mixed with urine and dew) to the quid of entire sun-dried leaves that are actually sucked rather than masticated. In the Amazon, the leaves, by contrast, are dried in ceramic vessels over a fire, then pounded in a large mortar and pestle. Dried *Cecropria* leaves are burnt to ash which is added to the coca powder. The final product is carefully filtered to the consistency of talc. Placed into the mouth, the powder is formed into a moist quid which is slowly consumed orally.

The related species *Erythroxylum novogranatense* differs from *E. coca* in a number of morphological features, but more significantly it has evolved distinctive chemical and ecological traits and has become genetically isolated from its parental relative. *E. novogranatense* consists of two well defined varieties adapted through human selection to grow in arid conditions where *E. coca* could not survive.

E. novogranatense var. *truxillense* is cultivated in the dry river valleys of the north coast of Peru at elevations between 200 and 1,800 metres. It is grown largely for consumption and as a flavouring for the soft drink Coca-Cola. *E. novogranatense* var. *novogranatense* is the coca of Colombia. Well adapted to xeric conditions, it is often grown in the arid inter-Andean valley, along the Caribbean coast and in the Sierra Nevada de Santa Marta.

In sharp contrast to the disruptive and convoluted phenomenon of cocaine use in modern societies, the use of coca leaves in the Andes and northwest Amazon is a force of stability and equilibrium, the purest expression of what it means to be human. In many parts of the Andes, distance is measured not in kilometres, but in coca chews.

When people meet on a trail they exchange leaves rather than handshakes. Those who divine the future by consulting the coca leaves are empowered to do so by having survived a lightning strike. Among the Kogi who dwell in the Sierra Nevada de Santa Marta in northern Colombia, a young man begins to chew the sacred leaves at the time of his marriage. The high priests, or *sacerdotes*, emerge from their isolated huts to officiate at the ceremony. The priest impregnates the bride, and then perforates the gourd, which will thenceforth contain the burnt seashells that will potentiate the coca leaves. Thus, as a man weds himself to a life of procreation, so he commits himself to a life of chewing the sacred leaves. For the Kogi, the societal ideal is to abstain from sleep, food and sex while devoting oneself to chewing the leaves and chanting to the ancestors. For the Kogi, as for so many tribes in Andean South America, the chewing of coca is the most profound expression of culture.

In addition to its crucial symbolic and religious function, coca plays a direct and critical role in the biological adaptation of Andean peoples. True, the leaves contain a small amount of cocaine, perhaps 1.0-1.5% dry weight, which acts as a useful and mild stimulant in a particularly harsh environment. But coca leaves also have valuable vitamins, more calcium than any other known plant (an invaluable asset for lactating women in a culture that traditionally lacked a source of milk), and enzymes that may enhance the body's ability to digest carbohydrate at high elevation (a useful feature in a mountainous society that lives on potatoes). Coca is the remedy of choice for altitude sickness and is such an effective medicine for stomach aches that Coca-Cola was long listed in western pharmacopoeia. Clearly, to judge the coca plant by the deleterious impact of cocaine is as inappropriate as to judge coffee by the physiological effects of ingesting or inhaling pure caffeine.

The bounty of the rainforest is by no means limited to medicinal drugs. Most of the common foods eaten in North America and Europe were first domesticated by indigenous peoples, and in many instances originated in the tropical rainforests. Indeed, if North Americans had to subsist only on cultivated plants native to the U.S. and Canada, our diet would consist of pecans, cranberries, Jerusalem artichokes and maple sugar. Without the agricultural contributions of Central and South American Indians, Switzerland would have no chocolate, Hawaii no pineapple, Ireland no potato, Italy no tomato, India no eggplant, North Africa no chili, England no tapioca and none of us would have vanilla, papaya or corn. The rainforests of Southeast Asia have given to the world sugar cane and, most recently, the kiwi fruit, which contains ten times as much vitamin C as oranges.

Of an estimated 75,000 edible plants found in nature only 2,500 have ever been eaten with regularity, a mere 150 enter world commerce, and a scant 20 (mostly domesticated cereals) stand between human society and starvation.

Yet in the Amazon there are wild trees that yield 300 kg of oil-rich seeds a year, a palm whose fruits have more vitamin C than oranges and more vitamin A than spinach, and another palm, known as a living oil factory, whose seeds contain 27% pure protein. There are trees in the Amazon that produce resins which, if placed raw into the fuel tank, will run an engine. There are shrubs whose fruits contain natural compounds 300 times as sweet as sucrose, leaves coated with industrial waxes, seeds covered by brilliant pigments and dyes, and lianas impregnated with biodegradable insecticides.

What economic value can we place on the potential of these plants? In certain cases the returns can be substantial and immediate. Brazil, for example, currently spends $95 million each year to import olive oil from Portugal. In the Amazon, however, a palm species, *Jessenia bataua*, produces an abundance of seed oil that is, in terms of taste and chemistry, indistinguishable from olive oil. The development of this plant through artificial selection could permanently free the nation of a chronic drain on its trade balance. Another equally dramatic example of the economic potential of wild plants is a simple discovery made by Hugh Iltis, a plant explorer from the University of Wisconsin. In 1962, on one of his botanical expeditions, Iltis collected the seeds of an apparently useless wild relative of the tomato. That plant, crossed back into cultivated tomatoes, proved to have certain genetic traits that vastly improved the domesticated varieties. Iltis later calculated that his efforts cost the government agency that funded his research about $40. The value of his discovery to the tomato industry, in contrast, has been about $8 million a year, or $80 million over the last decade.

These are the potential gifts of the rainforest — plants that heal, fruits and seeds that bring forth the foods we eat, magic plants that transport us to realms beyond our imaginings. Yet critically, in unveiling this indigenous knowledge, we must seek not only new sources of wealth, but also a vision of life itself, a profoundly different way of living with the forest.

A year ago I was in Sarawak, the Malaysian state in northern Borneo, living among the Penan, one of the last nomadic peoples of the tropical rainforest. In terms of species diversity, their homeland, a varied and magical landscape of forest and soaring mountains, dissected by crystalline rivers and impregnated by the world's most extensive network of caves and underground passages, is one of the most significant regions of the world. Today, after centuries of isolation, it suffers from the highest rate of deforestation in the world. In a clumsy attempt to rationalize the decimation of these forests, James Wong, the Sarawak Minister of the Environment, has gone on record as saying that it is a national embarrassment to have the Penan "running around the forest like animals". "They should be taught," he has said, "to be hygienic like us and to eat clean food." How could one who knew the Penan reconcile such a statement with an image of them bathing in their clear streams, or in the forest, manipulating their plants with a dexterity equal to that of a laboratory chemist?

I recall a morning when a group of visitors shared their "clean food" with Asik Nyelik, a nomadic Penan from beyond the headwaters of the Baram River. The night before, Asik had slept poorly in a bed, and that morning at breakfast, looking rather tired, he sat uncomfortably in a chair. He drank from a glass of water as would a deer, dipping his mouth to the surface. Then came breakfast, a depressing offering of cold canned beans, a sorry-looking fried egg, and a slice of tinned sausage. Asik politely looked around the table, then to his plate, then once again at the people eating this food. He rotated his plate, hunting perhaps for an angle from which the food might appear palatable. Backing away from the table with a look of sincere pity, he slipped out of the building and into the forest. An hour later smoke rose from the edge of the forest and Asik was found hunched over a fire, slowly roasting a mouse deer that he had killed with a blade.

Several nights later, there was a full moon. It reminded Asik of a story he had heard about some people who had travelled there and returned with dust and rocks. He asked if the story was true. Told that it was, after a moment of silence, he asked, "Why bother?"

It is a strange set of values that has dictated the course of economic development in the tropical rainforests. At a time when the annual worldwide sales of plant-derived pharmaceuticals totalled $20 billion, Daniel Ludwig spent $1 billion levelling a stretch of Brazilian forest the size of Connecticut in a failed attempt to grow pulp fiber. His costs soared to that level in part because so little fundamental research had been done in the tropics that his foresters and engineers had to begin from scratch, literally experimenting on a massive scale as they went along. Had they known at the onset what they discovered so painfully later on, it is unlikely the project would have ever begun.

Yet in 1980, just as Ludwig's Jari project was coming to a final crisis, the total funding for global research in the basic biological sciences of ecology, taxonomy and plant exploration was $30 million, a figure that represents less than the cost of a single F-16 fighter, and perhaps the amount spent in New York City's bars in two weekends.

E.O. Wilson, renowned Harvard biologist, has said that the 20th century will be remembered, not for its wars or its technological advances, but rather as the era in which men and women stood by and either passively endorsed or actively supported the massive destruction of biological and cultural diversity on the planet. Our prosperity has been purchased at a cost that may rightly fill our descendants with shame.

Sensitivity to nature is not an innate attribute of indigenous peoples. It is a consequence of adaptive choices that have resulted in the development of highly specialized perceptual skills. But those choices in turn spring from a comprehensive view of nature and the universe in which man and woman are perceived as but elements inextricably linked to the whole. It is another world view altogether, one in which humanity stands apart, that threatens the forests and our world with devastation.

Perhaps the greatest gift of the indigenous peoples will be their contribution to a dialogue between these two world views, such that folk wisdom may temper and guide the inevitable development processes that today ride roughshod over much of the earth. In an increasingly complex, bloodstained world we need the visions of the indigenous peoples even as we need that view of the earth from space described so eloquently by Story Musgrave, and the wild lands themselves, alive and intact, because they stand apart as symbols of the naked geography of hope.

References

Abelson, P.H. 1990. Medicine from plants. Science 274: 513

Balick, M.J. 1984. Ethnobotany of palms in the Neotropics. Adv. Econ. Bot. 1: 9-23

Bisset, N.G. 1989. Arrow and dart poisons. J. Ethnopharmacol. 25: 1-41.

Boom, B.M. 1985. Amazonian Indians and the forest environment. Nature 314: 324.

Caufield, C. 1984. In the rainforest. Heinemann, London, England.

Davis, E.W. 1985. Hallucinogenic plants and their use in traditional societies. Cultural Survival Quarterly 9: 2-5.

Davis, E.W., and J.A. Yost 1983. The ethnomedicine of the Waorani of eastern Ecuador. J. Ethnopharmacol. 19: 273-298.

Day, D. 1989. The encyclopedia of vanished species. Mclaren Publishing, Hong Kong.

Farnsworth, N.R. 1979. The present and future of pharmacognosy. Am. J. Pharmacol. Ed. 43: 239-243.

Farnsworth, N.R. 1988. Screening plants for new medicines. *In* Biodiversity, *edited by* E.O. Wilson. National Academy Press, Washington, DC. pp. 83-97.

Farnsworth, N.R., and R.W. Morris 1976. Higher plants — the sleeping giant of drug development. Am. J. Pharmacol. 148: 46-52.

Forsyth, A., and K. Miyata. 1984. Tropical nature. Scribners, New York, NY.

Grainger, A. 1980. The state of the world's tropical forests. The Ecologist 10: 6-54.

Head, S., and R. Heinzman. 1990. Lessons of the rainforest. Sierra Club Books, San Francisco, CA.

Myers, N. 1983. A wealth of wild species. Westview Press, Boulder, CO.

Myers, N. 1985. The primary source. W.W. Norton & Co., New York, NY.

Norton, B.G. (Editor) 1986. The preservation of species: the value of biological diversity. Princeton University Press, Princeton, NJ.

Plowman, T. 1984. The ethnobotany of coca (*Erythroxylum* spp., Erythroxylaceae). Adv. Econ. Bot. 1: 62-111.

Posey, D.A. 1983. Indigenous ecological knowledge in the development of the Amazon. *In* The dilemma of Amazonian development, *edited by* E. Moran. Westview Press, Boulder, CO. pp. 225-257.

Posey, D.A. 1984. A preliminary report on diversified management of tropical forests by the Kayapó Indians of the Brazilian Amazon. Adv. Econ. Bot. 1: 112-126.

Posey, D.A. 1985. Native and indigenous guidelines for new Amazonian development: understanding biological diversity through ethnoecology. *In* Change in the Amazon basin. Vol. 1. Man's impact on forests and rivers, *edited by* J. Hemming. Manchester University Press, Manchester, England. pp. 156-181.

Prance, G.T., and T. Lovejoy (Editors) 1985. Amazonia. Pergamon Press, Oxford, England.

Shoumatoff, A. 1978. The rivers Amazon. Sierra Club Books, San Francisco, CA.

Schultes, R.E. 1988. Where the gods reign. Synergetic Press, Oracle, AZ.

Schultes, R.E., and Hofmann, A. 1979. Plants of the Gods. McGraw-Hill, New York, NY.

Schultes, R.E., and Hofmann, A. 1980 The botany and chemistry of hallucinogens. Charles C. Thomas, Springfield, IL.

Schultes, R.E., and R. Raffauf. 1990. The healing forest. Dioscorides Press, Portland, OR.

Soejarto, D.D., and N.R. Farnsworth. 1989. Tropical rainforests: potential sources of new drugs? Perspectives in Biology and Medicine 32: 244-256.

Stone, R.D. 1985. Dreams of Amazonia. Penguin Books, New York, NY.

Wilson, E.O. 1984. Biophilia. Harvard University Press, Cambridge, MA.

Wilson, E.O. (Editor) 1988. Biodiversity. National Academy Press, Washington, DC.

Addressing the Retention of Biodiversity in National Forests in the United States*

Jack Ward Thomas

Forestry and Range Sciences Laboratory, USDA Forest Service, 1401 Gekeler Lane, La Grande, Oregon, U.S.A. 97850.

Abstract

Forest managers seem to be increasingly at odds with the public as to how forests should be managed: "for whom and for what". This situation must be remedied if forest managers are to be effective in maintaining the public's support or forbearance. Remedies for present forest management ills will, inevitably, include a combination of striving to adjust to public will while educating the public in the intricacies of forest management. To the extent that humans have dominion over Earth, we must retain and restore the systems and processes that, in turn, sustain the communities of which humans are but a part. Humans do not exist, except in our own minds, as entities apart from Earth's processes.

Some people think that the outcome of the way we treat the forests of the world is certain, inevitable and bleak. They have lost hope of living in harmony with the forests. *I choose* to disagree. Why? Knowledge is cumulative. Most scientists who ever worked are living today. Our cadres of natural-resource management professionals and experienced land stewards develop, synthesize and put new knowledge to work. These scientists and stewards can develop new and deeper insights into how to exploit the forest and keep it alive and well. They care deeply about sustaining the productivity of the forests. That is not

* This paper was first presented as the summary to a one-week workshop on considering biodiversity in the management of national forests in the United States held in Spokane, Washington, in November 1990. The specifics were aimed at natural resource-management professionals in the USDA Forest Service, but the ideas are presented here in the hope that the overall thrust and message may have meaning in the Canadian context as well. The thoughts expressed herein are solely those of the author. No endorsement by the USDA Forest Service is implied or should be assumed.

In *Our Living Legacy: Proceedings of a Symposium on Biological Diversity,* edited by M.A. Fenger, E.H. Miller, J.A. Johnson and E.J.R. Williams, pp. 47-63. Victoria, B.C.: Royal British Columbia Museum.

enough. Natural-resource professionals must always be ready to see anew through old eyes — to take a new perspective on how to exploit land responsibility. There is no shame and much honour in that.

The land-use planning process conducted by the U.S. Forest Service over the past decade may have been the most productive exercise in the history of forest management in North America. How can that be said? It took much longer and cost so much more than anticipated and is leading to outcomes that leave almost no one pleased. Our developing land ethic, a human concept, must include the needs and desires of people, which implies providing goods, products and services from the land. Aldo Leopold's vision of what such an ethic might entail is a good place to begin. Now we must ask how such an ethic may be modified to retain biodiversity, maintain economic stability, preserve productivity, and sustain provision of goods and services, all simultaneously. We are further down that trail, intellectually and technically, than ever before. But the path not yet taken stretches ahead.

The developing land ethic, however, must be applied with the question, "Forests, for whom and for what?" ringing in our ears. The most vexing of the problems to be faced will be linking the "for whom and for what" question with the biological capabilities of the land in determining forest policy and management.

Introduction

Biodiversity is the variety of life in an area, including the variety of *genes*, *species*, plant and animal *communities* and *ecosystems*, and the *interactions* of those elements.

The responsibilities to address biodiversity in forest planning can be "the building blocks of biodiversity", which are to:
- manage to recover and conserve threatened and endangered species;
- manage habitats to maintain viable populations of existing native and desired non-native wildlife, fish and plant species;
- conserve sensitive species;
- manage special plant and animal communities to achieve overall multiple-use objectives; and
- manage for selected management indicator species.

Change in the air

Several years ago, the Society of American Foresters held its annual convention in Spokane, Washington. Dr Hal Salwasser, Director of New Perspectives for the U.S. Forest Service, addressed a plenary session on "New Perspectives in Forestry". That talk included a discussion on maintaining biodiversity in managed forests and alterations in current forestry practices aimed at sustaining forest productivity over the very long term. This might include significant changes in current forestry approaches, philosophy and practices. At the next break in the proceedings, as the crowd milled about in the foyer of the conference centre, an executive of one of the largest forestry schools in the United States approached me. He was agitated and asked whether I had heard Salwasser's speech and, if so, what did I think of it?

Before I could answer, he began a lecture on the need and obligation of humans to subdue nature and to bring it under control to produce the food, fibre and wood needed to sustain civilization. The rise of humans to our present state was traced, he said, in increasingly successful efforts to bring order out of the chaos of nature. Animal husbandry, agriculture and forestry were examples; they had been precursors to and were now the underpinning of cities, states and nations. To suggest that foresters re-examine their tenets, thereby implying that current forestry practices might be less than perfect or less than acceptable to the public, seemed almost heretical to him.

I suggested that these were examples of what has too frequently resulted from instituting too much order on nature too rigidly for too long. Such order, in many instances, had dramatically reduced the capability of the land to produce forests and crops. Such order can lead to increasing simplification and homogenization of highly evolved communities. Simplified systems seem to require more and more inputs of energy, resources and human effort to sustain their ecological and economic productivity.

Tombstones and Proclamations

I repeated to the university executive a story told to me by a friend who visited Lebanon years ago. While sightseeing with someone who worked in that country, they walked out on a hill to view the Mediterranean. The city of Beirut lay within view. My friend saw few trees as he viewed the landscape and asked jokingly about the "cedars of Lebanon". His guide motioned him to follow further along the ridge to a monument that had been left by the Romans. The Latin letters on the marker said, "These forests are protected by the Emperor."

When I heard the story years later, the latest war to rage in that land had severely damaged Beirut, the Paris of the Mediterranean. Once again, as has happened many times over the centuries, armed people of several factions fought in that land to set things "right". But to what end? the storyteller asked. The land is depauperate. The forests are gone. Grazing is a marginal enterprise. Goatherds graze their flocks where forests once stood. The agricultural lands are much reduced in productive capability, or destroyed. No matter which faction wins the latest war — and the ones that follow — the land will never regain its former productivity. Most of the food, fibre and wood needed by the people living on this poor land must come from elsewhere.

If the Emperor had known what the ultimate fate of the forests protected by his edict and his empire would be, I wondered, would he have considered new perspectives? The university executive walked away shaking his head. We were of different minds.

We must now, taking a new perspective, realize that the chaos perceived in nature is not chaos but intricate, always evolving, order. The challenge in new perspectives in forestry is to see anew, to be more humble, and to strive intelligently to retain enough of the ecological systems accorded to us by evolutionary processes to assure that ecosystems and processes, eons in the making, can continue their evolution.

Are we doomed to repeat inevitably the errors and follies of our predecessors? The signs along the highways that say Entering the Whichever National Forest are but a later version of that monument in Lebanon. The two signs have the same intent — these forests will be sustained, for the Emperor tells me so. Will the proclamation, "These forests are protected by the Forest Service", mean as little in one, two or three centuries as did the Emperor's proclamation?

Conversely, must we leave our cities and return to being hunter-gatherers to avoid land degradation? Of course that is impossible. As a species, we are too far down the road of taking dominion over Earth for that. But, conversely, we cannot continue as we have in the past. We must bring human population growth and the increasing demand on natural resources in line with the ability of ecological systems to sustain their productivity. The hope of increasing joint production of various mixes of goods and services from the forest can only go so far. There are limits, a reality only now coming to bear in North America, and coming to grip with that reality is painful to a society unused to limits.

Forest managers seem more and more at odds with the public as to how forests should be managed: "for whom and for what" (Clawson 1975). This situation must be remedied if forest managers are to be effective in maintaining the public's support or forbearance. Remedies for present forest management ills will, inevitably, include a combination of striving to adjust to public will while educating the public in the intricacies of forest management. To the extent that humans have dominion over Earth, we must retain and restore the systems and processes that, in turn, sustain the communities of which humans are but a part. Humans do not exist as entities apart from Earth's processes, except in our own minds.

The Land Stewards

Some people think that the outcome of the way we treat the forests of the world is certain, inevitable and bleak. They have lost hope of living in harmony with the forests. *I choose* to disagree. Why? Knowledge is cumulative. Most scientists who ever lived are working today. Our cadres of natural resource management professionals and experienced land stewards develop, synthesize and put new knowledge to work. These scientists and stewards can develop new and deeper insights into how to exploit the forest and keep it alive and well. They care deeply about sustaining the productivity of the forests. That is not enough. Natural-resource professionals must always be ready to see anew through old eyes — to take a new perspective on how to exploit land responsibility. There is no shame and much honour in that.

We are a relatively new breed, we natural-resource specialists, we foresters, we wildlife biologists, we citizens of the world concerned with appropriate exploitation of natural resources. We were and are revolutionaries. Like all revolutionaries in the heat of battle we have, too often, been sure of our own righteousness. And some revolutionaries lose heart when they learn that in the struggle is no respite, and that many defeats accompany few victories.

We, like all other professions in their beginnings, have turned rudimentary knowledge into theory and then into dogma that is overturned only slowly by accumulating knowledge to the contrary. We are reluctant to face how much of our underpinnings is more akin to religious beliefs than the infallible, scientifically derived understanding that we claim as our foundation.

Management Plans

The land-use planning process conducted by the U.S. Forest Service (FS) over the past decade may have been the most productive exercise in the history of forest management in North America. How can that be said? It took much longer and cost so much more than anticipated and is leading to outcomes that leave almost no one pleased. When we natural-resource management professionals started the process more than a decade ago, there was little doubt in most of our minds that the mandated planning and resource allocation could be done efficiently, effectively and well. Results were anticipated that would attain at least a public consensus of acceptability. If those are the criteria used to judge the results, the effort fell short.

But, at the very least, the planning effort was educational. We learned a lot. We brought increased rigour to analysing land capability and resource allocation. Some professional dogmas trembled in the face of internal and external examination. It was found that, as Frank Egler said, "Ecosystems are not only more complex than we think, they are more complex than we can think" (Jenkins 1977: 43). Imposing political influence through public input and administrative oversight to justify desired outcomes led to an almost unintelligible cacophony of sound and fury.

In some cases, this planning led to ends difficult to defend technically, and politically and operationally impossible to carry out as advertised in many cases. Adjustments are being made. Planning for management of publicly owned land in a democracy is more technically and politically difficult than was imagined. In the process, what have traditionally been local or state issues, and traditionally treated as such, evolved into national issues. Public tastes changed and power shifted rapidly even as the planning took place over a decade, and court decisions altered, time after time, the anticipated forest management scenarios.

The forest planning process brought, at least to some professionals, a becoming modesty that replaced too much certainty too forcefully expressed for too long. Such humility, if not taken too far, is an admirable characteristic for natural-resource professionals. Best of all, the process had led society to more satisfactory ways of performing coordinated land-use planning. The multi-agency, multi-disciplinary, multi-ownership, multi-interest group approaches being applied by the State of California are pioneering that effort.

The Full Circle — Back to the Future

If we examine the history of the FS, a constantly evolving agency philosophy can be discerned. It seems that the FS has come almost full circle back to its beginnings and the original reasons for its existence. Are the national forests to be managed like other lands held by individuals and large corporations? If that were so the lands could, and perhaps should, be placed in the custody of private owners. This is not to fault management of privately owned forests per se, but private forests serve a different master — the owners of those lands. The Forest Service's master is the people of the United States — all the people —

and not just those of a single state or region, and not just those who receive economic benefit from exploiting the plants and animals produced on those lands.

But as perceptions change about for whom and for what these forests are managed, it is imperative that the social and economic effects of these shifts be recognized and softened. People and their welfare should also weigh in the balance. When Jerry Franklin (pers. comm.) speaks of a "kinder, gentler forestry" we must extend that same consideration to the people and communities that will be affected by the changes that come from changes in forestry practices.

As we think about that and about where the FS started as an agency (Pinchot 1947) and examine the roads travelled since, T.S. Eliot's observation in *Gerontion* has a special meaning: "We shall not cease from our exploration, and the end of our exploring will be to arrive where we started, and know the place for the first time." It is well to recognize, however, that while Pinchot and Roosevelt broke new ground, they stood on foundations built by their predecessors. As we consider reformation now, we have an even stronger foundation, built up over 100 years, on which to build. We must know where we have been in order to know where we want to go.

Natural-resource managers have, or should have, learned something along the way. Nature is not in chaos. Rather, nature is in ever-evolving order along an untold number of intertwined pathways. Yet it is essential to impose some degree of order on nature if the land is to yield enough to sustain humankind in anything close to present numbers and present standards. But, in working to impose the requisite degree of order, more care must be taken to retain critical natural processes if the land is to be forever fruitful. We have resurrected a little-used word from the forestry lexicon for our banner — sustainability. We now must demonstrate to the people owning the forests that we land managers are worthy to carry that guidon.

Natural-resource managers bear a burden that others do not, at least not yet. That burden is thinking not only in the terms of the next year or the next decade, but also in terms of a thousand years or more. The bottom line on our accounting ledgers must count more than dollars and cents in the sums. That requirement and mind-set puts us apart from others. Our first rule must always be to master "the art of handling the forest so that it renders whatever service is required of it without being impoverished or destroyed" (Pinchot 1914: 13). So we return, again, to the beginning, and know the place for what it is.

Keeping Cogs and Wheels

There may be much of value for human survival to be learned in tracing and understanding why the forests that once grew around the Emperor's monument in Lebanon no longer exist.

We know that the admonition to "go forth and subdue the earth" is bad advice when pursued too vigorously for too long. We have often interpreted subdue to mean "... to bring under, to conquer by force or superior power..." There is a second, more applicable meaning which is "... to overcome ... by persuasion, kindness and other mild means ..." Instead of being overpowering we must be gentle and understanding lovers that woo and court the earth to sustain the sojourners on this planet. To do that, we must retain the soil and the ecological processes developed over the millennia, including extinction and development of species and communities, for such are the engine of life.

This is, perhaps, what Aldo Leopold meant in saying, "To keep every cog and wheel is the first precaution of intelligent tinkering" (1953: 147). We now admit that we do not know, and probably never will know, the roles and assembly plan for all the "cogs and wheels" of any ecological system. But at least we now suspect that if we discard cogs and wheels, we are in perpetual danger of discarding the drive shaft of ecological systems, the engine of life.

Saving cogs and wheels — retaining ecological processes — is but a pragmatic admonition to retain biodiversity. It now seems beyond question that saving all the pieces is a good idea: we must struggle to figure out how. Saving all the pieces is difficult when the icon driving our decision-making process and our land-use planning is still almost solely economic analysis. But that is changing.

Aldo Leopold (1949: 225) is increasingly recognized as correct when he wrote, "The fallacy the economic determinists have tied around our collective neck, and which we now need to cast off, is the belief that economics determine *all* land use. This is simply not true...." We now know what he meant. That recognition has come clear in trying to save endangered species, the forerunner or perhaps the stalking-horse for concerns about retaining biodiversity. Our nation's actions in such situations as saving wetlands for waterfowl, sustaining Grizzly Bears, protecting what remains of anadromous fish runs, and reserving old-growth forests of very high economic value to save a threatened owl and its associates show that public morals are changing. Society is demonstrably ready to include factors other than economics in land-use decisions. It is a beginning.

Most people responsible for the public's forests know and willingly accept, beyond economic analysis or social consequence, the ethical and professional burden to pass on the land in as good or better state than when they became its stewards. For good or ill, humans do have some limited dominion over the earth. The key to knowing if the steward's trust is fulfilled will be measured in the land's ability to produce desired goods and services now and over the centuries. That hillside in Lebanon, with the ancient tombstone proclaiming that there a forest once lived, is stark testimony to what can and does happen when failure occurs.

Evolution in Law

The Endangered Species Conservation Act (ESA) of 1969 (U.S. Laws, Statutes, etc., Public Law 91-190) is now recognized as a first, primitive attempt to retain biodiversity coincidental with the continuing exploitation of the nation's natural resources. This was followed by the National Forest Management Act of 1976 (U.S. Laws, Statutes, etc., Public Law 94-588), and the regulations issued pursuant to that act requiring that viable populations of all native and desired non-native vertebrate species be maintained and be well distributed in the planning area. This is now recognized as another step in considering the biodiversity in managing national forests. There is no shortage of law to direct management of the national forests. What we do and how we do it is another matter (Thomas 1987).

From Owls to Biodiversity to Sustainability

The issues in forest management that arose from developing and instituting a credible and legally adequate conservation strategy for the "threatened" Spotted Owl, developed by the Interagency Scientists' Committee (ISC), have sharply focused the attention of officials and the citizens on just how great the direct and opportunity costs of maintaining biodiversity may be (Thomas et al. 1990). And certainly other species are yet to be designated as threatened or endangered.

The ISC strategy may provide the stimulus for change in the underlying concepts of the Endangered Species Act. The aftermath of publishing the ISC report, the statement that forest management by the Forest Service in the Pacific Northwest will be carried out "in a manner not inconsistent with" the ISC report, and partial adoption by the Bureau of Land Management of its recommendations as the interim owl management strategy have been dramatic. Recent court decisions may mean that even more stringent management rules will be necessary to comply with the law. It is increasingly obvious that conservationists must look beyond the welfare and sustenance of single species to the larger context of ecosystems and communities and ecological processes.

The ISC conservation strategy is the latest and most expensive in a series of proposed strategies (USDA Forest Service 1988, Thomas et al. 1990). These efforts are the only existing management effort that involves sustenance of ecosystems at a landscape level (Salwasser 1988). That achievement occurred by a serendipitous route.

The strategy was built around the welfare of a single subspecies of owl that was believed by the FS to be an "indicator species" of the old-growth ecosystem. The validity of the indicator species concept is being debated, but the emerging realization is that this focus is too narrow and that ultimately a broader consideration of ecosystem issues must be undertaken (Landres et al. 1988).

After the ISC report was released, immediate attention came from economists evaluating the direct and opportunity costs and the political and social disruption that might result from its adoption (Rasmussen 1989; Bueter 1990; Gilless 1990; Greber et al. 1990; Lee 1990; Olson 1990). The impacts, the estimates of which differ dramatically depending on the analysts' assumptions (job-loss estimates run from less than 10,000 to more than 100,000), may be dramatic enough to stimulate one of two courses of action. The first is retreat from the evolving land ethic typified by the philosophy embodied in the ESA. The second might be the acceptance in law of a broader concept of conserving biodiversity (Blockstein 1989).

Elected and appointed officials, land managers and the public may no longer tolerate the recurrent disruption of orderly natural-resource exploitation that results from addressing the status, listing and recovery of one potentially threatened or endangered species after another. The need for social, economic and political stability coupled with the political unwillingness to significantly modify the ESA may be adequate stimuli to take the next step. That step could be a requirement for planning at either the landscape or ecosystem scale, or both, with retention of biodiversity as a major objective.

The ISC strategy, expanded from the realm of biology and science into the realm of public debate and politics, is influenced by ongoing legal, economic and social considerations. Scientists have a continuing role, but it is the turn of the politicians and gladiators from various interest groups to stride upon the stage. And the lawyers, of course, will do what lawyers do. But no matter what happens to the ISC strategy, management of the public's forests in the west has changed — suddenly, dramatically and significantly. And, for good or ill, things will never be the same.

A Change in Management

This change in managing publicly owned forests in the west, so long as it lasts, is the first serious broad-scale effort at saving cogs and wheels at the regional level (Salwasser 1988). In turn, this is the first broad-scale attention to the conservation of ecosystems and their inherent processes.

By the end of this symposium you will have been exposed to concepts of what biodiversity is, and to its consideration in land-use planning and management. Some people may assume that they will then be prepared to go forth and ensure the retention of biodiversity as a part of forest management. Probably not.

Are you now suddenly confused? There is no wonder in that. Retaining biodiversity as part of land management may be the intuitively obvious, the scientifically sound and the intelligently cautious thing to do to ensure maximum options for future generations to ponder. It is something to consider while choices remain, though choices become fewer on a daily basis. Yet our present understanding of how to apply best the concepts of biodiversity in forest management can be likened to eating Jello with chopsticks: the objective is clear, but much must be mastered in the execution.

Some things forest managers can do, however, at the fine-filter, site-specific level, and now. Leaving "legacies" from stands where trees are to be cut is one. Leaving some large live trees, some standing dead trees, logs and small remnant patches of mature trees are among such possibilities. This can be accomplished by forest managers at the timber-sale level with few broader-scale considerations. These ideas, being promulgated, evaluated and spread by a group of foresters, ecologists and wildlife biologists, have been called "new forestry". Dr Jerry Franklin of the FS and the University of Washington is most closely identified with this revolution in forest practice. The concept has won both praise and criticism. No matter; Franklin and colleagues have dared to question, to re-examine and to see anew. The debate is on. In that alone is value.

Forest management at the landscape scale will increase the complexity of management, requiring a broadened ecological view and increased co-ordination among ownerships, disciplines, and the public. This new view will encompass thinking beyond current administrative boundaries that, too frequently, impose rigid constraints on thinking and action. The ISC's owl conservation strategy involved thinking simultaneously about the management of federal, state and private lands under administrative control of five agencies of the federal government, three states with six agencies and private lands owned by hundreds — perhaps thousands — of parties. The biology, complex as it was, proved easier for the ISC to come to grips with than lines on maps designating ownership boundaries or agencies assigned to various missions. The human-constructed barriers to a landscape approach to forestry may, in the end, be the biggest hurdle to overcome.

Ecosystem Research at the Landscape Scale

A growing literature on biodiversity incorporates the concepts of biodiversity and ecosystem function into natural resources management (e.g., Wilson 1988). Synthesis of that knowledge into a form suitable for use by land managers is inadequate. Such synthesis must be accomplished, sold to managers and the public, and incorporated into planning and management for these concepts to have any effect.

Research in biodiversity as it applies to land management lags far behind concept, theory and perception. Some such research is underway but not nearly enough. Such research into ecosystem function, particularly at the landscape scale, should be accelerated — now. Much lip service has been devoted to the need for such work and too few resources have actually been committed. Answers from research begun today are years away, while options are inexorably diminishing.

Landscape-level Research

Any speaker addressing biodiversity preservation seems obligated to say, in essence, that the time for 19th century reductionist, single-disciplinary research is waning. It is time for multidisciplinary research on ecosystem function at a landscape scale. Many who toil in the vineyards of natural-resources research intuitively agree but feel compelled to add caveats.

Such research will be conceptually and operationally more complex and much more expensive, in money, time and people than current research efforts. Future research will require shifting resources into a forestry research community, federal, state, academic and private, that has for some years seen steadily decreasing resources. This has occurred simultaneously with the humbling experience of forest land-use planning that made forest managers face the inadequacy of our understanding of how forest ecosystems work and what may be required to sustain the forests' function and process.

A prominent political figure said recently that he was uncertain whether to increase efforts in forestry-related research, because too frequently more research resulted in reductions in the timber supply which, in turn, caused economic and social problems. I commented that the annual quantity of timber sold is a short-term concern and less significant in the long term than sustaining the productivity of the forests. The official responded that changes in the gross national product occurred annually and elections occur every two, four or six years. Furthermore, neither he nor I nor the present voters would be around in a hundred or a thousand years to know or care.

And there lies the rub. Few among us can or want to think about sustaining ecosystems over the centuries. It is impossible for most people to conceive that the form of government in the year 3000 may be less important than still having the soils of earth with their ecological processes intact and producing biomass. If the earth and its ecology are felt to be important, many things are possible for humankind and for earth's communities, of which humans are but part. If not, there is little chance and little hope.

The political figure with whom I was speaking is a successful politician, elected time after time. Obviously he reads the electorate well. He shook his head and turned away. We were not of like mind. Yet generations of people in power change much more rapidly than the generations of trees in our forests. Perhaps his successor will understand and act differently in response to an increasing public acceptance of an evolving land ethic.

Back to the Beginning in the Forest Service

So too do the generations of power and influence in the FS and other agencies change. The FS has lived through eras dominated by founders who were dreamers and reformers. Then came periods dominated in turn by the custodians, the forest harvesters and the planters, the engineers, and the planners. And now, perhaps, the wheel comes again to the pragmatic dreamers and reformers. Good. The time is right.

Our recent FS emphasis on "new perspectives" seems as new to us as it was to our forebears in the time of Roosevelt and Pinchot. More accurately, it is a new, externally stimulated but self-cultivated willingness by the FS leadership to re-evaluate, rethink and re-examine where the agency and its employees have been and where they are going. We come again to the beginning. But we must move on from here. New perspectives is not a form of silviculture and not a new mystical revelation and not a *mea culpa* for paths followed to this point.

It is a statement of philosophy and of hope. We are able and willing to grow, to change, to cultivate additional constituencies, to realize that the will of the public shifts and to see anew. We can dare to have new perspectives in the FS because underneath it all lies a firm foundation that reaches to bedrock. Some say the FS has lost its vigour and its will to lead and that the agency is just one more old-line bureaucracy. I *choose* to think not. I believe the men and women of the FS *choose* to think not. I believe the agency leadership *chooses* to think not.

New perspectives will require personal, individual attention to the land by the very best people we have, for their entire careers. Increased emphasis will be placed on technical skills and the keenly developed "forester's eye" and heart that come after long experience in contact with the forest and the forest's response to management. Some way must be found to keep many, perhaps most, of the best and the brightest FS people in touch with the day-to-day business of forest management where the trees grow and fall, the elk bugles and the owl flies, and with the people who use their forests.

To do that we must make it possible for such professionals to have a career and make a decent living at the field level. Administration will remain no less important than it has always been. But our agency and our nation have followed a siren song that supposed, quite innocently, that the best and the brightest people using the best possible information would make a series of correct decisions that would lead, inevitably, to the best of all possible worlds. When these decisions are made using cold-blooded perusal of computer run after computer run, by technicians in sterile rooms far from where the trees grow, one senses danger. When the computer spews forth answers and working folk wearing Vibram-soled work boots think the answers are wrong, it is time to step back and reconsider. Perhaps the best computer of all, the human brain, and the best sensitivity barometer, the human heart, must also be brought to bear in land-management decisions. Too much trust in technology is just as bad as too much reliance on intuition. Balance is the key.

The best administrators in all creation cannot touch and smell and know and feel the land as they did in their youth. They do not exult in the triumphs of successful acre-by-acre management and suffer with the failures. We can no longer move most of those with exceptional talent into administrative roles. Administrators are important. Yet it is equally important to cater also to those who want to stay in touch with the soil and the plants and the animals and the human forest-users, getting better and more expert at the job of careful forest management with the passing years. They must always be there, these masters of the forest management art, to ensure that requisite skills and sensitivity to the living forest, and love of it, are passed on to apprentices who will succeed them.

Changes coming from new perspectives and new forestry will lead to more sophisticated and artful forestry practices that require a broader, longer-range vision than is common today. To clearcut, burn and plant new trees does not require nearly as much technical knowledge, art or professional expertise and institutional memory as managing a forest to retain productivity, biodiversity and intact, functioning ecosystems while producing goods and services. Management approaches will be much more difficult to model and apply with a cookbook approach.

Such management will require a succession of master foresters, wildlife biologists and those of many other disciplines followed by their apprentices in long, continuing lineages. Somehow, when our predecessors became enamoured of German forestry, they failed to appreciate this significant attribute of that system — the need for some Forstmeisters and associates to spend their professional lifetimes in intimate contact with the same piece of the earth.

But we have come again to the beginning with a chance and a need to change. It is time to be open to new perspectives and to know we have been here before. The details are not significant; they will come later. An exact definition of "new perspectives" or "new forestry" is not important. Perhaps that will never come in final form or will differ from place to place and change over time. The attitude implied by new perspectives is what is important; the attitude encompassed in new perspectives is everything.

Aldo Leopold (1933: 634) called for extending ethics to the treatment of land. Some 45 years later, Starker Leopold (1978) could discern no movement toward his father's "pious hope". Now, in 1991, a pattern in law and developing consciousness in American society and elsewhere that the earth should be treated with respect indicates such an ethic is evolving. The developing concern about retaining biodiversity and sustainability is, it seems to me, aimed at "... preserving the integrity, stability and beauty of the biotic community" (Leopold 1949: 224-225).

The developing land ethic, however, must be applied with the question, "Forests, for whom and for what?" ringing in our ears. The most vexing of the problems to be faced will be linking the "for whom and for what" question with the biological capabilities of the land in determining forest policy and management.

Our developing land ethic, a human concept, must include the needs and desires of people, which implies providing goods, products and services from the land. Leopold's vision of what such an ethic might entail is a good place to begin. Now we must ask how such an ethic may be modified to retain biodiversity, maintain economic stability, preserve productivity, and sustain provision of goods and services, all simultaneously. We are further down that trail, intellectually and technically, than ever before. But the path not yet taken stretches ahead.

We can think of ourselves as the Long Green Line — a succession of men and women holding to the dream of forests sustained in perpetuity with their communities and ecological processes intact. That heritage gives us strong roots and a foundation on which we can dare to consider new perspectives and the changes that may be required. We have been there before. We can *choose*. There is nothing to fear. There is, however, much to hope for.

Literature Cited

Blockstein, D.E. 1989. Washington watch — biodiversity bill update. BioScience 39: 677.

Bueuter, J.H. 1990. Social and economic impacts in Washington, Oregon, and California associated with implementing the conservation strategy for the Northern Spotted Owl: an overview. Mason, Bruce, and Girard, Inc., Portland, OR.

Clawson, M. 1975. Forests for whom and for what? Johns Hopkins Univ. Press, Baltimore, MD.

Gillis, J.K. 1990. Economic effects in California of protecting the Northern Spotted Owl. Mason, Bruce, and Girard, Inc., Portland, OR.

Greber, B.J., K.N. Johnson and G. Lettman. 1990. Conservation plans for the Northern Spotted Owl and other forest management proposals in Oregon: the economics of changing timber availability. Forest Research Laboratory, College of Forestry, Oregon State Univ., Papers in Forest Policy I. Corvallis, OR.

Jenkins, R.E. 1977. Classification and inventory for the perpetuation of ecological diversity. *In* Proceedings of the Classification Inventory, and Analysis Fish and Wildlife National Symposium, Phoenix, AZ, June 24-27, 1977, *edited by* A. Marmelstein. U.S. Dept. of Interior, Fish and Wildlife Service, Washington, DC. pp. 41-51.

Landres, P.B., J. Verner and J.W. Thomas. 1988. Ecological uses of vertebrate indicator species: a critique. Conservation Biology 2: 316-328.

Lee, R.G. 1990 Social and cultural implications of implementing "A Conservation Strategy for the Northern Spotted Owl." College of Forest Resources, University of Washington, Seattle, WA.

Leopold, A. 1949. A Sand County almanac and sketches here and there. Oxford University Press, Inc., New York, NY.

Leopold, A.S. 1978. Wildlife and forest practice. *In* Wildlife and America: contributions to an understanding of American wildlife and its conservation, *edited by* H. P. Brokaw. Council on Environmental Quality, Washington, DC. pp. 108-120.

Leopold, L.B. (Editor). 1953. Round River: from the journals of Aldo Leopold. Oxford Univ. Press, New York, NY.

Olson, D.C. 1990. Economic impacts of the ISC Northern Spotted Owl conservation strategy for Washington, Oregon, and northern California. Mason, Bruce, and Girard, Inc., Portland, OR.

Pinchot, G. 1914. The training of a forester. J.B. Lippincott Company, Philadelphia, PA.

Pinchot, G. 1947. Breaking new ground. Harcourt, Brace and Co., New York, NY.

Rasmussen, M. 1989. Timber output impacts of the Northern Spotted Owl conservation strategy proposed by the Interagency Committee of Scientists. Timber Data Company, Eugene, OR.

Salwasser, H. 1988. Managing ecosystems for viable populations of vertebrates: a focus for biodiversity. *In* Ecosystem management for parks and wilderness, *edited by* J. K. Agee and D. R. Johnson. Univ. Washington Press, Seattle, WA. pp. 87-104.

Thomas, J.W., E.D. Forsman, J.B. Lint, E.C. Meslow, B.R. Noon and J. Verner. 1990. A conservation strategy for the Northern Spotted Owl. USDA Forest Service, Bureau of Land Management, U.S. Fish and Wildlife Service and National Park Service, Portland, OR.

Thomas, J.W., and H. Salwasser. 1989. Bringing conservation biology into a position of influence in natural resource management. Conservation Biology 3: 123-127.

Thomas, J.W. 1987. Multiple-use forestry — moving from platitudes to reality. *In* Northwest forestry in transition: the 1987 Starker Lectures. College of Forestry, Oregon State University, Corvallis, OR. pp. 43-52.

USDA Forest Service. 1988. Final supplement to the environmental impact statement of an amendment to the Pacific Northwest regional guide. Spotted Owl guidelines. 2 vols. Pacific Northwest Region, Portland, OR.

U.S. Laws, Statutes, etc., Public Law 91-135. [H.R. 11363], December 5, 1969. Endangered Species Act of 1969. United States statutes at large, 1969. Volume 83. U.S. Government Printing Office, Washington, DC. [16 U.S.C. section 668 (1970).] p. 275.

U.S. Laws, Statutes, etc., Public Law 94-588 [S. 3091], October 22, 1976. National Forest Management Act of 1976. United States code congressional and administrative news, 94 Congress, 2nd Session, 1976. Volume 2. West Publishing Company, St. Paul, MN. [16 U.S.C. section 1600 (1976).] pp. 2949-2963.

Wilson, E.O. (Editor). 1988. Biodiversity. National Academy Press, Washington, DC.

The Importance of British Columbia to Global Biodiversity

Bristol Foster

131 Thomas, Ganges, British Columbia, Canada V0S 1E0.

Abstract

The glacial and tectonic events that have occurred in British Columbia have produced a wide range of environments that now support the highest biological diversity in Canada. However, many of B.C.'s animals, plants, and ecosystems are endangered. There are many reasons for maintaining biodiversity, including keeping options open for future generations, and the economic values in everything from medicine to pest control. Provincial biodiversity can be put in a global perspective, and in this paper I examine the success of different countries in protecting their biodiversity resources.

The principles relating to the ecology of islands ("island biogeography") may help us measure how we are doing since protected natural areas are often merely "islands" of naturalness in a sea of development. It is also necessary to examine government policies and practices. Maintaining viable representation of different ecotypes is more important than looking at area alone, something not yet being addressed by the B.C. Parks Plan 90 or the B.C. Ministry of Forests Wilderness Strategy.

The adversaries of biodiversity include peoples' attitudes (particularly toward consumption) and population growth. We must acknowledge the impact that human disrespect for the planet has had on biodiversity, and focus attention on the fact that biodiversity is our most valuable resource.

Introduction

Most of British Columbia's life has immigrated here since the glaciers began retreating 12,000 years ago, but the province is biologically diverse compared to the rest of Canada and other temperate areas. Many of B.C.'s plants, animals and ecosystems are threatened or endangered by human activity. Much of this province's life is found

In *Our Living Legacy: Proceedings of a Symposium on Biological Diversity*, edited by M.A. Fenger, E.H. Miller, J.A. Johnson and E.J.R. Williams, pp. 65-81. Victoria, B.C.: Royal British Columbia Museum.

only here, and some species and ecosystems have already become extinct. Populations of many species have also vanished, particularly those in forested areas, thereby significantly reducing biodiversity. Many other species are unknown to science as taxonomists themselves are an "endangered species". Our success in protecting biodiversity in B.C. is examined in light of the principles of island biogeography and experience in protecting species and ecosystems. The success of national parks, provincial parks and ecological reserves in protecting biodiversity is examined. Recent government initiatives such as "Parks Plan 90" and the B.C. Ministry of Forest's "Wilderness Areas" are also reviewed. In the past our government has mainly protected alpine areas for recreationalists rather than protecting the far more biologically diverse lowland areas for biodiversity. This trend is also apparent in plans of the Ministries of Parks and Forests. Until we make basic changes in our attitudes, our biodiversity is likely to continue to decline.

The Diversity of British Columbia

Today much attention is focused on ancient and rich tropical ecosystems that contain over 50% of the biodiversity of the planet. In B.C., most of the flora and fauna has immigrated into our province from the south over the last 12,000 years, when the Canada-wide glaciers finally began to release their grip.

In spite of the youth of the mantle of life that covers our land, we in B.C. have inherited an amazingly diverse array of species and ecosystems. We have a larger biological inheritance than any other province in Canada, largely as a result of the ocean floor sliding under western North America during the last 120 million years. This massive tectonic event crumpled the land to produce towering mountains, plateaus and valleys with concomitantly varying rainfall and soils. The resulting wide range of environments accommodated the wealth of plants and animals that eventually arrived.

Even on a world basis B.C. has a rich collection of major ecological areas. B.C. has four of the world's five major ecological domains, including Cool Oceanic, Humid Temperate, Dry and Polar. In contrast, the entire continent of Australia has only three such "Ecodomains" (Bailey 1989). B.C. also has seven "Ecodivisions", more than in all of eastern Canada. These Ecodivisions are further divisible into 10 Ecoprovinces, then 33 Ecoregions, further emphasizing the enormous diversity of B.C. (Demarchi et al. 1990).

In comparing biodiversity of Canada to other temperate areas, Bunnell (1990) chose a European area of similar size, extending from

Iceland to the Mediterranean islands. He found that Canada hosts 13% more species of mammals and, notably, 28% more large (> 1 kg) mammal species. B.C. has far more species of mammals (102) than any country in Europe.

Within Canada, most wildlife diversity is in B.C. The province is home to nearly 500 bird species, representing about three-quarters of the total species in Canada—by far the highest for any province (Bunnell and Williams 1980). B.C. has 72% of Canada's terrestrial mammal species, 49% of its amphibian species, and 41% of its terrestrial reptile species. The province has about 3,150 species of plants, or about 60% of those found in all of Canada.

Paralleling its abundance in biodiversity, B.C. also boasts rich cultural diversity, with seven of the eleven language families in Canada. Some languages have become extinct, while others are endangered (D.A. Abbott pers. comm.).

Despite its huge size (almost a million square kilometres) with only about three million people, many of B.C.'s plants, animals, ecosystems and gene pools are endangered. Furthermore, most human activities, from farming to logging to cities and towns, are concentrated in the most biologically diverse parts of the province.

Straley (1990) identified 816 species of rare plants in B.C., many of which are endangered. Ten of B.C.'s mammals are on the Red List of endangered or threatened species, ranging from the Sea Otter to the Spotted Bat to the Vancouver Island Marmot (B.C. Ministry of Environment 1991). The status of 21 other mammals is considered to be "sensitive" and they could be moved to the Red List. These include the Badger, the Dall Sheep and Trowbridge's Shrew.

Two species of birds have already become extinct in B.C.—the Sage Grouse and the Yellow-billed Cuckoo. Ten others are on the threatened and endangered list, including the Spotted Owl, the Marbled Murrelet and the Pacific Coast race of the Peregrine Falcon. Fifty-seven other species are classified as "sensitive".

While no amphibians are on the Red List, five are classified as "sensitive". Two reptiles are on the Red List (the Sharp-tailed Snake and the Short-horned Lizard) and four others are classified as "endangered".

These lists attest to the dramatic and deleterious impact of modern humanity on the environment.

Endangered Ecosystems

Some natural ecosystems are endangered as well, particularly those of high productivity such as coastal old-growth forests, rangelands and estuaries. We are well aware of the controversy over how much old-growth forest must be saved to protect our diverse biological legacy. About 20% of B.C.'s bird species and 29% of mammals rely on cavities in old trees that contain some rot (Bunnell 1990). Insect species are about three times as abundant in old-growth forests as on tree farms, and old-growth forests house a far wider range of the arthropod predators that help keep foliage eaters in check.

Some natural forested ecosystem types have already been obliterated in B.C., including areas on southeastern Vancouver Island, in the Kootenays and near Terrace. Moore (1991) has assessed the natural and developed status of primary coastal watersheds in B.C. He noted that, of the eighty-nine Vancouver Island watersheds greater than 5,000 hectares in size, only five are still in the natural state, and of these, only one is in a park (Strathcona) with its biodiversity protected. This is obviously a very small remnant.

It is difficult to protect natural rangeland because independent ranchers have less flexibility than large logging companies. As a result, grassland ecosystems are under threat. Even some supposedly protected grasslands are suffering from overgrazing, firewood cutting and motor vehicle use.

Natural estuaries are the most endangered of B.C.'s ecosystems. Flat land near tidal water has always been a choice area for development. Any industrial activity is bound to affect downstream estuaries. Few large estuaries in the province remain natural and the most valuable of all, the Fraser River delta, is threatened by numerous human activities from golf courses to pulp mills hundreds of kilometres away.

No area illustrates how important B.C. is to global biodiversity better than the Fraser River delta. As the largest estuary on the Pacific coast of North America (followed by the Copper River delta in Alaska and the wetlands of California), the Fraser River delta is a key link in a chain of vital bird habitats needed for breeding grounds in Canada, Alaska and Siberia, as well as for wintering areas of birds in the southern United States and Central and South America. The Fraser River delta supports, in winter, the highest density of waterbirds, shorebirds and raptorial birds in Canada (Butler and Campbell 1987). On average about 500,000 waterbirds use the delta each year, but in some years the number rises to 1.4 million. Virtually the entire world population of Western Sandpipers, sometimes up to 1.2 million strong, uses the delta as a stopover for a few days in the spring and autumn to feed and rest. Snow Geese, on their spring and autumn migrations across the continent from California to Siberia, feed and rest in the delta for the long journey. For many species, no satisfactory alternative stopover is available within flying distance.

The Need for Protection

Because we are responsible for their protection, the kinds of animals, plants and ecosystems that are unique to B.C. are of particular concern. Since they are in our jurisdiction exclusively, we must protect them by safeguarding their habitat. Until now our province's Wildlife Branch has confined itself to concern about biodiversity only at the species level, so geographical races or "subspecies" are not protected (see Appendix B in Miller, this volume). As Bunnell and Williams (1980) point out, this has far-reaching implications. B.C. is the home of 288 breeding bird species and 117 mammal species, including 35 birds and 28 mammals found nowhere else in Canada. When subspecies are included, B.C. is home to 286 kinds of mammals (Nagorsen 1990).

I agree with Bunnell and Williams (1980: 257) when they state that restricting ourselves to protecting nothing more specialized than species "represents an abrogation of responsibility to husband the rich diversity of wildlife and initiates a quasi-official sanction of the extinction of affected subspecies". We must take the protection of our biodiversity seriously. We must define and take the necessary steps to protect adequate samples of the entire known diversity of the province, whether it be plants or animals or even fungi or bacteria, realizing that even then many kinds of organisms will be overlooked.

Until now, some of us have believed that protecting biodiversity at the species and subspecies levels was sufficient. Many scientists agree that we must also protect different populations of species to try to preserve a wide range of genetic diversity within them. For example, to plant tree farms the seedlings from a few individual trees impart low genetic diversity, and such an impoverished gene pool may not be able to respond adequately to insect or fungal outbreaks, depleted soil, the greenhouse effect or other environmental vicissitudes.

Robson Bight is an example of another unique ecosystem that is endangered. This bight has become world-famous because of its large gatherings and conspicuous activities of Killer Whales. Logging, fishing or tourism could destroy the special relationships that Killer Whales have with the area. Many other ecological associations such as those occurring in old-growth forests are threatened with extinction; indeed, some may already be extinct. We know little about the endangerment of biodiversity within B.C.'s old-growth forests, partly because forest ecologists working there have also been rare.

The predator-prey system involving large ungulates in B.C. is unique in North America. Nowhere else does one find four kinds of predators preying on seven to eight kinds of ungulates. We have a particular challenge in protecting such natural systems, as the habitat required is large and some ungulates occur only in our province.

The Need for Knowledge

When we talk about biodiversity, we should remember that we probably know as much about the subject as Aristotle knew about ecology. Of the 10-30 million species of organisms in our world, only about 1.6 million have been given a scientific name. Most of the unknown species are tropical, but the Royal B.C. Museum has recently illustrated our sublime ignorance of the biodiversity of our province. Cannings (1990) reported on an invertebrate survey on the Brooks Peninsula. Using nothing fancier than butterfly nets and yogurt containers he and his associates collected 3,600 specimens of terrestrial invertebrates. After receiving identifications from specialists around the world, a list of 519 species in 190 families was made. Of the 519 species, 31 were new to science and an additional 34 were previously unknown in Canada. R.A. Cannings (pers. comm.) also reports that of 100 species of gall midges he collected near Sooke, 90 turned out to be new to science.

Taxonomists, the specialists who describe new species of plants and animals, are also an "endangered species", unacknowledged while scientists with huge budgets rush madly to decode the human genome, identify a new galaxy or build a new weapon. Yet taxonomists are the key to decoding biodiversity. I make a special plea on behalf of taxonomists. Absolutely nothing is more basic to the theme of this symposium than knowing what species we are dealing with. Many species, perhaps most of the organisms we are trying to save that make up the biodiversity of the province, do not yet have a scientific name.

We have recently learned that many organisms, such as the mycorrhizal fungi associated with plant roots, are important for healthy trees. Undoubtedly other soil organisms are important for soil health but we do not know the existence of many of them. Scientists cannot learn about them until they are named and identified. But many have no name. If taxonomy is not given higher priority, all we can do is protect samples of habitats and hope that the samples contain a good representation of the biodiversity of the province (whatever "good" is). This is known as groping in the twilight. We need more light and more taxonomists.

Reasons for Retaining Biodiversity

Perhaps it is redundant to ask why we should retain biodiversity. While there are many reasons, I shall briefly list seven major ones. First, protecting biodiversity keeps options open for future generations.

Second, it has economic values in everything from medicine to pest control. Third, it has aesthetic values — enjoying life for its own sake. Fourth, a species is gone forever once it has become extinct. We must save the blueprints of the pieces of the ecosystem. Fifth, we need "indicator species", meaning those species that are more sensitive than ourselves to environmental changes that could warn us of environmental degradation. Sixth, there are moral values: everything has a right to live. And seventh, people want it. Biodiversity is a hot topic today as confirmed by recent task forces, symposia and polls. As wealthy and well-educated citizens of the planet we must set an example for the rest of the world. It has often been noted that the destruction of tropical rainforests appalls us while parallels closer to home have been excused.

The Global Perspective

Assuming that the protection of biodiversity is a worthy goal, how have we done so far in B.C? Before answering this question for our province, I would like first to give a global perspective. In the last few years I have travelled to 17 countries examining the success of each country in protecting its biodiversity. Here is a brief summary.

In Queensland, environmentalists were fighting to save the last of an incredibly rich tropical rainforest. The Federal government overrode the State's wishes on using State land and created a World Heritage Site to protect the biodiversity. Further south in Tasmania a Canadian logging company, against decisive local opposition, was attempting to turn into pulp the tallest flowering plants in the world, 100-metre-tall eucalyptus. The greenies are alive and well down under.

In the Malaysian State of Sarawak, one of the last hunter-gatherer tribes in the world, the Penan, was having the remains of its ancient (130 million years) and biologically diverse rainforest cut to make a billionaire out of the Minister of Environment and to supply lumber for cement forms in Japan.

Thailand, until recently 80% clothed in forest, is now only about 18% covered. After massive and deadly landslides caused by logging, the Thai government banned all further logging in the country. Thai loggers have since moved on to destroy the rich forests of Burma and Laos.

In Nepal we heard people say jokingly that the country's chief export was topsoil. But it is no joke. Deforestation has removed protection from the soil to such an extent that the country is literally washing away into the sea. Nepal's rich biodiversity is making the same trip.

Kenya has the fastest population growth in the world, doubling every 16 years. Pressure on the natural resources is doubling at least as quickly. The government has a birth control program that is making some progress.

The tropical forests along the Panama canal are being cut with the result that Lake Gatun is filling with silt and soon there could be insufficient water for the canal to function. Needless to say, species are going extinct, perhaps daily.

In Tanzania we found one of the world's best park systems. Poaching, however, is out of control and some populations are suffering severely.

In many ways little Costa Rica is a model for the world to follow with about 20% of its biologically rich and diverse land protected from development. But refugees from nearby troubled countries and a recent visit by the Pope have resulted in soaring human numbers, and forest often has to be cut so poor people can merely survive.

Frequently loss of biodiversity is accompanied by loss of cultural diversity: the Penan and the African Pygmies are two examples.

While instances of saving biodiversity in the world are found, they are distressingly rare.

Island Biogeography

So how are we doing? In answering this question we usually examine government policies and practices and note those species and natural areas that are protected. It is also worthwhile to understand some evolutionary principles related to the ecology of islands ("island biogeography"), since protected natural areas are often merely "islands" of naturalness in a sea of development.

Islands invariably have fewer species than the adjacent mainland. The smaller the island, the fewer the number of species that inhabit it. An island one tenth the size of another will have about half the number of species, other things being equal. The farther islands are from each other or from the adjacent mainland, the fewer the number of species present. Extending these principles to B.C.'s system of mostly small, scattered protected areas, it is not difficult to see that we have inadequately protected biodiversity.

I will give three examples of how we can learn from the study of islands. When the Panama Canal was built, Lake Gatun was formed with Barro Colorado in the middle of it. As the lake level rose, what was once part of a huge Central American tropical forest became an island. Studies before and after the lake level stabilized showed that 45 of 209 species of birds disappeared. For the most part it could not be determined why these species became extinct. They just did. No doubt many other kinds of animals and plants became extinct as well. The island reached a new and lower level of equilibrium of species related its smaller area. For the same reasons, unless the public insists that B.C.'s protected areas be enlarged, they too are likely to lose species and ecosystems as they move towards lower levels of equilibria.

As a second example, the Channel Islands in California have been studied for many decades. Over 50 years about one third of the original bird species have vanished, but the total number of bird species on the island remained the same. New immigrants from the mainland maintained the equilibrium. About a third of the species would have been lost had there been no nearby source of immigrants. The concept of providing natural corridors to facilitate movement of animals and plants between protected areas has never been applied effectively in B.C.

A final example of the equilibrium that species maintain on islands is from Florida. An experiment was carried out on a few small islets, with all insects being killed. In a fairly short time insects immigrated back. While the islets then had about as many kinds of insects as they had originally, many of them were different. As for California's Channel Islands, an equilibrium was maintained, but only because there was a nearby source from where the insects came. Had there been no source, diversity of insects would have settled at a much lower level of equilibrium. Counterpart examples in B.C. are easy to find. Lowland old-growth forest is becoming increasingly rare, fragmentary and highly scattered. If more old-growth forest with easy access to other protected areas can be protected, then a species that dies out in one patch can be replaced through recolonization from another.

Because fewer species dwell on islands than on the nearby mainland, the factors that influence island species also differ profoundly, and evolution on islands proceeds differently. Thus island species and ecosystems that we try to protect are likely to change irreversibly. Change is generally hastened because protected areas are generally a small part of a much larger whole, so gene pools will become impoverished. One solution to these dilemmas is to make preserves as large as possible.

Because biodiversity on islands varies with island size, much debate has occurred over the years about whether it is better to protect biodiversity by having a few scattered preserves or a single large one of the same area as all the smaller ones. Some people have proposed that the minimum size of preserves may be estimated by first identifying a target or keystone species, such as Gray Wolves or Sea Otters, whose disappearance would significantly decrease the value and integrity of the reserve; then determining the minimum number of individuals in a population needed to guarantee a high probability of survival of these species; and finally estimating from this information the area needed to sustain the minimum number.

Grambine (1990) believes the current reserve network is inadequate to protect many species much beyond the next 50 years. The species-level approach must be augmented by landscape-level strategies that recognize ecosystem patterns and processes. An integrated system of large nature reserves is needed to preserve biodiversity.

Pickett and Thompson (1978) argue that simple applications of the principles of island evolution will not work, because recolonization sources are disappearing. Reserves must be based on minimum dynamic area: the smallest area with a natural disturbance regime that maintains internal recolonization sources.

The only satisfactory way of protecting forested ecosystems is to protect entire valleys because any upstream industrial activity affects the integrity of lower areas, and lower areas must be protected to preserve a watershed's greatest biodiversity. As a result of forest fragmentation through logging, the options for setting aside samples of forested ecosystems are decreasing rapidly. Besides the few remaining natural watersheds on Vancouver Island, only one primary pristine watershed over 5,000 hectares in size remains on the southern mainland coast (Moore, 1991).

Government Policy and Biodiversity

In our world, political considerations often play a greater role than do biological ideals in decisions to protect natural areas. In the past we have created mostly small, highly disjunct preserves, and the number of species in them moves to new and lower levels of equilibrium. To ensure the preservation of taxa and ecosystems we must fundamentally change our outlook. For a start, we must understand that biodiversity sustains us. Thinking that we sustain nature is as unrealistic as believing that the sun revolves around the earth.

In B.C. we have many ways of protecting wild lands and, consequently, biodiversity. They include national parks, provincial parks, ecological reserves, wilderness areas of the Forest Service, wildlife management areas and private land purchased by the Nature Conservancy or Nature Trust and leased to a government agency for management purposes. Beyond these formal methods are opportunities to minimize impacts of industrial activity on biodiversity by, for example, selective logging or more careful mineral exploration and development. But to protect biodiversity on unprotected land requires far better stewardship of the land than has occurred up to now. Pulp mills continue to poison the air and marine environment, mining effluent still leaches into waterways and cattle move into ever more marginal lands. So most people put their hopes for protecting biodiversity in formally protected areas. In B.C. only 6.6% of the land is given some protection by a government or agency.

Often the highest level of government provides the most secure protection. This is true in the case of Canada's national parks, but even here all is far from well. In the July-August 1990 issue of *Equinox* magazine, the article "Islands of Extinction" describes the sorry state of our national park system. National parks represent less than 2% of the area of Canada, placing Canada 22nd in the world in percentage of its protected land. Even in this small amount of protected land the government has had difficulty refusing developers, whether they be forest companies or ski tow operators. More money has been spent in national parks on road maintenance than on conservation, and on the development of park visitor facilities than on basic inventories of natural resources. Many kinds of natural areas are not yet represented in the park system and options to do so are running out. Obviously we must elect a federal government with more concern for the future of national parks.

The provincial park system may offer higher hope. It covers 5.3% of the province if one includes recreation areas in which industrial activity is permitted. The parks have been chosen for recreation and scenery, with little thought for biodiversity. Much thought, however, has been given to minimizing resource conflicts and, as a result, most of the protected area is bog or alpine. For example, of the 12% of the land in the East and West Kootenays protected by park status, 91% of the protected area is alpine or subalpine. In the past, park areas have been arbitrarily and drastically reduced, or have been changed through allowing mining, hunting and logging. Unless park areas are permanent, free of resource extraction, representative of the province's natural diversity and exempt from all but minimal tourist development, their contribution to protecting biodiversity will be negligible.

In B.C., the Provincial Parks Branch has recently announced "Parks Plan 90", a public process to study how and when we can best complete the park system. This initiative, although tardy, offers the hope that park planning will be less secretive and will involve a productive sharing of minds.

At the same time, the Ministry of Forests has announced another worthy initiative: a public review of their program for establishing wilderness areas in provincial forests. As in the case of provincial parks, these wilderness areas will be protected. Most of the controversial areas of recent concern to the public need to be included in planning by governmental agencies: lower Tsitika River, Carmanah-Walbran, Tahsish-Kwois, the Koeye, Khutzeymateen and Megin rivers, and Brooks Peninsula land including the East Creek and Klashkish, Battle and Power rivers. As well as containing valuable timber, all these areas contain a wealth of biodiversity and cultural significance.

Areas proposed for protection by the two provincial agencies contain few areas of high biodiversity, particularly those proposed by "Parks Plan 90". In the Kootenays, for example, Ministry of Forests proposals cover 5% of the land and include forests that have 4% low capability for logging, 14% moderate capability and 8% high capability, while the "Parks Plan 90" proposals cover only 1% of the land and include only forests of 4% low capability. The tradition that Parks protects alpine terrain for recreationalists continues. Protecting biodiversity apparently is still not a high priority. The Forest Service must be commended for its somewhat broader outlook.

Some key ecosystems, however, are omitted from both plans. No areas in the biologically diverse rangeland of B.C. are even recommended for study. The Parks Plan affirms that only the most important natural features will be protected, and then only if the features are not already protected in existing regional or national parks. This policy has the effect of placing all our biodiversity eggs in one basket. We must pressure the government to protect at least three examples of an ecotype (if they exist) to mitigate against inevitable natural disasters. Even with triplicates, much variation within species may be lost.

B.C. Parks, in its booklet "Striking the Balance", states that its intention is to raise park area from the present 5.3% to at least 6% in the next 20 years. Considering the widely supported movement in Canada to raise the protected area to at least 12% in the next 10 years, the B.C. Parks proposal fails to acknowledge the wishes of many of the public.

Invariably it has taken years from idea to realization to gazette a protected area. For Ministries of Parks and Forests proposals to be considered as serious by the public, moratoria on resource development must be in effect until protected status is either achieved or abandoned. The B.C. tradition of "chop while they talk" is absurd.

One reason that these two agencies have aired their respective plans is that doing so follows one of the recommendations of the Wilderness Advisory Committee. In the *Wilderness Mosaic* (1986), the Wilderness Advisory Committee states that, ideally, wilderness areas should exceed 5,000 hectares in size. Unfortunately, Parks now considers an area to be large if it is more than 1,000 hectares and Forests uses the figure of 1,000 hectares to define wilderness. It appears that until now Parks in particular has tried to do the minimum to satisfy the groundswell of public opinion that is demanding more parks, and that in a few more years the definition of large, or wilderness, may be reduced to an area of 500 hectares and eventually even smaller. We must reverse this direction.

It is puzzling that the Ecological Reserves program gets only passing mention in the Ministry of Parks literature even though this program belongs to the same ministry and is directly involved with activities similar to parks, namely protecting samples of our biodiversity. The Ecological Reserves program needs more attention.

This year, 1991, marks 20 years since approval of the Ecological Reserves Act. One of the Act's key objectives was to protect representative, unique, rare and endangered species and ecosystems for research and education.

For ten years I worked for the provincial government on our Ecological Reserves program. When I resigned in 1984 the government was boasting that we had the best ecological reserves program in the country. And we did. But this spoke poorly for the rest of Canada. B.C. had protected only one tenth of one per cent of the land base of the province, and that was mostly bog and alpine. Productive sites such as estuaries, old-growth forest and rangeland were and are almost impossible to secure.

When the Ecological Reserves Act passed the Legislature in 1971, then Liberal MLA Pat McGeer was recorded in Hansard as saying that he thought the Act was a lot of "window dressing" and that the Act would fool people into thinking that biodiversity was being protected and therefore resource use could go full speed ahead with a clear conscience. Unfortunately, time has proved McGeer to be largely correct. As the study of islands has shown, these mostly tiny, widely scattered protected areas are unlikely to make a significant contribution towards protecting biodiversity. Some important research has been done in a few reserves but most reserves have not had even a basic inventory. Management of reserves has improved thanks to a group of dedicated volunteer wardens.

The reasons for the mediocre success of the Ecological Reserves program are many. First, by nature it is a rather esoteric program. The reserves are created for research and education and not for recreation. The public and therefore politicians naturally identify with recreation, rather than with research and education. Second, the basis for selection of ecological reserves involves a full knowledge of the province's complicated biogeoclimatic zones and subzones, something that has until now been missing from the fingertips of most of the public and all politicians. Perhaps it would be easier for a natural scientist to acquire the skills of a politician than for politicians to become educated in the natural sciences. Third, ecological reserves do not allow the taking of resources except for research purposes. So the tiny Ecological Reserves Unit of two to four people ran headlong into the separate demands of all the resource departments. Any other agency could veto a proposal for an ecological reserve, and often did. As a biologist I expected other biologists in the Fish and Wildlife Branch to be supportive of the program, but I was naïve. Some biologists were very effective in blocking proposals as they did not want to lose the option to hunt in even a small part of their region. Clearly the need to protect samples of all our biodiversity, including samples of unhunted wildlife, must be championed more vigorously among the public. Fourth, the numbers of the Ecological Reserves staff were and remain small. There are only two full-time staff in Victoria. Regional Park staff help, but often they do not have the knowledge or time needed to adequately support the program.

Finally, and fundamentally, the public has not yet convinced politicians of the seriousness of the program. When I left the Ecological Reserves program seven years ago we had just finished the third version of the systems plan, a detailed rationale of how many ecological reserves were needed and where. Seven years later the plan is still not public despite a parallel systems plan being unveiled by the Parks Branch. The Ecological Reserves program remains mostly window dressing. (Information on the Ecological Reserves Program and existing reserves are available from an undated pamphlet and map produced by B.C. Parks and a B.C. Environment and Parks 1989 publication.)

The Adversaries of Biodiversity

The adversaries of biodiversity are easily identified. First and foremost are people's attitudes, particularly our attitude towards consumption. The second is population growth, particularly a growth in populations that propagate materialistic and spiritually bankrupt lifestyles.

As witnessed here in B.C., some people value a tree only for its monetary value as cut timber, totally blind to its values to culture and biodiversity. In addition, much tropical rainforest destruction has been in response to economic pressures created here in North America, in Europe or in Japan. The coffee, bananas and some hamburgers we all insist on having come from land where rainforest once stood. And the rainforest is still falling to make sure that we in the First World are never deprived. We could and should switch to herbal teas and apple juice while buying environmentally friendly tropical products such as Brazil nuts and natural rubber.

Paul Erhlich states that a baby born in North America has 50-300 times as much impact on biodiversity and the environment as a baby born in Costa Rica or India. Canada's pressure on world resources, therefore, causes our population of 26 million to be the Third World equivalent of at least 1.3 billion people. The 3 million of us in B.C. are ecologically equal to at least 150 million Third World citizens, and we place terrific economic pressure on highly productive areas such as old-growth forests. With economic pressure we lose habitat and biodiversity.

No doubt our material standard of living will eventually fall, but quality of life can be maintained. It's a matter of attitudes. Our quality of life needs to be independent of our high consumption of material wealth. Experience can teach us that happiness comes from inside us. Contrary to what we tell ourselves in the media, happiness does not come from a new TV.

One of the most unfortunate of modern attitudes is people demanding the maximum in short-term profits. It is daunting to imagine what our great-great-grandchildren will be thinking in a hundred years, but I suspect that they will be wondering how their ancestors could have been so thoughtlessly greedy as to have removed so many of their own descendants' options in just a few decades. They just might decide that preservationists would have been good ancestors.

Some people argue that extinction is a natural process so we need not be overly concerned with the loss of species. Of course extinction is natural, but the rate at which it is occurring now far exceeds the rate at which it occurred when the dinosaurs became extinct (see Scudder, this volume). Significantly, in past extinction episodes, it was mainly animals that went extinct. This time, 10% of the quarter million species of vascular plants in the world could go extinct in the next 10 years. This is a particularly horrifying trend because plants, through photosynthesis, are the key to sustaining life on our planet.

Part of the attitudinal problem includes population growth. Three people are added to the world every second. It is estimated that humans now use 40% of all terrestrial food production, an alarming statistic in light of the possibility of the human population doubling in the next 50 years.

Clearly we must acknowledge the impact that human disrespect for our planet is having upon biodiversity. We must also acknowledge that, until now, biodiversity has been our least appreciated resource. Therefore we must focus attention on collectively understanding that biodiversity is, in fact, our most valuable resource. Here in B.C. we must proceed with the utmost urgency in protecting at least 12% of the land base, a step recommended by the Brundtland Commission and endorsed nation-wide. Samples of all ecosystems must be protected. And outside of preserves we must treat the land with a kinder hand.

When Iraq attacked Kuwait there was instant world reaction. Unfortunately hundreds of Iraq fiascos of grave consequence to our descendants are going on in the world at this moment: human population growth out of control, 25 billion tons of top soil washing to the ocean yearly, thinning ozone layer, greenhouse effect and so on, all of which could nullify our efforts on behalf of biodiversity. The public has yet to react because, compared to the Iraq invasion, all these horrors are incremental, the change being gradual and therefore less noticeable. We must change our attitudes so that our descendants will enjoy at least most of the biodiversity we enjoy today.

Acknowledgements

I wish to acknowledge Dennis Demarchi and Mike Fenger of the Ministry of Environment and Wayne Campbell of the Royal B.C. Museum for their assistance.

Literature Cited

Bailey, R.G. 1989. Ecoregions of the continents (map). USDA Forest Service, Washington, DC.

B.C. Ministry of Environment and Parks. 1989. Guide to ecological reserves in B.C. B.C. Ministry of Environment and Parks, Victoria, BC.

B.C. Environment. 1991. Managing wildlife to 2001: a discussion paper. B.C. Ministry of Environment, Wildlife Branch, Victoria, BC.

B.C. Ministry of Parks. Undated. Ecological reserves in British Columbia (pamphlets and map). B.C. Ministry of Parks, Victoria, BC.

Bunnell, F.L. 1990. Biodiversity: what, where, why, and how. In Symposium Proceedings, Wildlife Forestry : A Workshop on Resource Integration for Wildlife and Forest Managers, edited by A. Chambers. B.C. Ministry of Forests, FRDA Report 160, Prince George, BC. pp. 29-45.

Bunnell, F.L., and R.G. Williams. 1980. Subspecies and diversity — the spice of life or prophet of doom. In Threatened and endangered species and habitats in B.C. and the Yukon, edited by R. Stace-Smith, L. Johns, and P. Joslin. B.C. Ministry of Environment, Victoria, BC. pp. 246-259.

Butler, R.W., and R.W. Campbell. 1987. The birds of the Fraser River delta: populations, ecology, and international significance. Can. Wildl. Serv., Occ. Pap. 65.

Campbell, R.W., N.K. Dawe, I. McTaggart-Cowan, J.M. Cooper, G.W. Kaiser and M.C.E. McNall. 1990. The birds of British Columbia, Vol. 1. Royal B.C. Mus., Victoria, BC.

Cannings, S. 1990. Endangered invertebrates in B.C. Bioline (Assoc. Prof. Biologists of B.C.) 9(2): 15-17.

Demarchi, D.A., R.D. Marsh, A.P. Harcombe and E.C. Lea, 1990. The environment. *In* The birds of British Columbia, Vol. 1, by R.W. Campbell, N.K. Dawe, I. McTaggart-Cowan, J.M. Cooper, G.W. Kaiser and M.C.E. McNall. Royal B.C. Museum, Victoria, BC. pp. 55-144.

Grambine, E. 1990. Protecting biodiversity through the greater ecosystem concept. Natural Areas J. 10: 114-119.

Moore, K. 1991. An inventory of watersheds in the coastal temperate forests of British Columbia. Earthlife Canada Foundation and Ecotrust/Conservation International, Vancouver, BC.

Nagorsen, D., 1990. The mammals of British Columbia: a taxonomic catalogue. Royal B.C. Mus., Memoir 4.

Pickett, S.T.A., and J.N. Thompson. 1978. Patch dynamics and the design of nature reserves. Biol. Conser. 13: 27-37.

Straley, G. 1990. Rare and endangered vascular plants in B.C. — an update. Bioline (Assoc. Prof. Biologists of B.C.) 9(3): 21-23.

Wilderness Advisory Committee. 1986. The wilderness mosaic. Vancouver, BC.

Principles of Biodiversity

Measuring Biodiversity:
Quantitative Measures of Quality

E.C. Pielou

RR #1, Denman Island,
British Columbia, Canada V0R 1T0.

Abstract
Laws to protect tracts of land with high biodiversity will have to contain quantitative specifications, that is, numerical measures of diversity (diversity indexes). It is impracticable to measure the diversity of organisms of all kinds; plants are a useful surrogate, since they both respond to and create the habitat diversity that affects all other life. Indexes of plant diversity are needed that are adequately descriptive and easily understood; they also must be computable from data that can be assembled quickly and cheaply, or that are already on file. Three indexes seem necessary and sufficient for any given tract of land: an estimate of S, the number of plant species; H, the habitat diversity; and R, a measure of the degree of isolation of species populations, as a rough indicator of genetic diversity. This paper describes methods of estimating S, H and R for a given tract of vegetation.

Introduction

No thoughtful person would deny the urgent need to conserve biological diversity by all means available. Biodiversity is threatened everywhere, by the population explosion in low latitudes and by technological explosion and land development in high latitudes. Prompt, effective action is needed on several scales and the forms action can take obviously depend on the scale concerned. In this paper I consider the provincial scale, and the problem is: what can be done to protect the biodiversity of British Columbia from the continuing erosion it is subject to? Protecting threatened and endangered species is only

In *Our Living Legacy: Proceedings of a Symposium on Biological Diversity*, edited by M.A. Fenger, E.H. Miller, J.A. Johnson and E.J.R. Williams, pp. 85-95. Victoria, B.C.: Royal British Columbia Museum.

part, albeit an important part, of the task. We must also protect areas of unusually high biological diversity even if they are not known to contain any particularly noteworthy species.

Many steps must be taken, without delay. The first consists in continuing to educate people about the subject and the need for action. The second step is legislation, and this is where a knotty problem immediately arises. How does a "Diversity Protection Act" specify precisely what people must and must not do? What actions would constitute law-breaking? How will the law be enforced? If previous experience is any guide, this third step will probably be the weakest link in the chain. I am concerned here, however, with a specific aspect of the second step: how do you indicate, quantitatively, what you are trying to conserve?

As a parallel, consider a law intended to prevent water pollution by pulp-mill effluent. The law specifies the maximum number of kilograms of designated chemicals permitted to be discharged, per week or per month, into the soil or into streams, rivers or the sea. Note the exactness of the specification. It admits of no argument. It specifies a *number*, the number of kilograms per unit of time. If the number is exceeded, the law has been broken.

A law to conserve biodiversity, if it is to do any good, must be equally specific. There needs to be a numerical measure (or measures) of diversity as a preliminary to legislation. We must be able to say that if the biodiversity of a tract of wilderness exceeds a certain specified value, then the tract should be preserved intact, because once diversity is destroyed, no remedy can restore it. This may come as a surprise to non-biologists, some of whom believe that damaged ecosystems can be wholly restored provided no species of animals or plants have been driven to extinction. That is emphatically not so, as other papers in this volume explain. My task, however, is to consider how diversity should be measured.

The Need for Simple, Useful Diversity Measures

Research ecologists have used a variety of so-called diversity indexes in the past, the first having been devised in the 1940s. The subject is not new. But none of these indexes is suitable in the present context, for three reasons:

1. The scientific indexes are too mathematically abstruse for use by non-specialists, who would not understand their purposes and derivations.

2. Each index was devised to measure the diversity of a rather narrowly defined ecological community (e.g., the breeding birds on a rocky island, the plants in a meadow or the freshwater plankton of a lake), and none is suitable as an inclusive diversity measure that takes into account all the living organisms in all the varied habitats in even a few hectares of land, let alone those in a big tract.

3. Most existing indexes are computed from data whose collection requires large amounts of time and effort, even when the community being studied is small both spatially and taxonomically.

Therefore, indexes are needed (a single one is insufficient as I shall explain) that are meaningful to non-ecologists, adequate for specifying the diversity of large ecosystems and easily calculated from readily obtainable data. Judging whether my proposed indexes meet these requirements can only be subjective. Therefore, for clarity, I present my proposals without defending them in detail; obviously opinions will differ on whether they meet the requirements just listed and on whether the requirements themselves are exhaustive. I consider terrestrial systems only.

Plants: a Surrogate for All Other Organisms

To begin, I propose that in measuring the diversity of a terrestrial ecosystem, observations be confined to plants. This is because it is impracticable to inventory other taxonomic groups, particularly arthropods and soil invertebrates generally. Taxonomists with the necessary expertise are too few, and the population densities of these organisms fluctuate so greatly, both from place to place and from time to time, that it is impracticable to collect the data needed for reliable inferences.

Considering only plants (vascular plants plus bryophytes) does not, however, lead to an unacceptably narrow view of diversity. Plants both reflect diversity and create it. A large number of plant species implies a diversity of habitats, home to (or fodder for) a correspondingly large number of animal species. Habitat diversity results in part from abiotic causes, chiefly geological and microtopographic, that affect both plants and animals; and in part from the plants themselves, which constitute a diversified array of animal habitats varying from tree-tops to moss cushions.

The Need for Three Indexes

No single index sufficiently describes the diversity of a large ecosystem, even if it is adequate for a circumscribed community. The minimum number of indexes needed in the conservation context is three: first, a measure of plant species diversity (the total number of plant species present in the ecosystem); second, a measure of habitat diversity; and third, a measure of local rarity, to be explained below.

It would be easy to devise additional indexes allowing a fuller description of ecosystem diversity, but one must resist the temptation to use them. Too many indexes would defeat the purpose. The three listed above seem to be the irreducible minimum in the present context. We consider them in turn.

1. The Number of Species

The number of species in an ecosystem is the most obvious index of its diversity. But the number of species cannot be determined exactly, except for very small areas. Unless it is feasible to closely examine every square metre of the area concerned, the number has to be estimated rather than determined.

This is done by making a complete list of all the plant species in each of a large number of small sample plots regarded as representative of a given tract of land, and counting the total number of different species found in all the plots taken together. Given enough plots, this number may closely approach the true total, but there is no guarantee that a few species have not been overlooked. Mathematical methods are required to infer the true total, with some slight but unavoidable inexactness, from the incomplete data. A mathematically more rigorous account of the procedure is given in Appendix A.

An observer will nearly always come across a few "extra" species that, by chance, are not included in any of the sample plots originally laid out. These species should not be ignored. The best way to cope with them is to create extra sample plots to contain them.

The estimated number of species, based on observations on all plots (including extra plots), is the first of the trio of indexes required. Denote it by the symbol S.

2. Habitat Diversity

No satisfactory measure of habitat diversity has so far been devised. An indirect measure can be obtained by the following procedure using the data already collected (a species list for each of a number of sample plots).

Take the species lists of any pair of plots and compare them. Specifically, count the number of mismatches between the two lists, that is, the number of species present in the first plot but not in the second, plus the number present in the second plot but not in the first. Do this for every possible pair of plots and compute the average.

It is intuitively clear that a high value for this average implies greater habitat diversity than does a low value. In an area of low habitat diversity, and thus with fairly uniform vegetation, the mismatches between the species lists from pairs of sample plots will be comparatively few. Conversely, in an area of high habitat diversity, for example a mosaic of coast forest, sand dunes and salt marsh, the number of mismatches between the species lists of plots from contrasted habitats will be great, and the average number of mismatches over all possible pairs of plots (similar pairs as well as contrasted pairs) will be higher than that from uniform vegetation with the same number of species.

The average number of mismatches depends in part on the number of species but it cannot, for mathematical reasons, exceed one-half that number. Therefore, to standardize the average (so that results from different areas will be comparable), it should be doubled and divided by the number of species. Call the result H, the index of habitat diversity. It cannot be less than zero or greater than one. Mathematical details are in Appendix B.

3. Local Rarity

So far we have considered species diversity and habitat diversity. What about genetic diversity? If our irreplaceable genetic resources are to be conserved, it is insufficient merely to prevent currently endangered species from going extinct. We must also ensure that every species retains its full within-species genetic diversity: without genetic diversity, a species loses its capacity to adapt to changing conditions and *ipso facto* becomes endangered. It is not understood nearly as widely as it should be that genetic impoverishment is as dangerous to a species as rarity (see Ledig, this volume). Conserving genetic diversity requires conserving as many as possible of the isolated "local populations" of each species.

All plants grow as local populations. The commonest, most wide-ranging species have local populations that cover big areas, have ill defined margins, and merge gradually with neighbouring populations. In other species, local populations tend to be more isolated from one another and, as a consequence, more genetically distinct because little gene exchange occurs among them. The more isolated a population, the more likely is it to contain unique genes or gene combinations.

For conservation purposes, the best measure of the "genetic potential" (not genetic diversity per se) of a tract of land would be an index (or indexes) based on data giving the numbers of local populations and the degree of isolation among them for every species.

This is obviously impracticable. All we can hope for is some rough and ready measure of the proportion of the plant species in the tract that are represented by only a single population. Then, if the tract is large, these single populations are likely to be isolated (not certainly, as any population near the border of the tract may be exchanging genes with populations outside the border; nor will the single populations be the only isolated ones). An estimate of the proportion, based on data already in hand, is given by the proportion of all the observed species that occur as singletons (a singleton species being one found in only one of the sample plots).

This proportion, which must be less than or equal to one, will be called the index of local rarity and be denoted by the symbol R.

Using the Indexes

The three indexes described, S, H and R, are all computed from easily obtainable data, namely a list of the species in each of a set of sample plots. The indexes are also easy to compute and their meaning is plain to anybody. The least satisfactory of them is R (index of local rarity), because its link with the average within-species genetic diversity (an unmeasurable quantity) is so tenuous. But for lack of a better alternative, it must do.

The next question, one that cannot now be answered, is this: how does one rate the "worth" of a tract of land, in the biodiversity sense, given its trio of diversity indexes (S, H, R)? What combination of values would lead one to decide that the tract should (or should not) be conserved? The decision is necessarily subjective and can only be made judiciously by experienced ecologists who have become familiar with the magnitudes of the indexes yielded by ecosystems they are familiar with. Gaining such familiarity should be fairly easy. The Ministry of Forests is said to have a large body of data obtained by sampling forest plots all over the province, and many more data can no doubt be found in the data banks of all manner of plant ecologists. New data, because of their simplicity, will be quick and easy to gather.

To conclude on a note of urgency, the continued destruction of this province's biodiversity must be halted without delay. But we cannot make demands without stating concisely and exactly what we are demanding. Numbers are essential; anything else is too vague. The indexes I propose could undoubtedly be fine-tuned in various ways, but we have no time for further discussion. Indeed, one must be wary of invitations to interminable round tables and discussions. Their chief accomplishment may be to postpone action.

Appendix A:
Mathematical Notes on the Index S

On a map or air photo, divide the tract to be evaluated into broadly defined sampling strata to separate such obviously contrasted vegetation types as forest, alpine tundra, salt marsh and the like. Determine the area of each stratum. Next, sample the whole tract systematically (or as systematically as is feasible), keeping sampling intensity the same in each stratum. Using sampling plots (quadrats) of one square metre, list all the species in each quadrat. List trees and shrubs having parts vertically above a quadrat.

Estimate the total number of species in each stratum separately and later pool the results to get the grand total. For the rest of this appendix, these notes refer to data from a single stratum.

Write S' for the true total of species in the stratum and s for the observed total. Let $q(s-j)$ be the number of quadrats that had been accumulated when the jth species was encountered for $j = 0, 1, 2$, etc. And let $t(j+1) = q(s-j) - q(s-j-1)$. Consider the values $t(1)$, $t(2)$, ..., $t(7)$. These are the lengths of the final 7 "treads" of the step function approximated by a collector's curve constructed from the data. Treads of length zero are possible; if a quadrat is collected which raises the accumulated number of species from k to $k+m$ with $m > 1$, then $m-1$ treads have length zero.

Randomly permute the ordering of the quadrats N times to obtain N different sets of sample values of the 7 "tread lengths" [i.e., the set $t(i)$ for $i = 1, 2, ..., 7$], with each set based on a different permutation. Make N as large as is computationally feasible, which will depend on the computer being used and on the number of quadrats sampled. The number of quadrats sampled will depend on the area of the tract being evaluated. From these N sets, find mean values of the $t(i)$ for $i = 1$ to 7. Fit a polynomial to the t values, using a polynomial of degree 2 (or 3 if it gives an appreciably better fit). Use the fitted polynomial to extrapolate the step-function to the right until its right-hand end corresponds with the area of the whole stratum (on the abscissa). The ordinate then gives S'(est), the desired estimate of the total number of species in the stratum.

Because extrapolation can introduce large errors, sample size (the number of quadrats examined) should be large enough to ensure that s/S'(est) ≥ 0.85. If, after the first round of computations, the proportion is found to be less than this, additional sampling will need to be done.

Finally, add S'(est) from all strata and subtract repetitions (species that have been counted more than once), to obtain S, the estimated grand total of plant species in the tract.

Appendix B:
Mathematical Notes on the Index H

This index, based on data from all strata, measures the habitat diversity both within and between strata. Before constructing a species x quadrat data matrix on which to carry out the computations, ensure that no one stratum is over-represented, as follows.

For each stratum, construct a collector's curve (actually a step function) by plotting the number of species accumulated (on the ordinate) against the number of quadrats examined (on the abscissa), using the data averaged over N permutations as described in Appendix A. From this graph, read off the maximum number of quadrats, say Q (on the abscissa) corresponding to a value of $s/S'(est) = 0.85$ (on the ordinate). Round up these values to the nearest integers. Then this stratum should be represented by Q quadrats in the data matrix. For each stratum in turn, determine the appropriate Q and discard surplus quadrats selected at random. This ensures that internally homogeneous strata (for which Q will be low) will not be over-represented at the expense of internally heterogeneous strata (for which Q will be high).

Let $N = \Sigma Q$ be the total number of quadrats (from all strata) retained. Let s be the total number of species in these quadrats.

Now construct an $s \times n$ data matrix $\mathbf{X}$ in which

$$x\,(i,j) = \begin{cases} 1 \text{ if species } i \text{ occurs in quadrat } j \\ 0 \text{ otherwise} \end{cases}$$

with $i = 1, 2, ..., s$ and $j = 1, 2, ..., N$

Next determine the number of mismatches, $h(j,k)$ between columns j and k of the matrix. That is

$$h(j,k) = \sum_{i=1}^{s} |x(i,j) - x(i,k)|$$

There are $2^S (2^S - 1)/2$ inter-column comparisons, and therefore

$$\overline{h} = 2 \sum_{j<k} h(j,k)/2^S(2^S - 1)$$

It can be shown that if there are 2^s quadrats, each with a unique species list, then Σh takes its maximum value for the given s, namely

$$\text{Max}\ (\Sigma h) = s\ 2^{2(s-1)}$$

Hence Max $(\bar{h}) \rightarrow s/2$ rapidly with increasing s.

Thus $2\ \bar{h}/s = H$ is a measure of habitat diversity in the range [0,1], and the higher the value of H, the higher the habitat diversity.

Statisticians will notice that h is the mean square dispersion of the data. Let the data be plotted (conceptually) as a scatter diagram in s-dimensional space. Then all the data points are at the apices of an s-dimensional hypercube and h is the mean of the distances squared from every point to the centroid of the whole set of points.

Appendix C:
Why Not Measure Species Quantities?

Some ecologists argue that diversity indexes should always be based on measurements of the quantities of each species in each quadrat rather than on simple binary (presence-absence) data. The arguments against this (in the present context) are as follows: (1) The plants concerned range in size from trees to mosses; measurements of their "amounts" would have to be artificially weighted to give meaningful results. (2) Data collection would be prohibitively laborious unless subjective estimates were allowed; I believe the latter to be so affected by personal bias as to be inadmissible. (3) Many plant species change enormously in size through the growing season (spring-flowering geophytes wither, other species expand) making observations in different months non-comparable. (4) Measuring plant quantities usually means measuring only the above-ground parts that form an unknown proportion of the whole.

Any one of these four reasons alone is enough to make quantitative observations not worth attempting, in the contexts of conservation.

Appendix D:
What About Standard Errors?

No attempt has been made to derive expressions for the sampling variances of S, H and R. To do so would entail making theoretical assumptions (about species abundance distributions and the like) whose correctness would take years of research to confirm. The alternative, namely to estimate the sampling variances empirically, would entail repeating all the field observations and subsequent analyses 20 or 30 times over, which is obviously impracticable.

Luckily, the lack of estimated standard errors for S, H and R is unimportant. Though it would be nice to have confidence intervals for the indexes, they are not strictly necessary. They would be indispensable only if significance tests were to be carried out. Non-statisticians have a tendency to test hypotheses indiscriminately. It would be absurd, for instance, to test whether the number of plant species on the Brooks Peninsula "differed significantly" from the number in an equivalent area in the Gulf Islands, for there is no reason whatever to entertain the null hypothesis that these numbers are the same, even if they did happen to be close. That is the essence of hypothesis testing: a null hypothesis that is not inherently reasonable should not be constructed and tested. Conservationists, because they deal with tracts of land which, on account of their size, are necessarily unique, have little use for such tests.

Some Spatial Aspects
of Biodiversity Conservation

Larry D. Harris

School of Forest Resources and Conservation,
University of Florida, Gainesville, Florida, U.S.A., 32611.

Abstract

Biodiversity is more than just variation among living organisms: it must be viewed in a variety of ways and at a variety of scales. We do not see genotypic variation, but rather phenotypic variation, as it has been shaped by ecological processes acting on individuals and populations. The size of an organism, its use of space and its length of life all vary greatly and demand a number of conservation strategies. The largest organisms (e.g., coral reefs, Giant Kelp, Redwoods) are sessile. Movement of communities or populations of such species is apparent only when seen in the context of millennia. Large size, longevity and slow movement are intercorrelated. Four examples illustrate the complexity that surrounds conservation of biological diversity and the importance of considering different levels of ecosystem hierarchy. A conservation strategy for the Monarch Butterfly, which moves great distances at low elevations across many states and through local jurisdictions, is contrasted to the relatively sessile White-footed and Deer mice (*Peromyscus*). A reserve would protect Deer Mice successfully, but a reserve for the congregation areas of the butterflies would be only one step in maintaining the species. The endangered Manatee's primary source of mortality is collisions with motorboats. Despite reserves, Manatees follow the water courses, so their fate is the responsibility of several states and local authorities. The Florida Panther is a rare subspecies of Cougar. Most of the remaining individuals are within a single reserve, with high mortality outside it. Inbreeding has led to maladies, and reserves have led to isolation of their populations.

In *Our Living Legacy: Proceedings of a Symposium on Biological Diversity*, edited by M.A. Fenger, E.H. Miller, J.A. Johnson and E.J.R. Williams, pp. 97-108. Victoria, B.C.: Royal British Columbia Museum.

Introduction

Biodiversity can be defined as variety and variability among living organisms and the ecological complexes in which they occur. Ecologists consider that the most important aspect of this definition is the phrase "the ecological complexes in which they occur", otherwise the diversity could just as well be kept in seed banks, botanical gardens, zoos and museums. I consider that a further criterion must also be considered, namely that of *all the levels of hierarchy in which they occur*. This means that the combinations of biochemicals that cause the diversity of fall leaf colours need to be considered along with the genetic diversity within and among individuals and the obligate ecological links between populations and species. But because biological diversity is shaped and maintained by ecological processes acting on individuals, it is the phenotypic and not the genotypic diversity that is generally perceived by humans and that is of selective advantage to species. The ecological and biogeographical complexes within which these phenotypes occur are so manifold that in some cases it is the gender of the organism that is determined by environment (such as in turtles), while in other organisms it is a complex of species (such as lichens: an alga and a fungus). Moreover, the organizers of this symposium were correct in concluding that the location and spatial arrangement of human activities across the landscape affects the occurrence and maintenance of biological diversity.

Biogeography and the Use of Space by Organisms

With few exceptions, living organisms require space for their existence and, as a generality, their use of space is intimately tied to their body size. Large organisms such as Redwood trees and elephants clearly require more space than small ones such as bugs and bacteria. Moreover, because the lifespan of smaller organisms tends to be infinitesimal, while the lifespan of large organisms may be immense, individuals of large organisms may occupy a million times more space for a million times longer than the smaller organisms. This relationship is made even more complex, however, by the fact that most organisms must — at some stage of life — move to live, acquire resources, or reproduce. In the case of higher plants such as trees, and in some animals such as oysters, this movement normally occurs during the gametic stage (as in pollen blowing across the landscape) or during the propagule stage (as in seeds being transported by animals). But one reason that ecology is such a rich science for those who study it and so frustrating for one who must contend with it, is because it is the long-lived adults of animals such as beetles, birds and bats that are necessary for the movement of short-lived pollen and propagules of plants.

To make matters worse, it is not the immortality of individuals that conservationists need to worry about, but rather the maintenance of viable populations and thus genetic diversity through space and through time. A single large Redwood tree might occupy a single site for a millennium, with natural selection operating on as few as ten generations since the Pleistocene. Contrast this with the Arctic Tern, with single individuals moving tens of thousands of kilometres annually in migration, while natural-selective processes of an entire hemisphere have operated on perhaps ten thousand generations over the same period. Ecologists and biodiversity managers must not only understand integral calculus but be able to understand how populations and species integrate environmental variance in both time and space over dozens of orders of magnitude.

The amount of movement and thus the amount of space used by organisms depends upon the length of observation time involved. Organisms that are only observed for minutes can not possibly be seen to use as much space as similar organisms that are observed for months or millennia. This leads to the obvious methodological principle that parameters such as territory or home-range size can only get larger with increased numbers or duration of observations. Similarly, while the distributional range of a species or a biological community may appear to be stationary when observed for time periods as short as weeks or years, paleoecologists know that entire plant and animal communities move through space when measured in terms of centuries or millennia.

This relation is best captured by considering an example such as the Africanized Honey Bee (*Apis mellifera*). A foraging bee may move only centimetres from one flower to the next when observed for periods of seconds, but the same bee may require several minutes to travel the tens or hundreds of metres necessary to return to the hive. The individual cannot exist alone, and it depends upon a population that behaves as a super-organism that disperses over kilometres during the swarming period. It also belongs to a species that is expanding its range over thousands of kilometres per century and is presently invading North America from the south (Fig. 1).

But relations between size and movement are not always simple (Fig. 2). Among vertebrates, it appears that the relation of body size to movement is positive in mammals, fish and reptiles, but is inverse for birds (at least of certain foraging types). Thus, whereas individual whales and elephants require more space than individual mice, smaller birds generally migrate farther than large birds. It follows that to

A: Common Units of Time

5×10^4 minutes	= 1 month
5×10^5 minutes	= 1 year
5×10^6 minutes	= 1 decade
5×10^7 minutes	= 1 century
5×10^8 minutes	= 1 millennium

B: Sea-level Encroachment

80 km inland movement in 10,000 years

$$= \frac{80 \text{ km}}{10,000 \text{ yrs}} \quad \text{x} \quad \frac{1000 \text{ km}}{1 \text{ km}} \quad = \quad \frac{8 \text{ m}}{\text{yr}}$$

C: Life-zone Migration

500 km movement in 50 yrs

$$= \frac{500 \text{ km}}{50 \text{ yrs}} \quad \text{x} \quad \frac{1000 \text{ m/km}}{52 \text{ wks x (168 hrs/wk)}}$$

$$= \frac{500,000 \text{ m}}{486,800 \text{ hrs}}$$

$$= \frac{1.14 \text{ m}}{\text{hr}}$$

Figure 1.

A: To compare movement phenomena of greatly different orders of magnitude, it is convenient to use numerical time units raised to different exponents (analogous to logarithms). Note that common English units such as minutes, hours, days, weeks, months or years are translatable to simple multiples of a common base number such as five seconds.

B: The sea level was nearly 100 metres lower during the Pleistocene and the western shoreline of peninsular Florida was much farther west than it is today. Concurrent with glacial melt and rising sea level, the shoreline has transgressed onto the peninsula about 80 km in a period of 10,000 years for an average rate of 8 m/yr.

C: Human activities are probably causing a rise in global temperature that may be 2 °C in the next 50 years. Bioclimatic simulation models suggest that in areas such as Florida where topographic relief is modest, current life zones would need to migrate several hundred kilometres northward to remain in equilibrium with climate. A movement of 500 kilometres over 50 years is approximately one metre per hour. Clearly, biodiversity conservationists must concern themselves with maintaining the opportunity for whole galaxies of fauna and flora to move across the landscape.

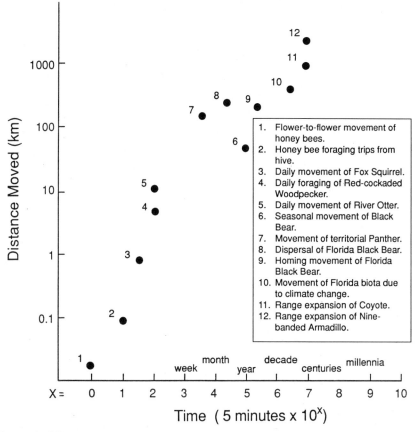

Figure 2. When logarithmic scales of time and distance are used, it is possible to map ecological phenomena of greatly different spatial scales onto a single graph. Movement, ranging from small-scale flower-to-flower foraging of honey bees, to large-scale seasonal migrations of large vertebrates, to range expansion of species across continents, and even to the shifts in life zones and vegetation communities in response to deglaciation and global climate change, seems to follow approximately the same pattern across many scales of time and space. Biodiversity conservationists who are concerned with maintaining the function of plant and animal communities must be aware of these spatial scales of change, and must design movement corridors with dimensions that are appropriate to the phenomena in question.

maintain viable populations of some species such as migrant birds will require at least a low level of protection over vast ranges (the extensive approach), while maintaining species such as Grizzly Bear will apparently require more isolated but more focused and perhaps more intensive approaches (the intensive approach) to protection.

Because movement requires energy and because different environments are subject to different amounts of energy, the medium in which the organisms occur and the mode of movement must also be considered before natural patterns of distribution can be understood. As a generality, organisms that occur on sites where energy and resources flow past them (such as streams) can develop more sedentary life-history traits than those that occur in environments where the resources are stationary and patchy, and where the organism must move about to acquire those resources. Organisms or organism complexes like corals and Redwoods that can remain sedentary and have resources presented to them by wind or water currents, may achieve very large size compared with those that must forage widely for dispersed resources. It is probably not a simple coincidence that the largest biological structures on earth (coral reefs or bioherms) occur in ocean currents, and that the tallest or most massive trees and forests of North America (e.g., Redwoods, kelp and cypress) occur in conjunction with currents of flowing air or water and nutrients.

The complex relations among the scales of hierarchy as they are perceived by humans in space and time are expressed in still other ways. While the chemical diversity of plants (e.g., pigments) can be observed and measured in biochemical terms by stationary scientists in their laboratories, most lay people appreciate the diversity of pigments only when they appear as fall colours arrayed in regional landscapes. In this case, pigments remain stationary while their regional patterns of colouration are caused by differences in soils, topography and the movement of cold air currents up and down valley channels. Their value as a tourist attraction is proportional to the distances travelled by tourists to view the kaleidoscope of colours that occurs across a region. The biochemical diversity of both the coral reef and the autumn-coloured foliage can be maintained in the short term by simple preservation techniques and strategies. But in realistic and useful terms that project significantly into the future, neither the molecular diversity that causes foliage leaf colours nor the rich species assemblages of the coral reef can be maintained by simple preservation techniques. Whereas designating a stationary patch of coral reef or kelp forest as a park or preserve may be necessary to protect it from people who visit and exploit it, this will not suffice in and of itself.

Because the medium and resources flow past a stationary coral reef or kelp forest, the quality of the water in the current determines the ultimate fate of the diversity. No matter what size the coral reef preserve and no matter what degree of protection is afforded within the preserve, the reef can be destroyed easily by a pollutant that enters the ocean current ten thousand kilometres away. The simple conclusion from the above is:

Maintenance or conservation of one or a few individuals over short periods of time such as years may require only small units of designated space such as hectares. The maintenance or conservation of viable populations or species through periods of time such as decades requires larger units of space, measured in kilometres. The maintenance of still higher levels of hierarchy, such as plant and animal communities, for periods of time such as centuries, requires attention to entire regional landscapes or seascapes. Either the organisms will need to move across entire landscapes and, in some cases, entire continents, or if they are expected to remain stationary, then the resource currents that flow past them must be maintained.

Applications of Scale to Biodiversity Conservation

In addition to the issues raised above, molecular and genetic diversity is distributed differently among individuals within species depending upon the life history of the taxon in question. Whereas the genetic diversity encompassed by some taxa would seem to be distributed homogeneously across the entire range of the taxon, the within-taxon diversity of other groups is highly segregated at the species and subspecies levels. The following two examples demonstrate extremes in spatial arrangement of within-taxon biological diversity for two widespread North American organisms.

Monarch Butterfly of Eastern North America
In certain taxa, the annual movements of individuals are large relative to the size of the taxon range and, in at least some of these cases, the full complement of genetic diversity is distributed homogeneously among individuals that move nearly the length of the continent (Fig. 3). Not only is the breeding range of the Monarch Butterfly distributed from Canada to Florida, but individual butterflies themselves migrate the full length of the continent annually. With essentially all individuals moving over the entire range annually there seems to be little opportunity for genetic isolation, differentiation or subspeciation to occur.

Because the migration distances are great relative to the small size

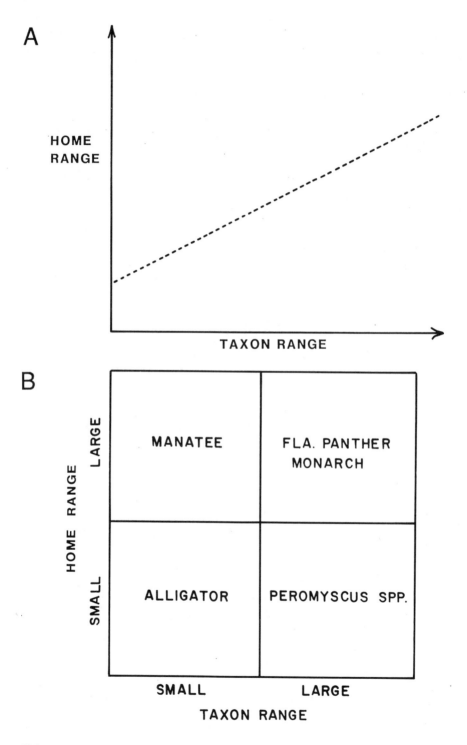

Figure 3. To reveal the full complexity of biodiversity conservation issues and to derive meaningful strategies it is useful to consider the extremes of movement among different kinds of organisms. When the normal home range size or movements of individuals are graphed against distributional range (A), at least four distinct groups result (B). In the case of the Monarch Butterfly of eastern North America, the range of individuals is the same as the species' range, and conservation efforts must be concerned with movement through the landscape context. In the case of mice in the genus *Peromyscus*, the genus's range is at least twice as large as that of the Monarch Butterfly, but because sedentariness prevails among individuals the conservationist must focus on capturing ecotypic diversity across scores of differentiated subspecies. Like *Peromyscus*, the geographic range of the Florida Panther (a subspecies of Cougar) is very large and, therefore, even though the home range size of individuals is large, there is ample opportunity for genetic differentiation. Unsustainably high levels of mortality result from movement outside of the conservation reserves while inbreeding depression results among those that are sequestered within isolated reserves. Because individual West Indian Manatees migrate over distances at least 50% as great as the geographic range of the species' range in North America, it is not possible for an isolated sanctuary approach to effectively conserve the species. Again, attention must be given to the landscape (seascape) context within which the sanctuaries occur.

---

and frailty of the individuals, the full migration cycle depends upon metamorphosis through several life stages of at least two generations. Individual butterflies that are born in Canada must migrate to, find and overwinter in very few remaining high-elevation, relictual fir forests in Mexico that they have never visited before. And, unlike the ecology of birds, there are neither any experienced veteran parents from whom to learn the route nor will these same individuals ever return all the way to Canada.

Because this species could be conserved easily in captivity and because it can be propagated with ease, there appears to be little threat of species extinction. Yet it would be trivial to merely save the species when an entire hemispheric biological phenomenon is at risk.

The phenomenon involves a small and frail invertebrate species that must metamorphose through multiple life stages in order to migrate by the millions over thousands of kilometres to arrive at a precise location on a distant mountain.

Simply saving the species and its genome does not seem to address the real issue. Rather, conservation of a truly great biological phenomenon, involving the movement of genes through time and space in ways not yet understood by humans, is the real issue.

Because the annual Monarch migration phenomenon involves movement near ground level, mortality during migration depends greatly upon human activities. Thus, for all practical purposes, the spatial scale that must be considered for conserving the genetic resource is the same as that required for conserving viable populations. This is identical to the scale necessary to conserve the entire migration phenomenon. An effective conservation strategy must consider not only land-use and land-management practices within the confines of habitat management areas positioned along the full length of the continent, but also the context within which these management areas exist. In the absence of planning, a single east-west interstate highway carrying high-density traffic on a single holiday weekend may eliminate a significant proportion of an entire year's production. A single local government decision to eliminate a 25-hectare stand of Mexican fir forest may destroy 10% of all remaining overwintering habitat. Successful Monarch Butterfly conservation will depend upon sensitization of people ranging from local farmers of at least three different nationalities to government agencies spanning the state and provincial hierarchy of three independent countries (Canada, United States, Mexico). In this extreme case, the ultimate conservation of (1) the species, (2) the unique genetic programming contained therein and (3) the intercontinental migration phenomenon are all one and the same issue.

White-footed and Deer Mice of North America.

Mice of the genus *Peromyscus* are distributed over the entire continent of North America but, unlike the Monarch Butterfly, individual mice do not normally move widely. Thus, even though the genetic diversity encompassed by this taxon may be intrinsically no greater than that of the genus of Monarch Butterflies, it has differentiated profusely. One authority designates well over 200 discrete species and subspecies in North America (Miller and Kellogg 1955). *Peromyscus* individuals are not known to migrate, and it is precisely their localized distribution and propensity for sedentariness that lead to their demise. Perhaps a quarter of the recognized species and subspecies have either already been or are being extirpated because of local land-use developments such as human building complexes on isolated coastal islands.

A conservation strategy that encompasses the genetic diversity among species and subspecies of *Peromyscus* must be very different than that dictated by the biology of the Monarch Butterfly. An ideal strategy for conserving the three levels of diversity of *Peromyscus* (genetic, species and continental) will involve many reserves strategically dispersed across the entire continent. But, because the animals do not move great distances, little if any attention will need to be given to the movement of individual mice among reserves. Emphasis can be focused on selecting refugia and preserves on the basis of their habitat content with little or no consideration of the surrounding context within which the reserves occur.

Other Examples

I now draw attention to two biodiversity conservation cases located on the southeastern corner of the North American continent in order to contrast additional patterns in the erosion of native biological diversity.

In North America, the range of the West Indian Manatee is virtually limited to inland and coastal waters of Florida. The Manatee is North America's largest inland mammal (> 1,000 kg) and, as is fitting such a large creature, individuals of this species live for decades and migrate hundreds of kilometres between seasonal refuges. Because they are surface-breathing, inshore mammals, Manatees migrate in shallow waters that are occupied or used heavily by humans. Consequently, motorboat collisions with Manatees have become the major source of mortality for this endangered species. Even though small, isolated Manatee sanctuaries function as effective seasonal refuges, Manatees rarely stay in them for extended periods of time. Such refuges provide little if any protection for Manatees when they migrate in open water. Thus, although human-use regulations and perhaps some management must occur within the sanctuaries themselves, the real challenge for conservationists involves the nearly lethal context within which the sanctuaries occur. Future conservation strategies will need to be aimed almost totally at the protection and maintenance of Manatee movement opportunities over hundreds of kilometres of near-shore seascape. In contrast to the Monarch Butterfly, where all countries of North America must be involved, the highly localized distribution of Manatees in North America means that only a few state and local authorities hold all options, and bear all responsibility for maintenance of an endemic species that is of great interest to many people.

The case of the Florida Panther dramatizes yet a different combination of spatial and diversity hierarchy issues. Florida Panthers represent a distinct subspecies of a species (Cougar) that is distributed throughout the Americas, a geographic range that spans much more distance and environmental diversity than all of the species mentioned above. Because of the species' wide range and environmental diversity, there are selective pressures for adaptation to local conditions. Because the day-to-day movement of individuals is small relative to the species' range, geographic variation and genetic differentiation have occurred. Unlike the Manatee, however, the distribution of the Florida Panther has become almost totally restricted to a single reserve complex at the southern tip of the Florida peninsula. Mortality resulting from impact by automobiles causes serious concern when wide-ranging males attempt to move among reserves, or when animals of either sex cross major roadways within reserves. The high mortality rate in the landscape matrix surrounding the reserves effectively isolates the reserves and the remaining Panthers to an area of approximately one million hectares. Several maladies that apparently result from close inbreeding now predominate and signal the occurrence of inbreeding depression. In the case of the Florida Panther, it appears that individuals that venture far outside the isolated reserves are doomed to unsustainably high probabilities of human-induced mortality, while those that occupy the more secure territory within the reserves suffer the long-term threats of inbreeding depression.

Conclusion

In total, the four cases presented above demonstrate that the diversity of life-history and biogeographical traits among species is too great to allow for any simple or single conservation prescription to be effective for all. The cases also demonstrate that any strategy aimed at conserving biodiversity spanning three or more levels of hierarchy, and a significant range of taxa for a significant length of time, must explicitly address landscape-level issues. These landscape-level issues derive not only from the intrinsic environmental heterogeneity encompassed within a region, but also from the diversity of environmental processes such as wind and water currents that shape and maintain biodiversity patterns as we know them. A full spectrum of human impacts ranging from intense but localized activities such as road building, clearcutting or site conversion to much less intense but more extensive impacts such as ocean pollution or acid rain must also be addressed. The full spectrum of diversity across all these dimensions must be considered in any realistic attempt to maintain B.C.'s biological diversity resource.

Literature Cited

Miller, G. S., Jr., and R. Kellogg. 1955. List of North American Recent Mammals. U.S. Nat. Mus., Bull. 205.

Biodiversity Over Time

G.G.E. Scudder

Department of Zoology, University of British Columbia, Vancouver, British Columbia, Canada V6T 1Z4.

Abstract

This paper traces the evolution of multi-celled organisms from 600 million years ago to the present. The average annual extinction rate has been about 5 million years per species. There has been a background extinction rate of 9% of species per million years; *mass extinctions* are defined as at least a doubling of this rate. There have been five major mass extinctions with 50% or more of species losses in some. Several theories for mass extinctions have been proposed, including climatic instability, reversal of the earth's magnetic fields, cosmic radiation from supernovae, asteriod impact and vulcanism. Species survival through an extinction is not linked to success during normal periods and nothing is adapted to survive periodic catastrophic events. Therefore, survivors are lucky, not adaptively superior. The large mammal extinctions in the late Pleistocene (30,000-11,000 years ago) in North America and Australia are linked to the hunting cultures. The large and growing human population is the major cause of current extinctions. Tropical deforestation is a major contributor; at current rates, by the year 2100 there will be a 50% extinction level, similar to the mass extinctions of 66 million years ago when the dinosaurs plus 60-80% of the world's other species were lost. Two significant differences from previous extinctions are the large loss of plants and the persistence of the causal agent (human intervention) following the extinctions.

Introduction

In this paper, I will review the temporal aspects of the loss of biological diversity. Specifically, I will compare current estimates of the loss of species diversity today, with data from the geological record. I

In *Our Living Legacy: Proceedings of a Symposium on Biological Diversity*, edited by M.A. Fenger, E.H. Miller, J.A. Johnson and E.J.R. Williams, pp. 109-126. Victoria, B.C.: Royal British Columbia Museum.

will end by stressing the significance of the magnitude of the currently estimated losses over the next 50 to 100 years.

The Geological Record: Extinctions in the Deep Past

Life evidently first appeared on Earth more than 3 billion years ago and, for about the next 1.5 billion years, only the simple bacteria and blue-green algae, or autotrophic prokaryotes, existed (Schopf 1979). The most striking and generally characteristic feature of the Early Proterozoic fossil record is the first abundant and wide spread occurrence of formations called stromatolites, which resulted from activities of these early life forms (Glaessner 1984).

The first fossils of organisms with cells, a nucleus and chromosomes (eukaryotes) are from rocks about 1.4 billion years old (Schopf and Oehler 1976; Cloud 1978; Schopf 1978, 1979; Horodynski 1980; Vidal 1984). These are thought to have arisen from prokaryotes through symbiotic associations (Margulis 1970).

About 640 million years ago, there was a sudden appearance of a diverse assemblage of multicellular eukaryotic organisms (Glaessner 1961, 1971, 1984; Glaessner and Wade 1966). This fauna of soft-bodied forms in the Ediacaran Period did not survive very long, and was replaced by a new assemblage of forms in the Cambrian (starting about 570 million years ago; Seilacher 1984).

We know that the Cambrian fauna was very diverse and contained many forms with skeletons. Like the Ediacaran fauna, it contained many species that cannot be placed in modern phyla. The diversity of the Middle Cambrian Burgess Shale fauna is now well documented (Conway Morris 1979, 1985, 1989; Whittington 1985; Whittington and Briggs 1985). The fact that much of this diversity did not survive into succeeding periods is now a matter of popular knowledge (Gould 1989).

Since the Cambrian, there has been continuous evolution and extinction of life. Over a half billion species may have existed over the past 600 million years (Myers 1988). Current estimates are that there are between 5 and 30 million species [Erwin 1983; Stork 1987; Wilson 1988; but see May (1988)].

Calculations by Raup (1981, 1984) show that for organisms inhabiting shallow marine waters, the longevity of species is usually less than 10 million years (6 million in echinoids, 1.9 million in graptolites, 1.2 - 2 million in ammonites). Paul (1988) records the average survival for blastoid echinoderms as 5 million years. The mean species duration for all angiosperm plant taxa is around 5 million years (Niklas et al. 1985).

If the average survival of a species is about 5 million years and eukaryotic life has been around for over 600 million years, then there could have been at least 120 complete changes of species during the time of metazoan existence.

The uniformitarian model in geology that was promoted by Lyell (1833) and adopted by Darwin (1859) suggests gradual and cumulative change over time. We now know that the geological record does not demonstrate such gradual change: life on earth does not show a constant rate of change (Eldredge & Gould 1972). Neither do we see a constant rate of extinction over time. Over the past 700 million years there have been periods of major or mass extinction that have created biotic crises (Flessa 1979). These mass extinctions have occurred over a few million years (Simberloff 1986), and in the case of Cretaceous pelagic organisms, over a few tens of thousands of years (Hsu et al. 1982); in a few cases they may have been nearly instantaneous (Lewin 1983).

Jablonski (1986a) defined "mass extinction" as a doubling of the extinction rate over background levels for a large number of ecologically disparate taxa over a period that is brief relative to the average duration of the taxa involved.

Raup (1978) calculated that, since the Cambrian began, the average species extinction rate has been about 9% per million years. Raup (1988) translated this into a loss of about one species every five years in a biosphere of two million living species, but noted that this number could be at least a factor of ten higher, because palaeontologists in general cannot detect local endemic faunal change. Thus, the inflated rate would be about two species every year on average.

In contrast, mass extinctions over the same period are estimated to have eliminated at least half of the animal species each time (Raup 1988). Some mass extinctions resulted in much higher losses.

The earliest known mass extinction occurred about 650 million years ago, when a worldwide event eliminated as much as 70% of the plankton diversity on earth (Vidal and Knoll 1982). Seilacher (1984) believes that the Ediacaran fauna represents an entirely separate experiment in multicellular life, with no elements surviving into the Cambrian; Conway Morris (1989) believes there may have been a few survivors.

The other main mass extinctions in the deep past occurred during the Cambrian, Late Ordovician, Late Devonian, Late Permian, Late Triassic and Late Cretaceous (Flessa 1979):

1. The Late Cambrian event had a severe impact on the sponge-like archaeocyathids, several families of inarticulate brachiopods, and many families of trilobites.
2. The mass extinction in the Late Ordovician involved trilobites,

    nautiloid cephalopods, sessile echinoderms, graptolites, ostracods, inarticulate and articulate brachiopods, and bryozoans.

3. The Late Devonian extinction affected the tabulate and rugose corals, many families of articulate brachiopods, the ammonoid cephalopods, trilobites, many forms of conodonts, and the early placoderms and crossopterygian fishes.

4. The Late Permian mass extinction was the most destructive. It caused major extinction in the tabulate and rugose corals, the remaining families of trilobites, all the fusulinid foraminifera, the remaining eurypterid arthropods, two orders of bryozoans, the blastoid echinoderms, many families of brachiopods, a few amphibia, almost all of the synapsid reptiles and some anapsid reptiles. According to Raup (1988) this mass extinction resulted in the loss of 77 - 96% of the marine animal species of the time.

5. Many families of ammonoid cephalopods, all of the labyrinthodont amphibia and almost half of the existing orders or suborders of reptiles were extinguished in the mass extinction in the Late Triassic.

6. The Late Cretaceous mass extinction is perhaps the best known, as it caused the demise of the dinosaurs, as well as the ammonoid cephalopods, many echinoids and many genera of foraminifera and calcareous phytoplankton.

It is improbable that a single cause was responsible for all of these major extinctions over the past 600 million years (Flessa 1979). Both terrestrial and extraterrestrial events have been cited as causes.

In the terrestrial category, Crowley and North (1988) have hypothesized that terrestrially induced climate instability is a possible cause of rapid environmental change and biotic turnover. Since a persistent theme of mass extinction has been the decimation of tropical marine biotas, with less severe losses occurring at higher latitudes, temperature change could be expected to cause mass extinctions heavily biassed towards low latitudes (Stanley 1984).

Reversals of the earth's magnetic field correlate with mass faunal extinctions (Fig. 1). Thus, these reversals directly, or through the climate change they might induce, have been invoked as causal factors (Simpson 1966; Hays 1971; Reid et al. 1976).

Cosmic radiation from supernovae has been cited as a possible extraterrestrial cause of mass extinction (Terry & Tucker 1968). Alternatively, Russell & Tucker (1971) suggest that a nearby supernova explosion could produce climatic effects drastic enough to cause mass extinction.

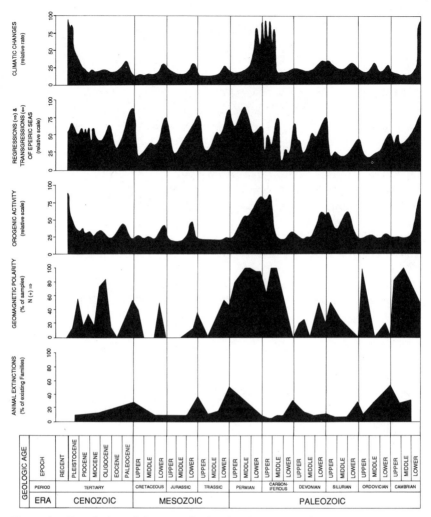

Figure 1. Correlation of the rate of animal extinctions with geomagnetic polarity, orogenic activity, epeiric seas and climate change (after Simpson 1966).

Alvarez et al. (1980) produced strong evidence for an extraterrestrial cause of mass extinctions. They documented high levels of the rare-earth element iridium around 65 million years ago, supporting the thesis that the mass extinction at that time was caused by a large (10 km diameter) asteroid impact. There has been much debate on this topic, with many variations on the impact scenario (Weisburd 1986, 1987). A cyanide-containing comet was suggested by Hsü (1980), while Whipple (1976) noted that a comet could act as a heavenly broom collecting cosmic dust to form the nucleus of a "dirty snowball".

A terrestrial volcanic cause of the Late Cretaceous extinction has also been strongly promoted (Hallam 1987; Officer et al. 1987). However, recent research stressing shocked mineral (tectite) occurrence has revived the extraterrestrial causation (Kerr 1988), with the Caribbean near Haiti being the latest place cited as the impact location (Sigurdsson et al. 1991; Smit 1991). Multiple causes for the Late Cretaceous mass extinction are perhaps nearer to the truth (Ehrlich and Ehrlich 1982; Walliser 1986).

Raup and Sepkoski (1984) analysed the temporal distribution of the mass extinctions of fossil families of marine vertebrates, invertebrates and protozoans over the past 250 million years, and showed a periodicity of 26 million years (Fig. 2). Only an extraterrestrial cause is thought to be able to account for such a pattern. Not all workers in the field agree with this periodicity, although they do find the mass extinctions to be episodic (Larwood 1988).

Although the occurrence of these mass extinctions has been recognized for some time, it has now been discovered that they were both quantitatively and qualitatively different to the background extinction that occurred during intervening intervals (Jablonski 1986b;

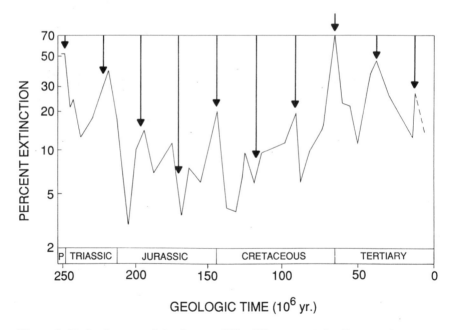

Figure 2. Extinction record for the past 250 million years (after Raup and Sepkoski 1984).

Raup 1986). Background and mass extinctions differed qualitatively in that they affected different groups of organisms (Lewin 1986a).

Jablonski (1986b) compared the evolutionary patterns among Late Cretaceous marine bivalves and gastropods during times of normal background levels of extinction, and over the period of the mass extinction at the end of the Cretaceous. He found that it was the mass extinction that shaped the composition of the biota, and that success during background extinction times had little relationship to survival through mass extinction. Lineages can be lost during mass extinction for reasons unrelated to the survival value of species during background times. Nothing becomes adapted to periodic catastrophic events and there is nothing adaptively superior about the survivors.

As Gould (1989: 305) has aptly remarked, "mass extinctions can derail, undo and re-orient whatever might be accumulating during the normal times between". Survivors owe their existence to good fortune, rather than adaptive superiority: dinosaurs did not do anything wrong in a Darwinian sense (Raup 1986)!

The other fact that we learn from the geological record is that although mass extinctions create opportunities for biotic change, the period of recovery takes many millions of years (Jablonski 1986b). Following the mass extinction in the Late Permian, the marine assemblages were clearly depauperate for at least 5 million years (Raup 1986). After the dinosaur crash, 5-10 million years went by before there were bats in the skies and whales in the seas (Myers 1988).

Late Pleistocene Extinction:
The Influence of Early Humans

Beginning in the Late Pleistocene, about 11,000 years ago, two-thirds of the large mammal species became extinct in America (Martin 1984a, 1984b, 1986). Martin (1984a, 1984b, 1986) has documented that an 86% extinction of large mammals in Australia preceded the megafauna extinction in the Americas; these extinctions were followed by extinctions on continental and oceanic islands (Fig. 3; Martin 1984a, Martin and Klein 1984).

Because most phyla did not suffer more than background extinction at this time, this selective loss of the large mammals appears to qualify as a mass extinction event (Martin 1986). However, it is out of phase with the 26-million-year cycle of mass extinctions in the marine fauna documented by Raup and Sepkoski (1982) and is discordant with any model of globally contemporaneous extinction.

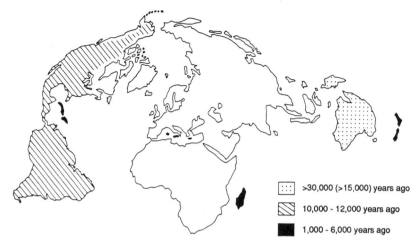

Figure 3. Late Pleistocene extinction of mammals and birds in different parts of the globe (after Martin 1984).

Martin (1967) has shown that Africa and Asia, with a long history of human evolution, suffered a less precipitous extinction of the megafauna than land masses reached by humans relatively late. Although archaeological evidence linking prehistoric societies and the extinction of the megafauna in America is sparse, the first unmistakable evidence of hunting culture coincides with the time of maximum extinction in the late Pleistocene (Martin 1967). It appears that modern humans made an initial onslaught on the biosphere some 11,000-30,000 years ago. The wave of late Pleistocene extinction in the Americas was clearly caused by the overzealous hunting activity of the newly arrived human population (Martin 1973). This explanation also seems to hold for other areas of the world (Martin and Wright 1967).

Recent Extinctions: Present and Future

Modern humanity has continued to have major impacts on planet Earth. The advent of herd hunting by Upper Palaeolithic *Homo sapiens* was responsible for what may have been the first hominid population explosion (Tullar 1977).

The second population explosion probably began about 10,000 years ago, as the agricultural revolution was beginning. By AD 1, the world population very likely had reached 250 million (Dorn 1962). The population has been rising at an increasing rate ever since. It required 1,650 years for the world population of AD 1 to double itself. By 1975 the world population was doubling every 45 years (Dorn 1962).

116

The doubling rate is now in the order of 25 years, with the majority of the increase in the tropical areas of the world. The estimated increase in the population of Asia from AD 1950 to 2000 will be roughly equal to the population of the entire world in 1958. Increases in population of this magnitude stagger the imagination (Dorn 1962). It is surpassed only by the magnitude of the destruction that this population can bring about.

The impact that this ever increasing human population has had and can have on life, ecosystems and the biodiversity on earth, is just now coming into focus. A classic example is the deforestation of the Mediterranean region (Thirgood 1981). No other part of the world so strikingly drives home the story of man's dramatic impact on biological diversity (Franklin 1988). Nowhere more than in the Levant can we find evidence that the deterioration and destruction of the environment have resulted from human impact on natural resources, an impact that has been no less destructive for being gradual and in ignorance of cumulative effects (Thirgood 1981).

Humanity has, unfortunately, not learned a lesson from the Mediterranean. Throughout the world, loss of forest cover is contributing to myriads of environmental problems, including species extinctions (Raloff 1988). The problems exist in both temperate and tropical forests.

Temperate forests, including the old-growth forests of the Pacific Northwest, are endangered habitats (Booth 1989; Dayton 1990). Biodiversity is abundant in the temperate zone and is worth saving (Franklin 1988). In British Columbia, species and populations that are at the limits of their geographic range are especially vulnerable to extinction and should be conserved (Terborgh and Winter 1980). Marginal populations have high evolutionary significance to species as a whole and are essential for the maintenance of genetic diversity and versatility (Scudder 1989).

The tropical rain forest is now the scene of a major massacre. The underlying cause is deforestation because of rapid population growth in tropical moist-forest countries (Caufield 1982). Clearing land for farming is the main cause of tropical forest loss today (Eckholm et al. 1984), but timber harvesting is also a main contributor (Caufield 1982). Progress, the free market, economic growth, technology and the systematic discrimination by humanity against everything that is not human, complete the seven major causes of tropical rain forest destruction (Jacobs 1988). Although people need wood for fuel, and this is an ever increasing problem in the Third World, the demand for firewood is not a prime cause of deforestation, except perhaps in southeast Asia (Eckholm 1975; Eckholm et al. 1984; Monastersky 1990).

The tropical rain forest is of utmost concern today for two main reasons. First, although rain forests cover only 7% of Earth's land surface, they contain more than 50% of the species in the entire world biota (Wilson 1988). Second, the tropical rain forests are being destroyed so rapidly that they will mostly disappear within the next 100 years, taking with them thousands of species to extinction (Wilson 1988).

At least 40% of all of the rain forest that ever existed is no longer present (Jacobs 1988). Current estimates are that about 1% of the tropical rain forest biome is being destroyed each year (Myers 1988; Monastersky 1990). The "Global 2000" report, drafted at the request of then U.S. President Jimmy Carter, estimated in 1980 that one million tropical moist-forest species could be extinct by the end of the century if deforestation continues at the present rate (Caufield 1982).

The situation in Madagascar is an example of the scale of recent human impact. A high proportion of the animal and plant species in Madagascar are endemic. In plants, the endemicity ranges from 89% to 95% in the various groups, and in animals it ranges from 88% in braconid wasps to 100% in tenrecs, cricetid rodents, carnivores and lemurs (Millot 1972). Many of these endemic species are now threatened with extinction as a result of human impact (Griveaud and Albignac 1972).

In Madagascar, numerous habitats have been destroyed since the arrival of humans about 2,500 years ago (Battistini & Verin 1972). Over the years, subsistence needs and cutting for fuel have severely depleted the tropical rain forests of the eastern coastal plains. Green and Sussman (1990) have used satellite imagery to measure the loss in recent times. They show that, by 1985, 66% of the original rain forest had been removed. In the 35 years between 1950 and 1985, 50% of the forest was removed, representing an average clearance of 1.5% per year. If cutting of this forest continues at the same pace, most of the remaining areas will be cleared within the next 35 years.

The rate of deforestation varies from country to country, and region to region, as do the forces driving forest loss (Monastersky 1990). African deforestation ranges from 0.2% a year in Zaire to 10% a year in Nigeria and the Ivory Coast (Caufield 1982). Many Asian forests are on the verge of commercial extinction after just 20 years of intensive logging (Caufield 1982). In the Amazonian rain forest, estimates are that the Brazilian portion is being cleared at the rate of 25,000 - 50,000 km^2 per year (Shukla et al. 1990). Countries such as Peru, Colombia, Ecuador and Venezuela, occupying the upper tributaries of the Amazon system, are losing forests even faster (Caufield 1982). If deforestation continues at the current rate in Amazonia, most of the tropical rain forest will disappear in 50 to 100 years. Once destroyed, it is unlikely to reappear (Shukla et al. 1990).

This broad-scale destruction of tropical rain forests is undoubtedly causing a rapid extinction of species. Although there are no actual data to show that any plant or animal species has yet gone extinct in the New World tropics (Simberloff 1986), there is circumstantial evidence indicating that neotropical bird species have gone extinct because of forest destruction.

We do not actually have to go to see the death of the last individual of a species to prove that species extinction must be occurring on a vast scale today. Species loss can be calculated on the basis of island biogeography theory and species-area relationships (MacArthur and Wilson 1963, 1967; Connor and McCoy 1979).

Actual data from area-species studies on island systems have been used to estimate decay constants (Diamond 1972, 1984; Terborgh 1974; Willis 1974). These theoretical estimates have been supported by empirical studies on areas of subtropical and tropical rain forest (Leck 1979; Willis 1979; Lovejoy et al. 1983). The studies show that about 50% of the species go extinct when a habitat is reduced by 90% in area.

Wilson (1988) used a conservative estimate to determine possible current species loss. He calculated that the annual rate of extinction in tropical forests is now about 17,500 species per year.

These are very rough figures, based on a number of assumptions, but they give an idea of the magnitude of the problem. Island biogeography theory also predicts that more species will disappear during the late stages of the current disaster than in the early stages.

Simberloff (1986) has calculated that 66% of all plant species in the area and 69% of all birds in the Amazon will disappear sometime in the next century if tropical forests in the New World are reduced to only those currently projected as parks and refuges.

Estimates by the Food and Agriculture Organization and the United Nations Environment Program lead to the conclusion that by the year 2135 tropical forests will have been completely destroyed if current rates of destruction are continued (Lewin 1986b). With the forest gone or largely disrupted, up to half of all the species on earth will disappear. A figure of 50% extinction will be close to the mass extinction 66 million years ago, when the dinosaurs finally disappeared together with 60 - 80% of the rest of the world's species (Lewin 1986b). The imminent catastrophe in tropical forests is commensurate with all the great mass extinctions, except for that at the end of the Permian (Simberloff 1986). It characteristically will eliminate species with no regard for their adaptive evolution. However, there are two differences between the current extinction and the mass extinctions of the past (Lewin 1986b). First, unlike previous events, current losses involve a large number of plant species. Second, the agent that is causing extinction — namely, human intervention — will persist.

Mass extinctions usually take about five million years to recover. We should realize that *Homo sapiens* will thus have to live without this biodiversity for a very long time and will have to manage without the genetic resources, medicinal plants, etc., contained therein (Ehrlich and Ehrlich 1982; Oldfield 1984; Abelson 1990).

In addition, with the World Bank estimating that the global human population will plateau at 11 billion (Lewin 1986b), it is unlikely that the forces that have led to tropical forest destruction will change and allow global biodiversity to rebound.

We now have a mass extinction at a time when there should normally be only background extinction. There is no precedent for what is happening to the biological fabric of the earth at this time (Soulé and Wilcox 1980). Instead of two species going extinct every year on average during background times, we now have a rate 1,000 - 10,000 times that because of human intervention (Wilson 1988). We are permanently altering the course of evolution in ways we cannot perceive (Lovejoy 1980). What will be the result if we have yet another terrestrially or extraterrestrially induced mass extinction?

We cannot predict when this will occur, or what will cause it, but there is one immediate ominous possibility. That is global climate change.

Most scientists now acknowledge that Earth's climate is changing (Rizzo 1988). The primary mechanism of this climate change is the increased concentration of carbon dioxide in the atmosphere (Barnola et al. 1987). The recent increase is human-induced, and results from deforestation causing an inbalance between the absorption and release of carbon dioxide by vegetation, and from the combustion of fossil fuels extracted by humanity from Earth's interior (McBean and McEwan 1990).

Climate modellers generally agree that a doubling of atmospheric carbon dioxide will lead to a mean global temperature increase of about 4°C (Schlesinger 1984). We know that this increase will translate into much larger increases at high latitudes and smaller increases near the equator (Harrington 1987). There will be a 6-10°C increase between 50° and 60°N; this may occur as soon as the year 2030 and is highly probable by 2050 (Bolin et al. 1986; Manabe and Wetherald 1986; Harrington 1987; McBean and McEwan 1990).

If we do not plan our conservation strategies correctly and provide large enough areas for species survival (Wilcove et al. 1986), and corridors for movement (Harris and Gallagher 1989; Harris, this volume), we could be on the verge of another mass extinction, for the first time concentrated in the temperate areas of the world.

A biodiversity crisis in British Columbia may not be far away. The year 2030 is only 37 years from now.

Literature Cited

Abelson, P.H. 1990. Medicine from plants. Science 247: 513.

Alvarez, L.W., W. Alvarez, F. Asaro and H.V. Michel. 1980. Extraterrestrial cause for the Cretaceous-Tertiary extinction. Science 208: 1095-1108.

Barnola, J.M., D. Raynaud, Y.S. Korotkevitch and C. Lorius. 1987. Vostok ice core provides 160,000 - year record of atmospheric CO_2. Nature 329: 408-414.

Battistini, R., and P. Verin. 1972. Man and the environment in Madagascar. Past problems and problems of today. *In* Biogeography and ecology in Madagascar, *edited by* R. Battistini and G. Richard-Vindard. Dr. W. Junk, Publishers, The Hague, Netherlands. p. 311-337.

Bolin, B., B.R. Döös, J. Jaeger and R.A. Warrick (editors). 1986. SCOPE 29: the greenhouse effect, climate change and ecosystems. John Wiley and Sons, New York, NY.

Booth, W. 1989. New thinking on old growth. Science 244: 141-143.

Caufield, C. 1982. Tropical moist forests. The resource, the people, the threat. Earthscan, London, England.

Cloud, P. 1978. Cosmos, Earth, and Man. A short history of the universe. Yale Univ. Press, New Haven, CT.

Connor, E.F., and E.D. McCoy. 1979. The statistics and biology of the species-area relationship. Am. Nat. 113: 791-833.

Conway Morris, S. 1979. The Burgess Shale (Middle Cambrian) Fauna. Ann. Rev. Ecol. Syst. 10: 327-350.

_____ 1985. The Middle Cambrian Metazoan *Wiwaxia corrugata* (Matthew) from the Burgess Shale and *Ogygopsis* Shale, British Columbia, Canada. Phil. Trans. R. Soc. Lond. 307B: 507-586.

_____ 1989. Burgess Shale faunas and the Cambrian explosion. Science 246: 339-346.

Crowley, T.J., and G.R. North. 1988. Abrupt climate change and extinction events in earth history. Science 240: 996-1002.

Darwin, C. 1859. On the origin of species by means of natural selection, or the preservation of favoured races in the struggle for life. John Murray, London, England.

Diamond, J.M. 1972. Biogeographic kinetics: estimation of relaxation times for avifaunas of south-west Pacific islands. Proc. Natl. Acad. Sci. USA, 69: 3199-3203.

_____ 1984. "Normal" extinctions of isolated populations. *In* Extinctions, *edited by* M.H. Nitecki. Univ. Chicago Press, Chicago, IL. p. 191-246.

Dorn, H.F. 1962. World population growth: an international dilemma. Science 125: 283-290.

Eckholm, E. 1975. The other energy crisis: firewood. Worldwatch Institute, Washington, DC.

Eckholm, E., G. Faley, G. Barnard and L. Timberlake. 1984. Fuelwood: the energy crisis that won't go away. Earthscan, Washington, DC.

Ehrlich, P., and A. Ehrlich. 1982. Extinction. The causes and consequences of the disappearance of species. Victor Gollancz Ltd., London, England.

Eldredge, N., and S.J. Gould. 1972. Punctuated equilibria: an alternative to phyletic gradualism. *In* Models in paleobiology, *edited by* T.J.M. Schopf. Freeman, Cooper and Co., San Francisco, CA. p. 82-115.

Erwin, T.L. 1983. Tropical forest canopies: the last biotic frontier. Bull. Entomol. Soc. Am. 29: 14-19.

Flessa, K.W. 1979. Extinction. *In* The encyclopedia of paleontology, *edited by* R.W. Fairbridge and D. Jablonski. Dowden, Hutchinson and Ross, Inc., Stroudsburg, PN. p. 300-305.

Franklin, J.F. 1988. Structural and functional diversity in temperate forests. *In* Biodiversity, *edited by* E.O. Wilson. National Academy Press, Washington, DC. p. 166-175.

Glaessner, M.F. 1961. Pre-Cambrian animals. Sci. Amer. 204 (3): 72-78.

_____ 1971. Geographic distribution and time-range of the Ediacara Precambrian fauna. Bull. Geol. Soc. Amer. 82: 509-514.

_____ 1984. The dawn of animal life. A biohistorical study. Cambridge Univ. Press, Cambridge, England.

Glaessner, M.F., and M. Wade. 1966. The late Precambrian fossils from Ediacara, South Australia. Palaeontology 9: 599-628.

Gould, S.J. 1989. Wonderful Life. The Burgess Shale and the nature of history. W.W. Norton and Co., New York, NY.

Green, G.M., and R.W. Sussman. 1990. Deforestation history of the eastern rain forests of Madagascar from satellite images. Science 248: 212-215.

Griveaud, P., and R. Albignac. 1972. The problems of nature conservation in Madagascar. *In* Biogeography and ecology in Madagascar, *edited by* R. Battistini and G. Richard-Vindard. Dr. W. Junk, Publishers. The Hague, Netherlands. p. 727-739.

Hallam, A. 1987. End-Cretaceous mass extinction event: argument for terrestrial causation. Science 238: 1237-1242.

Harrington, J.B. 1987. Climate change: a review of causes. Can. J. For. Res. 17: 1313-1339.

Harris, L.D., and P.B. Gallagher. 1989. New initiatives for wildlife conservation. The need for movement corridors. *In* In defense of wildlife: preserving communities and corridors, *edited by* G. Mackintosh. Defenders of Wildlife, Washington, DC. p. 11-34.

Hays, J.D. 1971. Faunal extinction and reversals of the Earth's magnetic field. Bull. Geol. Soc. Amer. 82: 2433-2447.

Horodyski, R.J. 1980. Middle Proterozoic shale-facies microbiota from the Lower belt supergroup, Little Belt Mountains, Montana. J. Paleontol. 54: 649-663.

Hsü, K.J. 1980. Terrestrial catastrophe caused by cometary impact at the end of Cretaceous. Nature 285: 201-203.

Hsü, K.J., Q. He, J.A. McKenzie, H. Weissert, K. Perch-Nielsen, H. Oberhänsli, K. Kelts, J. LaBrecque, L. Tauxe, U. Krähenbuhl, S.F. Percival Jr, R. Wright, A.M. Karpoff, N. Petersen, P. Tucker, R.Z. Poore, A.M. Gombos, K. Pisciotti, M.F. Carman Jr, and E. Schreiber. 1982. Mass mortality and its environmental and evolutionary consequences. Science 216: 249-256.

Jablonski, D. 1986a. Causes and consequences of mass extinction: a comparative approach. *In* Dynamics of extinction, *edited by* D. K. Elliott. John Wiley and Sons, New York, NY. p. 183-229.

_____ 1986b. Background and mass extinctions: the alternation of macroevolutionary regimes. Science 231: 129-133.

Jacobs, M. 1988. The tropical rain forest. A first encounter. Springer-Verlag, Berlin, Germany.

Kerr, R.A. 1988. Huge impact is favoured K-T boundary killer. Science 242: 865-867.

Larwood, G.P. (Editor). 1988. Extinction and survival in the fossil record. Clarendon Press, Oxford, England.

Leck, C.F. 1979. Avian extinctions in an isolated tropical wet-forest reserve, Ecuador. Auk 96: 343-352.

Lewin, R. 1983. Extinctions and the history of life. Science 221: 935-937.

_____ 1986a. Mass extinctions select different victims. Science 231: 219-220.

_____ 1986b. A mass extinction without asteroids. Science 234: 14-15.

Lovejoy, T.E. 1980. Foreword. In Conservation biology. An evolutionary-ecological perspective, edited by M.E. Soulé. Sinauer Associates, Inc., Sunderland, MA. p. ix-x.

Lovejoy, T.E., R.O. Bierregaard, J.M. Rankin and H.O.R. Schubart. 1983. Ecological dynamics of tropical forest fragments. In Tropical rain forest: ecology and management, edited by S.L. Sutton, T.C. Whitmore and A.C. Chadwick. Blackwell, Oxford, England. p. 377-384.

Lyell, C. 1833. Principles of geology. Vol. 3. John Murray, London, England.

MacArthur, R.H., and E.O. Wilson. 1963. An equilibrum theory of insular zoogeography. Evolution 17: 373-387.

_____ 1967. The theory of island biogeography. Princeton Univ. Press, Princeton, NJ.

McBean, G.A., and A.D. McEwan. 1990. Global climate change. A scientific review presented by the World Climate Research Programme. World Meteorological Organization, International Council of Scientific Unions, Geneva and Paris.

Manabe, S., and R.T. Wetherald. 1986. Reduction in summer soil wetness induced by an increase in atmospheric carbon dioxide. Science 232: 626-628.

Margulis, L. 1970. Origin of eukaryotic cells. Evidence and research implications for a theory of the origin and evolution of microbial, plant and animal cells on the Precambrian Earth. Yale Univ. Press, New Haven, CT.

Martin, P.S. 1967. Prehistoric overkill. In Pleistocene extinctions. The search for a cause, edited by P.S. Martin and H.E. Wright Jr. Yale Univ. Press, New Haven, CT. p. 75-120.

_____ 1973. The discovery of America. Science 179: 969-974.

_____ 1984a. Prehistoric overkill: the global model. In Quaternary extinctions. A prehistoric revolution, edited by P.S. Martin and R.G. Klein. Univ. Arizona Press, Tucson, AZ. p. 354-403.

_____ 1984b. Catastrophic extinctions and Late Pleistocene blitzkrieg: two radiocarbon tests. In Extinctions, edited by M.H. Nitecki. Univ. Chicago Press, Chicago IL. p. 153-189.

_____ 1986. Refuting late Pleistocene extinction models. In Dynamics of extinction, edited by D.K. Elliott. John Wiley and Sons, New York, NY. p. 107-130.

Martin, P.S., and R.G. Klein (Editors). 1984. Quaternary extinctions. A prehistoric revolution. Univ. Arizona Press, Tucson, AZ.

Martin, P.S., and H.E. Wright Jr. (Editors). 1967. Pleistocene extinctions. The search for a cause. Yale Univ. Press, New Haven, CT.

May, R.M. 1988. How many species are there on Earth? Science 241: 1441-1449.

Millot, J. 1972. In conclusion. *In* Biogeography and ecology in Madagascar, *edited by* R. Battistini and G. Richard-Vindard. Dr. W. Junk, Publishers. The Hague, Netherlands. p. 741-756.

Mitchell, J.F.B. 1983. The seasonal response of a general circulation model to changes in CO_2 and sea temperature. Quat. J. R. Meteorol. Soc. 109: 113-152.

Monastersky, R. 1990. The fall of the forest. Tropical tree losses go from bad to worse. Science News 138: 40-41.

Myers, N. 1988. Tropical forests and their species. Going, going...? *In* Biodiversity, *edited by* E.O. Wilson. National Academy Press, Washington, DC. p. 28-35.

Niklas, K.J., B.H. Tiffney and A.H. Knoll. 1985. Patterns in vascular land plant diversification: an analysis at the species level. *In* Phanerozoic diversity patterns. Profiles in macroevolution, *edited by* J.W. Valentine. Princeton Univ. Press, Princeton, NJ. p. 97-128.

Officer, C.B., A. Hallam, C.L. Drake and J.D. Devine. 1987. Late Cretaceous and paroxysmal Cretaceous/Tertiary extinctions. Nature 326: 143-149.

Oldfield, M.L. 1984. The value of conserving genetic resources. U.S. Department of the Interior, Washington, DC.

Paul, C.R.C. 1988. Extinction and survival in the echinoderms. *In* Extinction and survival in the fossil record, *edited by* G.P. Larwood. Clarendon Press, Oxford, England. p. 155-170.

Raloff, J. 1988. Unravelling the economics of deforestation. Science News 133: 366-367.

Raup, D.M. 1978. Cohort analysis of generic survivorship. Paleobiology 4: 1-15.

_____ 1981. Extinction: bad genes or bad luck. Acta Geol. Hisp. 16: 25-33.

_____ 1984. Evolutionary radiations and extinction. *In* Patterns of change in Earth evolution, *edited by* H.D. Holland and A.F. Trendall. Springer-Verlag, Berlin, Germany. p. 5-14.

_____ 1986. Biological extinction in earth history. Science 231: 1528-1533.

_____ 1988. Diversity crises in the geological past. *In* Biodiversity, *edited by* E.O. Wilson. National Academy Press, Washington, DC. p. 51-57.

Raup, D.M., and J.J. Sepkoski Jr. 1982. Mass extinctions in the marine fossil record. Science 215: 1501-1503.

_____ 1984. Periodicity of extinctions in the geological past. Proc. Natl. Acad. Sci. USA 81: 801-805.

Reid, G.C., I.S.A. Isaksen, T.E. Hozer and P.J. Crutzen. 1976. Influence of ancient solar-proton events on the evolution of life. Nature 259: 177-179.

Rizzo, B. 1988. Climate change - a global perspective. Canadian Committee on Ecological Land Classification Newsletter 17: 1-2.

Russell, D., and W. Tucker. 1971. Supernovae and the extinction of the dinosaurs. Nature 229: 553-554.

Schlesinger, M.E. 1984. Climate model simulations of CO_2 -induced climate change. Adv. Geophys. 26: 141-235.

Schopf, J.W. 1978. The evolution of the earliest cells. Sci. Amer. 239 (3): 110-138.

_____ 1979. Precambrian Life. *In* The encyclopedia of paleontology, *edited by* R.W. Fairbridge and D. Jablonski. Dowden, Hutchinson and Ross, Inc., Stroudsburg, PN p. 641-652.

Schopf, J.W., and D.Z. Oehler. 1976. How old are the eukaryotes? Science 193: 47-49.

Scudder, G.G.E. 1989. The adaptive significance of marginal populations: a general perspective. *In* Proceedings of the National Workshop on Effects of Habitat Alteration on Salmonid Stocks, *edited by* C.D. Levings, L.B. Holtby and M.A. Henderson. Can. Spec. Publ. Fish. Aquat. Sci. 105: 180-185.

Seilacher, A. 1984. Late Precambrian Metazoa: preservational or real extinctions. *In* Patterns of change in Earth evolution, *edited by* H.D. Holland and A.F. Trendall. Springer-Verlag, Berlin, Germany. p. 159-168.

Shukla, J., C. Nobre and P. Sellers. 1990. Amazon deforestation and climate change. Science 247: 1322-1325.

Sigurdsson, H., S. D'hont, M.A. Arthur, T.J. Bralower, Z.C. Zachos, M. Van Fossen and J.E.T. Channell. 1991. Glass from the Cretaceous/Tertiary boundary in Haiti. Nature 349: 482-487.

Simberloff, D. 1986. Are we on the verge of a mass extinction in tropical rain forests? *In* Dynamics of extinction, *edited by* D.K. Elliott. John Wiley and Sons, New York, NY. pp. 165-180.

Simpson, J.F. 1966. Evolutionary pulsations and geomagnetic polarity. Bull. Geol. Soc. Amer. 77: 197-204.

Smit, M. 1991. Where did it happen? Nature 349: 461-462.

Soulé, M.E., and B.A. Wilcox. 1980. Conservation biology: its scope and its challenge. *In* Conservation Biology. An evolutionary-ecological perspective, *edited by* M.E. Soulé and B.A. Wilcox. Sinauer Associates, Inc., Sunderland, MA. p. 1-8.

Stanley, S.M. 1984. Marine mass extinctions: a dominant role for temperature. *In* Extinctions, *edited by* M.H. Nitecki. Univ. Chicago Press, Chicago, IL. p. 69-117.

Stork, N.E. 1987. Arthropod faunal similarity of Bornean rain forest trees. Ecol. Ent. 12: 219-226.

Terborgh, J. 1974. Preservation of natural diversity: the problem of extinction-prone species. BioScience 24: 715-722.

Terborgh, J., and B. Winter. 1980. Some causes of extinction. *In* Conservation Biology. An evolutionary-ecological perspective, *edited by* M.E. Soulé and B.A. Wilcox. Sinauer Associates, Inc., Sunderland, MA. p. 119-133.

Terry, K.D., and W.H. Tucker. 1968. Biologic effects of supernovae. Science 159: 421-423.

Thirgood, J.V. 1981. Man and the Mediterranean forest. A history of resource depletion. Academic Press, London, England.

Tuller, R.M. 1977. The human species. Its nature, evolution, and ecology. McGraw-Hill Book Co., New York, NY.

Vidal, G. 1984. The oldest eukaryotic cells. Sci. Amer. 250 (2): 48-57.

Vidal, G., and A.H. Knoll. 1982. Radiations and extinctions of plankton in the late Proterozoic and early Cambrian. Nature 297: 57-60.

Walliser, O.H. (Editor). 1986. Global bio-events. A critical approach. Springer-Verlag, New York, NY.

Weisburd, S. 1986. Extinction wars. Science News 129: 75-77.

_____ 1987. Volcanoes and extinctions: round two. Science News 131: 248-250.

Whipple, F.L. 1976. Background of modern comet theory. Nature 263: 15-19.

Whittington, H.B. 1985. The Burgess Shale. Yale Univ. Press, New Haven, CT.

Whittington, H.B., and D.E.G. Briggs. 1985. The largest Cambrian animal, *Anomalocaris*, Burgess Shale, British Columbia. Phil. Trans. R. Soc. Lond. 309B: 569-618.

Wilcove, D.S., C.H. McLellan and A.P. Dobson. 1986. Habitat fragmentation in the temperate zone. *In* Conservation biology. The science of scarcity and diversity, *edited by* M.E. Soulé. Sinauer Associates, Inc., Sunderland, MA. p. 237-256.

Willis, E.O. 1974. Populations and local extinctions of birds on Barro Colorado Island. Panama Ecol. Mon. 44: 153-169.

_____ 1979. The composition of avian communities in reminiscent woodlots in southern Brazil. Pap. Avulsos Zool. 33: 1-25.

Wilson, E.O. 1988. The current state of biological diversity. *In* Biodiversity, *edited by* E.O. Wilson. National Academy Press, Washington, DC. p. 3-18.

Secret Extinctions:
The Loss of Genetic Diversity
in Forest Ecosystems

F. Thomas Ledig

Institute of Forest Genetics, Pacific Southwest Forest
Research, Forest Service - USDA, Box 245, Berkeley,
California, U.S.A. 94701.

Abstract
The biodiversity crisis is usually equated with species extinctions, but much
more common are the loss of genetic diversity through the extirpation of locally
adapted populations and the reduction of genetic diversity within species.
These local losses are the secret or hidden extinctions. They result in the erosion
of genetic resources and affect the recovery of damaged ecosystems. In general,
conservation efforts are justifiably focused on low latitudes, because genetic
diversity among species, populations within species and individuals within
species all tend to increase from high latitudes to the tropics. Likewise, loss of
diversity from overexploitation and habitat destruction increases along the same
gradient. Nevertheless, these losses, particularly tropical deforestation, will
have severe impacts at northern latitudes because they contribute to the
Greenhouse Effect and global warming. Conservation through the
establishment of reserves and through seed banks is important, but reserves and
seed banks in themselves are insufficient to halt the loss. Conservation must be
a factor in all land-management activities.

Introduction

According to some estimates, the rate of extinction is now about
100 species per day, exceeding any previous period in Earth's history
(Myers 1984). As drastic as this seems, the loss of species is only the tip
of the iceberg. Loss of populations and of genetic diversity within

In *Our Living Legacy: Proceedings of a Symposium on Biological Diversity*, edited by
M.A. Fenger, E.H. Miller, J.A. Johnson and E.J.R. Williams, pp. 127-140. Victoria,
B.C.: Royal British Columbia Museum.

populations may be even more critical than species extinctions. Genetic diversity is central to biological diversity at all levels and scales (Fig. 1). It leads to the uniqueness of individuals and populations or races. It is the genesis of forest structure, communities, ecosystems and ecological processes. Yet loss of genetic diversity is cryptic: it goes unnoticed. For every species that goes extinct, probably a score of secret extinctions occurs, the unheralded loss of locally adapted gene complexes. This crisis — this global threat to genetic diversity — is worse than any in Earth's geologic history because it will take place in a very brief span of time.

In this article, genetic diversity will be used to mean genetic diversity in the broad sense: the entire continuum between different forms of the same gene within species, through the many genetic differences that characterize different populations or races, up to the

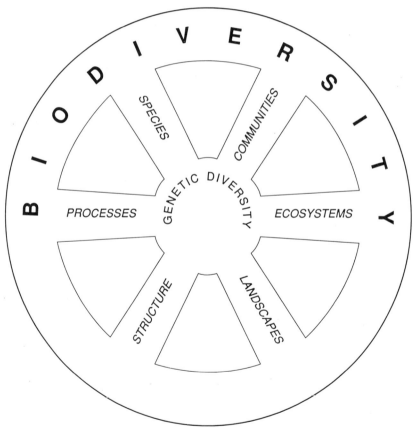

Figure 1. Genetic diversity is basic to biological diversity at all levels (from Keystone Center 1991).

whole libraries of genes that distinguish different species. The discussion will range over the entire latitudinal gradient from the boreal to the Amazonian forest, with British Columbia, California and Mexico as specific examples.

In the first section I review the distribution of genetic diversity which, in general, increases from north to south. The second section covers some threats to diversity including overexploitation, habitat destruction and environmental change. Overexploitation and habitat destruction, like diversity itself, increase toward the tropics. But environmental change, particularly global warming, is all-pervasive. In the third section, I discuss the importance of genetic diversity, especially for B.C. I finish with a brief consideration of conservation, which will not be accomplished simply by establishing a system of preserves. This paper is an overly brief treatment of a complex subject, and the reader is referred to a series of papers from which this one is drawn (Ledig 1986, 1988a and especially Ledig 1988b).

The Geographic Distribution of Genetic Diversity

Genetic diversity is not evenly distributed around the globe. Diversity tends to increase from the polar regions to the tropics. For example, B.C. has 44 tree species (or perhaps fewer, depending on the definition of tree; Hosie 1969). The taxonomy of trees and shrubs is stable and well defined, though some (e.g., birches *Betula* spp.) offer some problems. Most plant genera, both of trees and of herbs, include only a few species.

California has about twice as many tree species (86) as does B.C. and these are also well defined taxonomically (Griffin and Critchfield 1972). California, however, still has some botanical surprises: 5 - 10 new species of herbs are described every year (Shevock and Taylor 1987)!

In Mexico, there are 107 species of conifers alone (Martinez 1963) and over 2,000 species of hardwoods (J. Rzedowski pers. comm.). New species and new range extensions are discovered every year and taxonomically the situation is in flux, even in woody genera. For example, does Mexico have four species of Douglas-fir (*Pseudotsuga* spp.) or just one?

But Mexico is largely a temperate country and to really see a proliferation of species, head to the tropics. In some forests of South America, 300 different tree species per hectare are common and many species are still unknown to science (Gentry 1986). Species diversity increases by an order of magnitude from Canada to the tropics.

Among populations within species there may also be a tendency

129

towards greater diversity in the south than in the north, at least in species affected by glaciation. Little genetic differentiation exists among sites or elevations in populations of Jeffrey Pine (*Pinus jeffreyi*) or Western White Pine (*Pinus monticola*) in the northern parts of their ranges, but substantial differentiation occurs in California (Steinhoff et al. 1983; Furnier and Adams 1986). Presumably too few generations have elapsed since the northern latitudes were colonized by these species to permit evolution of finely tuned genetic adaptation of populations to the local environmental mosaic. Only 6,000 years have passed since the climate warmed, the glaciers receded, and coniferous forest migrated into the Puget Sound Basin, while the California Sierran mixed conifer forest has been in place for 10,000 years (Brubaker 1989).

Genetic diversity within populations follows an equally interesting pattern, probably reflecting genetic drift rather than selection. Here too, diversity increases from north to south. Coulter Pine (*Pinus coulteri*) is a good example. Diversity is twice as great in relict, southern populations as in more recently established northern ones (Ledig 1987). Diversity decreases northward along the mountain ranges from Baja California to Mt Diablo. A similar situation is found in many other conifers, including Giant Sequoia (*Sequoiadendron giganteum*), Jeffrey Pine, Eastern White Pine (*Pinus strobus*), and Western White Pine (Fins and Libby 1982; Ryu and Eckert 1983; Steinhoff et al. 1983; Furnier and Adams 1986).

Why do these trends exist? Perhaps because many species were eliminated from the northern parts of their ranges during glacial periods and populations were reconstituted by colonization after the glaciers melted. Southern stands remained in place for hundreds of thousands of years maintaining fairly sizable populations, which is a situation conducive to high levels of genetic variation. As the glaciers melted, new habitat opened up in the north and was colonized by long-distance dispersal from the south. Each migration or colonization event may have involved only one or a few seeds carried by the wind, a chance hurricane or birds. They came first from southern refuges and, in later generations, from the small advance colonies founded by early migrants. A few seeds, however, can carry only a sample of the parental genes. Therefore, each step northward was an opportunity for genetic loss because of successive subsampling from increasingly limited pools (Fig. 2).

In summary, a tendency toward higher levels of genetic diversity in southern latitudes and lower diversity in higher ones may characterize species, populations within species and genes within populations. These broad trends should guide conservation efforts, whether the goal is to maintain a representative sample of the biota or to conserve genetic resources for their potential in breeding.

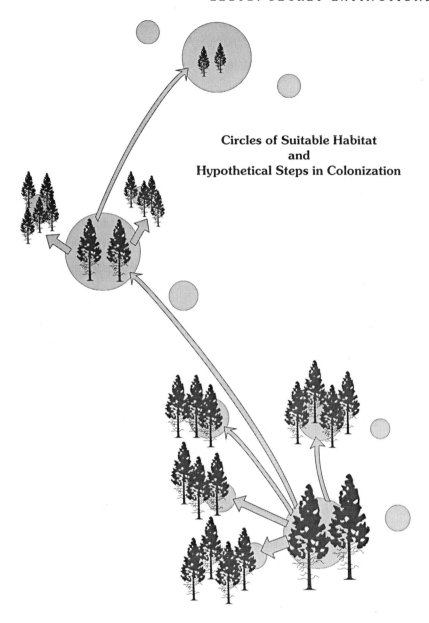

**Circles of Suitable Habitat
and
Hypothetical Steps in Colonization**

Figure 2. A model for dispersal northwards at the end of glaciation. Colonization proceeds in steps across generations. The chance of colonization decreases with distance to suitable habitat and increases with size of the habitat patch. The further the distance, the fewer seed that are likely to establish the colony and, therefore, the greater the opportunity for genetic drift and loss of rare alleles.

131

Extinction and Loss

John and William Bartram, father and son, were two of the earliest botanists in colonial America. In 1765 the Bartrams made a collecting expedition to Georgia (Kastner 1977). One of their discoveries was a grove of small flowering trees, the Franklin Tree (*Franklinia alatamaha*). William collected seeds from the trees in 1776. He and his father visited them again in 1803. The Franklin Tree has never been seen since — it is one of North America's first documented extinctions after colonization by Europeans. There have been others.

But the biggest threat to genetic diversity is not species extinction. It is loss of populations, loss of local genetic diversity. It is most unfortunate that we all concentrate so much on endangered species, because it detracts from this more serious problem. In virtually all economically important tree species, populations were eliminated as land was converted to agricultural and urban uses. And as agriculture moves to higher elevations, particularly in developing countries, more populations and more genes are lost, even though the "species" in the popular sense seems to be in no danger.

The loss can occur very quickly. In 1968 Bill Libby from the University of California (Berkeley) and Ken Eldridge from CSIRO in Australia visited Guadalupe Island off Baja California. Guadalupe Island had an isolated population of 368 Monterey Pine (*Pinus radiata*). Today, slightly more than 20 years later, only 45 remain, and the goats of Guadalupe may soon destroy those. The loss of the Guadalupe pines is particularly unfortunate because they have proven quite resistant to Western Gall Rust and Red Band Needle Blight, diseases that cause some concern in countries of the world that grow Monterey Pines (Cobb and Libby 1968; Old et al. 1986). But there is no great public outcry about the loss of the Guadalupe pines because the "species" is not in danger.

In other cases, the genetic resource has been degraded even without losing populations. Mahogany (*Swietenia mahogani*) has been seriously overexploited in the Caribbean. In many areas, Mahogany is no longer found as a tree. The species has been reduced to a multi-stemmed shrub (Styles 1972). In this case, genes for tree form may have been lost even though the populations themselves are still in existence.

In summary, the impending crisis is not simply one of species extinction, though that is a serious problem. The crisis includes loss of the genetic diversity that makes populations and individuals unique and enables them to adapt to new environments or evolve resistance to new challenges from pests and pathogens.

Threats to Diversity

As the United States and Canada developed westward, exploitation undoubtedly took its toll of genetic diversity. In California, low-elevation populations of Ponderosa Pine (*Pinus ponderosa*), White Fir (*Abies concolor*) and other species were eliminated a century ago because they were the most accessible timber sources for construction and mining props. Genes that enabled Ponderosa Pine to adapt to droughty, low-elevation sites may have been lost when those stands were cut. Where forests stood in 1849, brush-fields of manzanita (*Arctostaphylos* spp.) and other chaparral species now grow (Bolsinger 1980; Shoup 1981).

Today, overuse of forests continues predominantly in the developing countries. For example, La Ceiba (*Ceiba pentandra*) supported five major plywood mills and ten small sawmills along the Amazon of Peru just a dozen years ago. By 1983 two of the plywood mills were already shut down, the other three were in the process of folding, and the ten smaller mills were closed — all because the *Ceiba* were gone (Gentry and Vasquez 1988).

Habitat destruction is another threat to genetic diversity. Much forest land in Canada and the United States was lost to agriculture in the first three centuries of European colonization. At present, forest is being converted mainly to urban development, but the threat to wildland species is the same. An example is MacNab Cypress (*Cupressus macnabiana*), a species of limited occurrence endemic to California. A few years ago, Connie Millar of our Institute of Forest Genetics went to collect seed from Grass Valley, near the southern extreme of the species in the Sierra Nevada. To her dismay, she found the site had been bulldozed for condominiums. Most of the cypress were piled for burning. Only a few trees remain as landscaping between the buildings.

Biological diversity increases from north to south and, likewise, the amount of forest land being lost increases southward. The estimated loss of commercial forest to roads, railways and residences in B.C. was 0.04% in recent years (B.C. Ministry of Forests 1984). In California it was 1.2% in the 25-year period from 1952 to 1977 (Bolsinger 1980), but when statistics are available for the last 10 years, the figure will probably be higher. To anyone who visits Mexico frequently, widespread destruction of forests is obvious. Deforestation from 1981 to 1984 was 5.2% (Office of Technology Assessment 1984), exceeding in only four years the losses over 20- and 25-year periods in B.C. and California. Mexico's population is increasing at an annual rate of about 2.5% (Riding 1984). That equates at present to about two million people annually, many born to subsistence farmers, meaning that more land must be put under cultivation. You can watch the fields climb the mountains!

The worst destruction of habitat is in the tropical forest. Much of the destruction in the tropics is a result of swingle, or slash and burn, agriculture. The life of new agricultural fields is brief — a couple of years — but the damage may be forever. In many cases, genetic resources are lost and some of the soils will not soon support tall forest, at least not for hundreds of years. Norman Myers (1984) was accused of exaggeration when he estimated that tropical forests were disappearing at the rate of 7.3 million hectares per year. That was six years ago. But recent data from the Brazilian Institute of Space Research, still unpublished, show that 20 million hectares of Amazonian forest — *20 million hectares* — were burned in 1987 alone, a particularly bad year (Roberts 1988). That's the equivalent of clearing and burning *one-quarter* of B.C. in a single year! Eight million hectares were virgin forest. Thus habitat destruction in Brazil alone exceeds what we thought was happening in the entire tropics of South America, Africa and Asia.

Large-scale disturbance in the tropics is much more likely to disrupt communities and lead to extinctions than in Canada or the United States. Our northern communities, made up of weedy species, are more resilient. They are the survivors that adapted to repeated cycles of disturbance as the glaciers waxed and waned (Critchfield 1984). In addition, the level of endemism is much greater in the tropics than in most temperate forests. A few years ago the tropical botanist Alwyn Gentry (1986) found 90 new species of plants on Centinela Ridge in Ecuador, apparently endemic to an area only 5-10 km^2. Today they are gone. The last patch of forest on Centinela Ridge has been cleared.

The loss of tropical forest is a threat to us — a time bomb that will explode here in B.C. as well as in the Andes and the Amazon Basin. In part, that is because today's most insidious problem is environmental change. Though environmental change includes toxic air pollution and the destruction of the atmospheric ozone layer, the most serious threat is global warming (Harrington 1987).

Global warming results, in part, from an increase in atmospheric carbon dioxide. The evidence that atmospheric carbon dioxide has increased is now indisputable. Increased levels of carbon dioxide result from burning fossil fuels, coal and oil and (note the connection) from destroying tropical forests. The destruction of tropical forests contributes to increased levels of carbon dioxide in two ways. First, conversion of tropical forests by burning releases carbon that has been stored in wood — about 20% of the recent increase in carbon dioxide is a result of burning tropical forests. Second, burning or cutting the forest destroys a potential sink for atmospheric carbon dioxide; that is, it reduces the amount of leaf cover, thereby reducing carbon dioxide uptake by photosynthesis. But increased levels of carbon dioxide by themselves might be a blessing; low carbon dioxide limits plant growth. Forest productivity should therefore increase with increases in carbon dioxide. The catch is that carbon dioxide is a so-called greenhouse gas. It allows solar radiation to pass into the atmosphere but absorbs long-wave radiation that Earth reflects back toward space. Largely because of the increase in atmospheric carbon dioxide, climate models predict an average global increase in temperature of 2.5°C by the year 2050. The increase in temperature will be more pronounced in Canada than in the tropics. As a result, forest decline will be common throughout the high latitudes. I believe that it has already begun. Species with broad ecological ranges or wide ecological amplitude may survive, in part. Others may colonize newly favourable habitat northward or higher in elevation. Many species, however, will become extinct as the climate changes, because the change will be rapid and because migration routes are already closed by agricultural and urban development (Peters 1990).

The Rationale for Conserving Diversity

As the popular media tell us, the processes of species extinction and genetic loss are greatest in the tropics, in the developing countries. Does that have anything to do with B.C.? It has everything to do with B.C. As John Muir (1916: 157) said, "When we try to pick out anything by itself, we find that it is hitched to everything else." Neither B.C. nor California can afford to be provincial. Why should the people of this province worry about conserving genetic diversity? For the same reasons that should concern all of society — the four Es: economics, ecology, (a)esthetics and ethics.

Economic reasons sell better, so I will consider them in detail. B.C. has a stake in the genetic resources of the world. It depends on an immense variety of genetic resources that are non-native and this dependency will increase in the future. Agricultural production in this province is led by dairy products, fruit (mainly apples) and grains (wheat and barley) (B.C. Ministry of Agriculture and Fisheries 1988). *None* of these is native. Dairy cattle are derived from wild cattle that were domesticated in the Middle East and Europe; apples originated in the Caucasus Mountains of western Asia; wheat was originally from the mountains of Asia Minor and the Sudan of Africa; barley is from Abyssinia or Tibet (Janik et al. 1969). B.C. is not unique in relying on exotic crops. In the entire United States, only five native crops are important agricultural commodities: sunflowers, pecans, cranberries, blueberries and Jerusalem artichokes (Witt 1985). In brief, agriculture has been characterized by movement or importation of crop plants and animals on a global scale.

If predictions of climatic warming come true, B.C. may be pushed to consider importing new *forest* crops. If it does not import new species, at least it may need new provenances of the present species. After a doubling of atmospheric carbon dioxide and a projected increase in global temperature, southern boundaries of the cool temperate conifer forest that now lie south of the Klamath Mountains in California will be relocated northward (Bolin et al. 1986). In only 60 years B.C. may be at the southern extreme of the zone. The trees of local seed source now planted here may never reach rotation. The correct provenances of Douglas-fir (*Pseudotsuga menziesii*) for southern B.C. may be similar to the ones now growing in northern California, but we Californians may find it better to grow Mexican Douglas-fir. Therefore, the people of this province have a real stake in how we protect genetic resources in California, while Californians might be wise to focus their attention further south.

Let's turn to ecological reasons for conserving genetic diversity. Ecosystem function is important to us all. It provides free life-support services — for example, the production of oxygen. Yet humanity's insults to the environment are leading to changes on a global scale, changes that may endanger all life on the planet, even ours. Removing a few species here and there or destroying a few populations — with their more limited genetic diversity — may have no great effect on ecosystem function. But then again, it may. Can we predict? Eventually we should expect a cascade of extinctions, particularly if keystone species are lost. Suffice it to remind ourselves that we all have an interest in the vast changes that are occurring, even changes in the distant tropics, because they have an impact on *global* climatic and nutrient cycles.

Aesthetics and ethics are other important reasons for conserving biodiversity but, in the space allotted, I can't do those topics justice (see Ledig 1988b).

Conservation

Conservation practices can be divided into those operating *in situ*, meaning in place, and those that are *ex situ*, out of place (Ledig 1986). A large literature has developed on *in situ* methods: establishing reserves, minimum viable population sizes, whether a few large or many small reserves is best and the interconnection of reserves via migration corridors. But the truth is, even in the rich, developed countries of the world, the area put aside for reserves can never be sufficient to assure the persistence of all species, let alone all populations. National parks, wilderness areas and other protected areas occupy only 7% of the United States' territory (Fig. 3). In developing countries the area in reserves is often smaller and population pressure makes reserve boundaries meaningless. Fifty years ago, there were just over two billion people in the world; today, there are five billion (Johnson 1986). That is the key to the extinction crisis. Starving people will not respect reserve boundaries. Nor does atmospheric pollution, and in parts of Europe and southern California forest decline could make *in situ* conservation impossible.

Ex situ conservation in seed banks is insurance against loss of *in situ* reserves. Many tree species, however, particularly tropical species, have seeds that cannot be stored. These species can be maintained in arboreta or outplantings for several decades or even centuries, but the cost is high and saving a few specimen trees in seed banks or arboreta preserves only a fraction of the total diversity in a species. The process of regenerating these seed banks or arboreta will be plagued by problems of inbreeding and by biased selection of particular genetical characteristics, which will further reduce genetic diversity.

To be effective, conservation of genetic diversity must be an

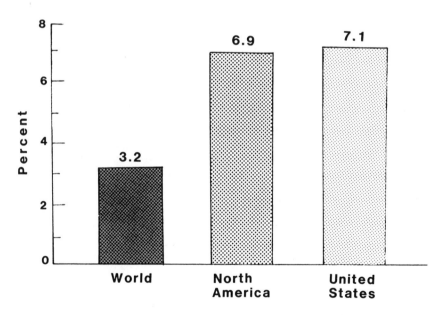

Figure 3. Bar diagram showing per cent of territory in major protected areas in 1985 (after data in World Resources 1988).

integral part of multiple-use management (Salwasser 1988). It must be integrated with other land uses such as timber harvesting, grazing, recreation, agriculture and even urban development. And, what's more, we must all help as citizens, not simply in our limited, professional roles as foresters or biologists.

Conclusion

In summary, I have argued that:
- Diversity is threatened by the loss of species, but the loss of populations is the great, unseen wave of extinctions — the secret extinctions.
- Genetic diversity is greater in the tropics than in the temperate regions.
- Like diversity itself, the direct threats to diversity through overexploitation and habitat destruction are also greatest in the tropics.
- On the other hand, environmental change through toxic atmospheric pollutants, destruction of the ozone layer and global warming will have their greatest effect in the high latitudes. Global warming is one reason B.C. must be concerned about tropical deforestation. All the world, however, shares significant economic, ecologic, aesthetic and ethical reasons for stopping the loss of genetic diversity.
- Finally, establishing reserves and seed banks is worth doing. It is business that all the nations of the world should get on with. Truly national or international *systems* of conservation are needed. But, by themselves or in combination, national parks and seed banks are insufficient to conserve our legacy of genetic diversity. Conservation must be made a part of our lives. It must be a consideration in every land-use decision and in our every action.

Literature Cited

Bolin, B., B.R. Döös, J. Jager, and R.A. Warrick (Editors). 1986. SCOPE 29: the greenhouse effect, climate change and ecosystems. John Wiley and Sons, Chichester, England.

Bolsinger, C. L. 1980. California forests: trends, problems, and opportunities. U.S. Forest Service Resource, Bull. PNW-89.

B.C. Ministry of Agriculture and Fisheries. 1988. 1987 farm, fish and food statistics. B.C. Ministry of Agric. and Fisheries, Victoria, BC.

B.C. Ministry of Forests. 1984. Forest and range resource analysis, 1984. B.C. Ministry of Forests, Strategic Studies Branch, Victoria, BC.

Brubaker, L. B. 1989. Climatic change and the origin of Douglas-fir/Western Hemlock forests in the Puget Sound lowlands. *In* Symposium Proceedings, Old-growth Douglas-fir Forests: Wildlife Communities and Habitat Relationships, *edited by* K.B. Aubry, L.F. Ruggiero, and M.H. Huff. Portland, OR. p. 13.

Cobb, F. W., Jr., and W. J. Libby. 1968. Susceptibility of Monterey, Guadalupe Island, Cedros Island, and Bishop pines to *Scirrhia (Dothistroma) pini*, the cause of Red Band Needle Blight. Phytopathology, 58: 88-90.

Critchfield, W. B. 1984. Impact of the Pleistocene on the genetic structure of North American conifers. *In* Proceedings, Eighth North Am. Forest Biology Workshop, *edited by* R.M. Lanner. Logan UT. pp. 70-118.

Fins, L., and W. J. Libby. 1982. Population variation in *Sequoiadendron*: seed and seedling studies, vegetative propagation, and isozyme variation. Silvae Genet. 31: 101-148.

Furnier, G. R., and W. T. Adams. 1986. Geographic patterns of allozyme variation in Jeffrey Pine. Am. J. Bot. 73: 1009-1015.

Gentry, A. H. 1986. Endemism in tropical versus temperate plant communities. *In* Conservation biology: the science of scarcity and diversity, *edited by* M. E. Soulé, Sinauer Assoc., Sunderland, MA. pp. 153-181.

Gentry, A. H., and R. Vasquez. 1988. Where have all the *Ceibas* gone? A case history of mismanagement of a tropical forest resource. For. Ecol. Manage. 23: 73-76.

Griffin, J. R., and W. B. Critchfield. 1972. The distribution of forest trees in California. U.S. Forest Service, Res. Pap. PSW-82.

Harrington, J. B. 1987. Climatic change: a review of causes. Can. J. For. Res. 17: 1313-1339.

Hosie, R. C. 1969. Native trees of Canada. 7th ed. Can. Forest Service, Dep. Fish. and For., Ottawa, ON.

Janik, J., R. W. Schery, F. W. Woods, and V. W. Ruttan. 1969. Plant science: an introduction to world crops. W.H. Freeman, San Francisco, CA.

Johnson, O. (Editor). 1986. Information please almanac. 40th edition. Houghton Mifflin Co., New York, NY.

Kastner, J. 1977. A species of eternity. Alfred A. Knopf, New York, NY.

Keystone Center. 1991. Final consensus report of the Keystone Policy Dialogue on biological diversity on federal lands. Keystone, CO.

Ledig, F. T. 1987. Genetic structure and the conservation of California's endemic and near-endemic conifers. *In* Conservation and management of rare and endangered plants, *edited by* T.S. Elias. California Native Plant Soc., Sacramento, CA. pp. 587-594.

Ledig, F. T. 1986. Conservation strategies for forest gene resources. For. Ecol. Manage. 14: 77-90.

Ledig, F. T. 1988a. The conservation of diversity in forest trees. BioScience 38: 471-479.

Ledig, F. T. 1988b. The conservation of genetic diversity: the road to La Trinidad. Univ. British Columbia, Leslie L. Schaffer Lectureship in Forest Science. Vancouver, BC.

Martinez, M. 1963. Las Pinaceas Mexicanas. Universidad Nacional Autonoma Mexicana, Mexico City, Mexico.

Muir, J. 1916. My first summer in the Sierra. Houghton Mifflin Co., Boston, MA.

Myers, N. 1984. Genetic resources in jeopardy. Ambio 13: 171-174.

Office of Technology Assessment. 1984. Technologies to sustain tropical forest resources. OTA-F-214. U.S. Congress, Off. Technol. Assessment, Washington, DC.

Old, K. M., W. J. Libby, J. H. Russell, and K. G. Eldridge. 1986. Genetic variability in susceptibility of *Pinus radiata* to Western Gall Rust. Silvae Genet. 35: 145-149.

Peters, R. L. 1990. Effects of global warming on forests. For. Ecol. Manage. 35: 13-33.

Riding, A. 1984. Distant neighbors: a portrait of the Mexicans. Vintage Books, New York, NY.

Roberts, L. 1988. Hard choices ahead on biodiversity. Science 241: 1759-1761.

Ryu, J. B., and R. T. Eckert. 1983. Foliar isozyme variation in twenty seven provenances of *Pinus strobus* L: genetic diversity and population structure. *In* Proceedings, Twenty-eighth Northeast. Forest Tree Improv. Conf., *edited by* R. T. Eckert. Univ. New Hampshire, Durham, NH. pp. 249-261.

Salwasser, H. 1988. Editorial. Conserv. Biol. 2: 275-277.

Shevock, J., and D. W. Taylor. 1987. Plant exploration in California, the frontier is still here. *In* Conservation and management of rare and endangered plants, *edited by* T.S. Elias. California Native Plant Soc., Sacramento, CA. pp. 91-98.

Shoup, L. H., with S. Baker. 1981. Speed power, production, and profit: railroad logging in the Goosenest District, Klamath National Forest, 1900-1956. Prepared in fulfillment of U.S. Forest Service Contract No. 00-91W8-0-1911. Klamath National Forest, Eureka, CA.

Steinhoff, R. J., D. G. Joyce, and L. Fins. 1983. Isozyme variation in *Pinus monticola*. Can. J. For. Res. 13: 1122-1132.

Styles, B. T. 1972. The flower biology of the Meliaceae and its bearing on tree breeding. Silvae Genet. 21: 175-182.

Witt, S. C. 1985. Biotechnology and genetic diversity. California Agric. Lands Project, San Francisco, CA.

World Resources Institute. 1988. World resources 1988-89. A report by the World Resources Institute and the International Institute for Environment and Development in collaboration with the United Nations Environment Programme. Basic Books, New York, NY.

Biodiversity Research in Museums:
A Return to Basics

Edward H. Miller

Research and Public Programs, Royal British Columbia
Museum, Victoria, British Columbia, Canada V8V 1X4
and Biology Department, University of Victoria,
Victoria, British Columbia, Canada V8W 2Y2.

Abstract

Museum collections of biological specimens are the fundamental reference
material that documents the world's biological diversity. They are essential also
for identification purposes, invaluable in educational programs and necessary
for countless investigations in environmental biology, ecology, evolution and
other fields. Museum collections need to grow and diversify rapidly because of
the accelerating loss of the world's species and habitats. Collections growth
should be based on: (a) biological inventories of threatened and disappearing
habitats; (b) opportunistic acquisition of specimens of rare or endangered kinds
of animals and plants; (c) voucher specimens from biological surveys; and (d)
diversification, including frozen-tissue collections and audio-visual archives.
Museum collecting and research must emphasize poorly known and diverse
groups of organisms, as well as groups that are ecologically important and have
high scientific, cultural and aesthetic value.

Museums are the major contributors to the science of systematics, which
embraces classification, taxonomy, evolutionary relationships and evolutionary
processes. Basic systematics research and training are declining seriously. It
needs to be increased, and museums should take the lead in doing so. Museums
also need to educate non-systematic biologists and the general public about the
essential role of systematics in the study of biological diversity and in
conservation biology generally.

In *Our Living Legacy: Proceedings of a Symposium on Biological Diversity*, edited by
M.A. Fenger, E.H. Miller, J.A. Johnson and E.J.R. Williams, pp. 141-173. Victoria,
B.C.: Royal British Columbia Museum.

Introduction

As a teenager, I trapped muskrats to earn money. One day while skinning one I noticed a dusting of brown and orange inside its ears. I peered closely and saw tiny specks, some moving. Fascinated, I swept several into an empty aspirin bottle using a fine brush, then added formalin. Surely somebody will be able to tell me what they are, I thought. It was not to be that simple. An entomologist at the Department of Agriculture transferred the specks to a vial of alcohol using an eye-dropper. He told me that they were mites, then he asked questions: Where had I captured the muskrat? When? What was its sex and age? Exactly where on the body were the mites found? He printed the information in jet-black ink neatly on a small label then dried it, immersed the label in the vial and inserted a rubber stopper firmly into the top of the vial to seal it. "We have nobody here who can identify these," he informed me. "We'll send them to somebody who can." Several months later I received a letter. Some of the specimens had been damaged (by my "fine" brush, I presume). The undamaged specimens represented at least two species, one of which may be new to science. May the Department of Agriculture retain the specimens for its collections?

I learned several lessons from this incident. Obviously, very specific information has to be recorded with scientific specimens and detailed procedures must be used to preserve them. Unless specimens are properly documented and preserved, it may be impossible to identify them. Even then, specimens might not be identifiable, for no specialists may be working on that group of animals, or the specimens may represent forms that are unknown to science. This last point astonished me, as I assumed that science had long ago found and named most kinds of animals. It has become a commonplace, however, that even now we neither know nor have names for most organisms with which we share this planet (e.g., Kosztarab and Schaefer 1990). By extension we obviously also know virtually nothing else about them, such as their ecological roles and relationships.

The purpose of this paper is to outline the responsibilities of museums in documenting biological diversity, in organizing information about it, and in making that information available to people. These responsibilities have become increasingly important as humanity strives to understand and preserve the world's rapidly disappearing natural ecosystems (Nicholson 1991).

"If You Can't Name It, You Don't Know What You're Talking About"

Many kinds of specimens are housed in museums and many uses are made of each kind. Stuffed birds are used to investigate distribution, geographic variation and seasonal movements. Pickled lizards are used to study anatomy, growth and evolutionary relationships. Biological collections include tape recordings of animal sounds, which help to document where vocal animals like frogs and crickets occur; microscope slides of pollen grains, needed to identify pollen in soil samples thousands of years old; plus an astonishing array of pickled, pressed, frozen and dissected animals and plants, with an equally astonishing array of uses (Miller 1984a). Museum collections are also diverse just for taxonomic and reference purposes, since organisms are so different from one another (see Taxonomic and Reference Collections are Diverse, below).

Collections in museums continue to grow and diversify as conceptual and technical advances find new uses for old specimens, and as modern science demands new kinds of specimens for previously unforeseen uses. Unimaginable in the past, for example, there exist now major collections of animal and plant tissues that are maintained at extremely low temperatures to halt chemical changes in them; these collections provide vital research material for research in genetics and environmental toxicology (Dessauer and Hafner 1984; see Collections Must Grow, and Fast!, below). Conceptual and technical advances have led to a flowering of research opportunities for biological collections. In the face of these opportunities, museum biology has become refocused on describing and analysing biodiversity, with a growing emphasis on biological inventories, field surveys and taxonomy.

The reasons for the revitalization of biological systematics are clear: vast and unique areas of the natural world are disappearing rapidly, yet the biological diversity in most areas is undescribed, and this diversity includes many organisms that are unknown to science[1] (Fig. 1A). A first step in environmental research is to clarify ecological relationships and functions, but such work cannot even be started unless people can identify and name the organisms present. The accelerating breakdown in our planet's ecological integrity gives particular urgency to the task of getting, organizing and disseminating basic information about biological diversity (Stuessy and Thomson 1981; May 1988; Roberts 1988). As well, there is a growing appreciation of the intrinsic value of the world's organisms and of their immense but largely unknown potential uses in industry, medicine and agriculture (Myers

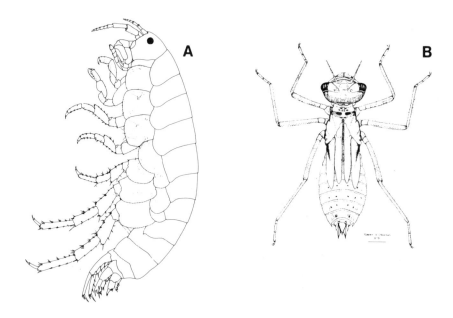

Figure 1. A basic task of museum curators is to describe new forms of animals such as new species (A) and undescribed life stages (B). (A) New species of "beach flea" (amphipod), *Platorchestia chathamensis*. Beach fleas are abundant and widespread animals on B.C.'s marine beaches. The female illustrated here was collected from Chatham Island, near Victoria, in 1959. It is the only specimen discovered to date and is stored in the National Museum of Nature in Ottawa. (After Fig. 11 of Bousfield 1982.) (B) Previously undescribed life stage of the dragonfly species *Sympetrum madidum*. Dragonfly larvae are important components of freshwater ecosystems in B.C. The specimen shown here was one of a series collected in 1978 and 1979 at Riske Creek and Langford. They are stored in the Spencer Entomological Museum at the University of British Columbia. (After Fig. 1 of Cannings 1981.)

1979; Stuessy and Thomson 1981; Wilson 1988a). Therefore, many museums have begun major initiatives to do biological inventories of threatened and biologically poorly known parts of the globe, such as tropical rainforests (e.g., Erwin 1988; Remsen and Zink 1988; Stap 1990).

Conserving biological diversity and ecological integrity in British Columbia, as in other parts of the world, is intimately connected to basic taxonomic knowledge. It is impossible to carry out biological inventories of threatened ecosystems such as coastal old-growth forests, southern interior grasslands or freshwater ponds and rivers, without being able to identify and name the distinctive and ecologically critical organisms that are present. Without such inventories of "name-able" organisms people cannot communicate, values of the areas cannot be assessed, necessary research cannot be done, and management and protection plans cannot be developed. Natural history museums have the responsibility to make well-documented field collections; to preserve them; to study them and to publish the results of research; to make them available for examination or loan; to serve as repositories for specimens and associated data; to provide expert information on taxonomy, identification and distribution; and to educate the general public about biological diversity and its values.

Taxonomy and Systematics[2] : Cornerstones of Biodiversity Research

Basic Curatorial Tasks

A museum curator sorts through a collection. The collection may have come from a recent field trip or may represent part of the larger museum collection that has not yet been studied. It may also have come from an inquiring scientist, museum, environmental consultant, government agency or member of the public. Whatever the collection's source, it will have been subject to some preliminary sorting: the curator is likely to be looking at just lichens, or just flies, or just roundworms, for example. Through knowledge of the group at hand and by reference to identification keys and previously identified specimens, a curator can assign names to most of the specimens. Others may be damaged or may represent life stages that cannot be identified (e.g., eggs or larvae are often difficult to identify; Fig. 1B). If few reference specimens are available at the curator's museum, it may be necessary to borrow some from other museums. After all this work, the curator may decide that part of the collection includes organisms that are not known to science. It is the curator's responsibility to formally describe and name them (or to send them to a specialist who can), and to place them within an existing classification (in some cases, as for poorly known groups, existing classifications may be inadequate and may need to be overturned). When a new form is described, its description, naming and classification must be done formally through original publication in a scientific journal or book. The published description of a new fish species from the Queen Charlotte Islands is given as an example of these conventions in Appendix A.

Usually, at least some of the specimens in a collection can be identified to the species level. The resulting information contributes to our knowledge about where and when species occur in B.C. Often, it is possible to obtain more detailed information also, such as about subspecies, life stage, gender or physiological state (e.g., moult) (Allen and Cannings 1984). After it has been sorted and identified, a collection becomes invaluable, for it can be re-examined, re-evaluated and re-studied for innumerable purposes and it serves as a permanent record.

Taxonomic and Reference Collections are Diverse

Some museum collections exist to aid identification specifically. Examples at the Royal British Columbia Museum are microscope slide collections of sea cucumber ossicles and plant pollen grains, all identified to species. Another contains bones organized by different bone types (e.g., pelvis, femur), which enables identification of individual bones, such as are obtained in a study of food habits of a predatory mammal or bird species (e.g., Nagorsen et al. 1991).

Most reference collections in museums, however, are organized taxonomically. An investigator thereby can have immediate access to an entire collection of beetles (order Coleoptera) within which he or she can have access to all ground beetles (family Carabidae) and within that family all species in the genus *Carabus*. In large museums, collections are often broken down further geographically or by subspecies[3] to give workers rapid access to appropriate specimens.

Taxonomically organized collections differ according to the kinds of plants and animals in them, because different attributes are used to distinguish species in different groups. A field guide to birds provides information on plumage, size, beaks, legs and general behaviour, enabling bird watchers to distinguish one species from another. A manual on plant identification relies on features of leaves, flowers, bark, buds and general habitat. For insects, usually it is necessary to examine or dissect entire specimens under a microscope, and with other organisms it may be necessary to make special microscope preparations of body parts (Fig. 2). Because identifying features vary among different kinds of organisms so greatly, museum collections include many kinds of collections, so specimens that are brought to a museum for identification must be prepared accordingly. Skulls are necessary to identify most shrews, for example, so a dried shrew skin without a skull is usually inadequate; in contrast, a dried bird skin by itself is generally identifiable to species. Thus, taxonomic collections of different organisms differ greatly from one another, as dictated by the important characteristics of each group.

Curators must use collections in both their own and other museums. Very few museums are so large that they contain enough specimens to support thorough research projects by themselves, so loans from other institutions are necessary. For example, in a research project

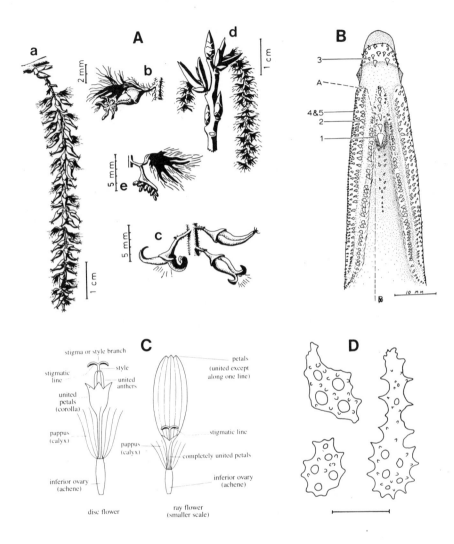

Figure 2. Fine details must be studied to identify different kinds of plants and animals. (A) Trembling Aspen (*Populus tremuloides*): (a) female catkin in flower, (b) female flower, (c) opening capsules, (d) twig with opening male catkins, and (e) male flower (after Fig. 4 of Brayshaw 1991). (B) A rare deep-sea eel species (*Xenomystax atrarius*): tooth pattern on roof of mouth, as seen from below (after Fig. 4 of Peden 1972). (C) Sunflowers (family Asteraceae): stylized diagrams of flowers in "cut away" views from the side (after Fig. 1 of Douglas 1982). (D) Sea cucumbers (*Cucumaria fisheri*): outline drawings of ossicles (calcified bits in the skin; scale bar = 0.1 mm; after Fig. 4 of Lambert 1990).

on the Upland Sandpiper (*Bartramia longicauda*), I examined 532 clutches in 53 museums. Most specimens, however, were in just a few collections. The 10 largest collections held 137, 69, 62, 28, 26, 17, 14, 11, 10 and 9 clutches or 77% of the total. Obviously, collections in large institutions are essential resources for research. For bird eggshells, these institutions include the Western Foundation of Vertebrate Zoology (Los Angeles), the National Museum of Natural History (Smithsonian Institution, Washington, D.C.) and the Field Museum (Chicago) — these have the three largest collections of specimens listed above. It is also apparent that small collections, especially those in small or remote museums, often are not worth an investigator's efforts to track down and study. In this study I found single — but old and valuable — clutches in small community museums in Connecticut, Illinois, Colorado, North Dakota and other places. Curators in charge of such small isolated collections may wish to donate them to large institutions that are more accessible to investigators, or to exchange them for material that is more appropriate to a small museum's needs.[4]

The Royal B.C. Museum's mandate in biology is to acquire, preserve and carry out research on collections pertaining to the province's natural history, and to communicate the findings to the general public. However, it is necessary for the museum's curators to maintain or examine specimens from much broader geographic areas, in order to interpret provincial information meaningfully (Fig. 3). As well, many specimens from B.C. are in other museums. Nagorsen (1990a) lists 12 museums in three countries that hold type specimens of mammal subspecies that occur in the province (see Appendix A). Similarly, most type specimens of B.C. mammals are in only a few places (half are in the National Museum of Natural History in Washington, D.C.!). On a local level, it is necessary, even just within B.C., to use resources in various collections, such as in the province's ten important herbaria: in Burnaby, Castlegar, Kamloops, Smithers, Vancouver, Victoria (four) and Williams Lake (Holmgren et al. 1990). These herbaria range in size from 3,000 (Kamloops) to 500,000 (University of British Columbia) and total nearly a million specimens!

How Many Specimens are Enough?

Countless biological specimens are housed in museums and the reasons are simple: millions of animal and plant species exist, and great variation occurs within each owing to gender, age, geography, season and other factors. Therefore, reference collections of even single species must be large enough to identify specimens and to support taxonomic and systematic research. But what is "large"? The answer to this question depends on how much variation exists within a species, which

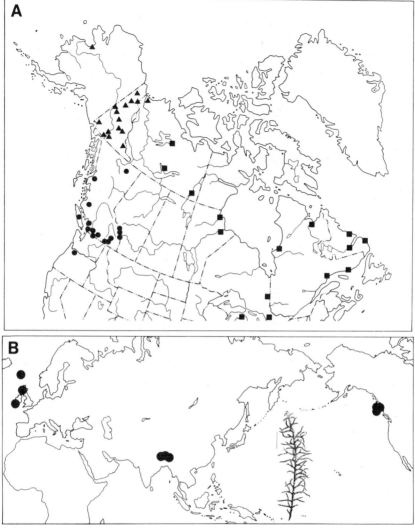

Figure 3. Museum research on B.C.'s biological diversity must be set in a meaningful geographic context. (A) Research on B.C. robber flies (*Rhadiurgus variabilis*) requires specimens from throughout the North American range (after Fig. 20 of Cannings 1992). (B) Research on the hepatic *Mastigophora woodsii* requires specimens from throughout the Northern Hemisphere (after Figs. 12 and 15 of Schofield 1989).

depends in turn on a species' geographic and ecological breadth. These points can be illustrated with an example:

The Hairy Woodpecker (*Picoides villosus*) is a widely distributed resident of suitable forest habitats throughout Central and North

Figure 4. How many specimens are enough? Museum collections must encompass many kinds of natural variation. Specimens of Hairy Woodpeckers (*Picoides villosus*) illustrate variation in back markings in this species on Vancouver Island.

America (Jackson 1970; Ouellet 1977; Short 1982). It varies greatly in size and plumage over this range: in male body weight alone the species varies nearly two-fold over 40 degrees of latitude (from about 40 to 80 grams). Further, size and plumage differ between males and females, and between adults and juveniles. Feather wear and moult of feathers cause other variations that must be taken into account, and extensive individual variation occurs also (Fig. 4). In light of these factors, an investigation of geographic variation in the species demands large numbers of specimens from throughout the range (Fig. 5).

Studies of geographic variation within a species are fundamentally important for documenting species' ranges and for documenting species' characteristics in different environments. Such information is essential for inferring the evolutionary pathways through which species have occupied and adjusted to their present ranges. Understanding these evolutionary patterns is needed to identify the elements of genetic diversity within species, such as distinctive island populations, or populations at the margins of a species' range that may merit special protection or research (Baker 1984; Ledig, this volume; see Research

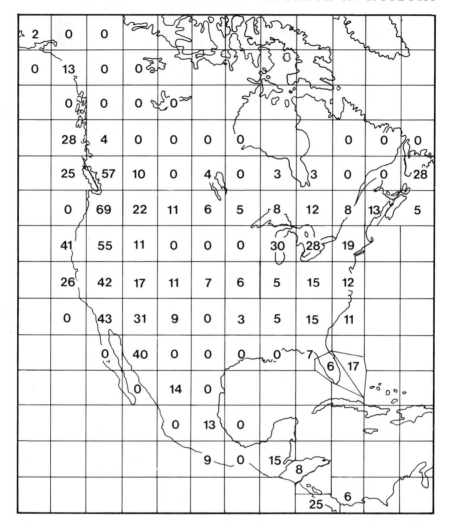

Figure 5. How many specimens are enough? Museum collections must be large enough to support comprehensive studies of geographic variation. Here, the numbers of available museum specimens of male Hairy Woodpeckers (*Picoides villosus*) in spring plumage, used in a study on geographic variation, are shown over the species' range. (After Fig. 2 of Jackson 1970.)

Needs to be Focused and Innovative, below, and Appendix C).

The Tailed Frog (*Ascaphus truei*) is an example of the importance of museum specimens in documenting distribution.[5] This curious species inhabits cold cascading streams. When such streams are exposed, as when forest cover is removed, water temperature can increase to fatally warm levels. Harmful silt loads can result also (Orchard 1990). We

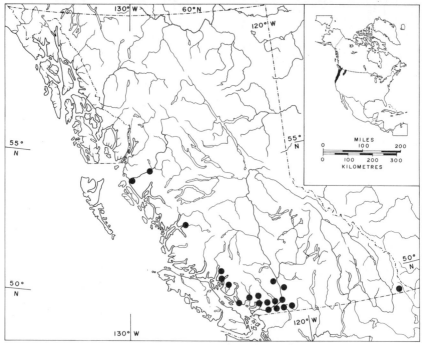

Figure 6. Documentation of a species' range depends upon solid evidence: the Tailed Frog (*Ascaphus truei*) in B.C. The localities marked are based on confirmed museum specimens and other information, cross-referenced with original data sources (see Appendix B). (After Orchard, ms. in prep.)

know little about the species' ecology and distribution. Its fragmented distribution in B.C. and its vulnerability to logging practices that remove streamside cover, however, make understanding its provincial distribution crucial for management and conservation (see Fig. 6, inset). An early distribution map showed two records at extremes of the range, one near Kitimat and one near Quesnel Lake in the Cariboo Mountains (Green and Campbell 1984). The Kitimat record was based on an accessioned museum specimen that had been discarded, so the record could not be verified. The Cariboo record was visual. Subsequently, Museum field work has confirmed the north-coastal distribution for this species, but acceptable evidence of its occurrence in the Cariboo (e.g., photographs or specimens) is still lacking (Fig. 6).

Getting the Word Out:
"If It Isn't Published, It Was Never Done"

A fundamental obligation of museum curators is to publish results of their original research fully and promptly (Mayr 1969). Prompt publication and wide dissemination of results are particularly important now for research on biological diversity and biological systematics (Stuessy and Thomson 1981). Publication in so-called "primary literature" is the basic accepted standard in most science and applies also to taxonomy and systematics. The publication may be a peer-reviewed journal, a monograph or other scholarly publication. It is no longer adequate to publish research results just anywhere: there simply is too much published literature for people to cope with. Thus vehicles that are widely distributed and widely available should be chosen. Increasingly, such vehicles are large English-language journals published in Europe or North America. Obscure publications with restricted distributions will reach only dedicated specialists, often after considerable delays. Large monographs, while highly desirable, are extremely expensive to publish and are subject to long delays in publication. Less expensive forms of publication that can be published quickly should also be considered and electronic publishing likely will become one important way to do this (Stuessy and Thomson 1981).

Research results must be published to make information available. But who is the information for? Ultimately it is for the general public, both directly, through museum exhibits, popular publications, etc., and indirectly, through impacts of "primary users" of research results like educators, environmental consultants, foresters and ecologists. Because of the needs and demands of such audiences, the global environmental crisis and museums' limited resources, priorities must be set in research, acquisitions, public programs and publications. Museums cannot support academic freedom to the extent of just "doing taxonomy for taxonomists" (Mound 1991). Instead, taxonomic and systematic research should focus on groups of animals and plants for which strong external needs for information have been expressed, for which conservation is an issue, that have notable ecological significance or that have high cultural or scientific value. When such criteria are used to set research priorities, the intended audiences for the completed research must also be carefully defined and a plan must be put in place for reaching them effectively through books, papers, reports, etc. In many cases the results must be made available for use in government policies or management decisions (e.g., in forestry or wildlife management) or for educating the general public directly. Consequently, "user-friendly"

153

publications, such as identification manuals, keys, checklists, lists of names and popular books, should be priorities for museums.

Trends and Priorities

How should museum resources be allocated to research and education on biological diversity in light of pressing environmental issues? I will answer this question by discussing collections growth, research, interagency collaboration and education.

Collections Must Grow, and Fast!

Museums should undertake or continue major programs of collections growth, especially for species and areas that are threatened or poorly known. To fulfil their mandate, museums must collect specimens, and such activity is often direct, through trapping, netting or shooting. The social acceptability of these activities will not be debated here, though I acknowledge that killing wild animals and plants rouses strong opposition from some people. Collecting by museums is done humanely and ethically, and with explicit reference to a species' conservation status. For groups like insects, large-scale field collecting is usually acceptable ethically because of their abundance. For birds and large mammals, on the other hand, collecting may have to be limited or opportunistic (e.g., road kills or donations of poached material from the Ministry of Environment). Natural mortality and donations of existing collections are other sources of specimens. Important recent additions to the biological collections of the Royal British Columbia Museum are: two collections of insects totalling 20,000 specimens donated by private citizens; 600 squids and related species from the estate of a world authority; 5,000 plant specimens from the Ministry of Environment; and 11,000 lots of "sea shells" (including bivalve molluscs and chitons) purchased from a retired scientist.

Collections should also diversify in respect to how specimens are preserved. Traditionally, ornithologists skinned birds and stuffed them to resemble corpses. Along with the skin, only parts of the leg bones, wing bones and skull were retained, all bound within the stuffed bird. It is difficult to know why this tradition prevailed for so long, since it conceals many important features of specimens. Wings and tails cannot be properly examined, for example, and bones are not accessible. In many museums this traditional form of preparation began to change a few decades ago to one in which more parts are saved and more features made accessible: flat skins, spread wings and tails, complete or nearly complete skeletons, etc. (Fig. 7). The resulting specimens are more

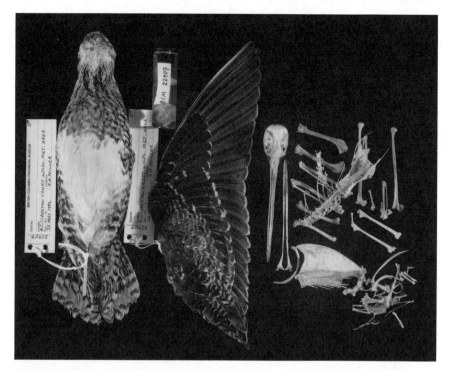

Figure 7. Example of a modern way to prepare museum specimens of birds. This non-traditional specimen includes a nearly complete skeleton, spread wing, stuffed body skin with spread tail plus one leg and tissue samples for genetical analysis (preserved in fluid in vial).

valuable scientifically as they contain more information. Fuller use of the specimen is also desirable on ethical grounds. Thus, in general, traditional ways of preparing museum specimens should be critically reviewed (Edwards 1984).

New kinds of collections also need to be established. In B.C., high priority should be given to establishing frozen-tissue collections and natural-history archives. Deep-freezing of tissues provides long-term protection against decomposition (Dessauer et al. 1988; Catzeflis 1991; Seutin et al. 1991). Important uses of such tissues are currently in environmental toxicology and biochemical genetics, but other uses are sure to be discovered (Becker et al. 1990; see next two sections). Archival collections are badly neglected and need to be established. Such collections should include photographs, audio-tapes and videotapes of habitats and organisms, along with field diaries, all of which will enrich our own and future generations. We have lost all opportunity to experience and appreciate the old-growth stand of cedars that was illegally destroyed near Monashee Provincial Park, as no images or information about it are publicly available (Anon. 1991). Many species of Amazonian plants may remain known to us only through paintings (Knox 1988; Mee 1988). A natural-history archive could be a repository and resource for such materials. One model is the ornithological archive at the National Academy of Sciences (Philadelphia), where about 65,000 photographs of birds, covering about half of the world's species, are archived (S. Holt pers. comm.; see Miller 1984b, 1988, 1991).

Research Needs to be Focused and Innovative

Museum research must become focused on critical habitats, poorly known groups of organisms, poorly known areas of the province that are likely to exhibit high biological diversity and particularly important questions (Roberts 1988; see Fig. 8). For habitats and species, the most important criteria by which to set research priorities are degree of endangerment, ecological significance, level of knowledge, scientific value (e.g., richness, uniqueness; see May 1990) and social values (e.g., aesthetic, educational). Some habitats in B.C. where museum research and collecting are badly needed are coastal old-growth forests, freshwater ecosystems and interior grasslands. The most important museum work in these areas is biological inventorying by collecting and identifying species in key groups (Kim 1987). Biological diversity in well-known groups of organisms may be a good indicator for diversity in other groups (e.g., some plant groups; see Pielou, this volume) and in such cases surveys built on known high diversity within good indicator groups should receive high priority (Roberts 1988). Essential parts of the collecting and research are to describe new forms and to prepare faunal and floral lists, along with identification keys or manuals (Stuessy and Thomson, 1981). However, the immediate needs are for inventories: "There is no time for exhaustive studies, elaborate phylogenies or science as usual" (Roberts 1988:1759; Stuessy and Thomson 1981).

Many new species await discovery and description, particularly in diverse and poorly known groups of small organisms like roundworms, mites and fungi (Stuessy and Thomson 1981; Wilson 1988b; Kosztarab and Schaefer 1990). Many poorly known groups play vital roles in ecosystem function (e.g., soils; BSC 1982; see Marshall, this volume) and research emphasis on these groups should be on large-scale collecting and basic taxonomic work. For better-known groups like vertebrates, emphasis needs to be on geographic variation (anatomical, genetical) and on distribution and life history, with particular reference to unique or threatened areas within the province (Fjeldså 1987; Ledig, this volume). Do Marbled Murrelets (*Brachyramphus marmoratum*) vary across their breeding range from California to Alaska? Are any

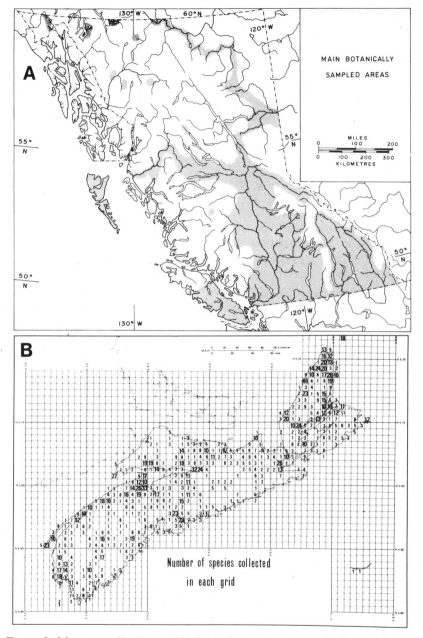

Figure 8. Museum collecting and biological inventories must be focused where information is lacking. (A) Map of B.C. showing approximate areas where botanical collections are reasonably complete ("higher" plants only; after Fig. 5 of Douglas 1982). (B) Detailed distribution of known museum fern specimens collected in Nova Scotia (after Fig. 3 of Aderkas 1986).

157

populations unique or strikingly different? Answers to such questions may identify important, distinct or isolated populations that merit special protection or further research. For important differentiated forms, formal subspecies descriptions should be refined or published anew (see O'Brien and Mayr 1991). The Peregrine Falcon (*Falco peregrinus*) is an example of a geographically varying species with distinct races or subspecies throughout its range, such as Peale's Peregrine Falcon (subspecies *pealei*), which has a range from Washington State to the Aleutian Islands (White 1988; White and Boyce 1988). This subspecies differs from others in distribution, plumage, size, habitat and life history, and is afforded special protection in B.C. (see Munro, this volume).

In other cases, new evidence can reveal unsuspected new species that can be validated with museum specimens. The abundant and widespread forest mouse *Peromyscus*, for example, was recognized as two species on Vancouver Island, rather than just one as previously thought, on the basis of their chromosomes: the Deer Mouse, *Peromyscus maniculatus*, and the Columbian Mouse, *P. oreas* (Nagorsen 1990a). On the basis of their chromosomal characteristics, voucher specimens (Appendix D) of the two species were studied, and diagnostic features of their skulls and study skins were identified. Using these distinguishing features, the thousands of existing museum specimens can now be identified and a detailed picture of where the two species are distributed in the province can be constructed. These mouse species, intimately bound up in a cycle that links soil ecology, tree growth and dispersal of fungal spores, are important ecological elements of forest ecosystems (Maser 1988, 1989). The species *Peromyscus maniculatus* and *P. oreas* likely have different roles in this cycle, so it is essential to identify and distinguish between them.

Basic taxonomic research relies on good collections and on sound, up-to-date understanding of concepts about how species arise and become different from one another. Therefore, curators must keep abreast of conceptual developments in taxonomy, systematics and evolution. Their knowledge must also be up-to-date in relevant methods and practices, such as statistical data analysis and genetics. The latter area includes research on chromosomes, proteins and other genetical elements. Such subjects may sound esoteric and remote from concerns about biological diversity, but they are fundamental to assessing how (and how much) species and populations differ from one another. They permit rigorous evaluation of genetic diversity both within and across species, and provide a sensitive indicator of whether something even is a species (Ledig, this volume). Incorporating biochemical genetics into research programs has high priority at the Royal B.C. Museum.

In recent years, technological advances that provide ways to extract new kinds of information have extended the uses of museum specimens: isotope analysis points to geographic areas of origin of elephant ivory (Merwe et al. 1990; Vogel et al. 1990); and a chemical amplification technique permits the investigation of relationships among species and subspecies using only tiny samples of tissue (e.g., hair or skin), even from museum specimens of long-extinct forms (Vigilant et al. 1989; Diamond 1990; Thomas et al. 1990; Catzeflis 1991; Pääbo et al. 1992; for other examples see Barrowclough 1984, Gibbons 1991, and Li and Graur 1991). The importance of these techniques can be illustrated by a freshwater worm with the scientific name *Capitella capitata*, a worm widely used as an indicator species in pollution monitoring. The "species" is actually a complex of species, not just one, and undoubtedly they differ from one another ecologically. These cryptic species can only be distinguished from one another biochemically, not morphologically (Parker 1990). Another example is the roundworm *Pseudoterranova decipiens*, a common parasite that has a severe economic impact on North Atlantic fisheries (Bowen 1990). Recent genetical research points to three separate worm species that differ geographically and ecologically from one another (Paggi et al. 1991). Other cases of recently recognized "cryptic species" are *Leuroglossus* fishes in the northeast Pacific Ocean and the unique New Zealand Tuatara (*Sphenodon*) (Peden 1981; Daugherty et al. 1990). These examples emphasize how important it is for museums to harness technological advances in their research programs and identification responsibilities, often through collaborative projects with other agencies and institutions.

Inter-agency Collaboration: Nobody Can Work Alone

The work of museums needs to be planned and carried out in conjunction with research and survey work of other organizations. By doing so, the expertise of diverse specialists is strengthened, and synthetic descriptions, analyses and evaluations are possible. Major provincial agencies that should be involved in broad collaborative efforts to study and document biological diversity are the Ministry of Forests, Ministry of Environment, Lands and Parks (especially the Wildlife Branch), the Royal B.C. Museum, environmental consulting firms, universities and environmental interest groups. Collectively, these agencies and organizations offer a wealth of talent and resources through specialized technical knowledge and expertise, including habitat description, biochemistry, economics, taxonomy and field observations.

Museums should be responsive to other organizations' needs for biological surveys and for the identification of specimens. In some cases this will mean strengthening certain collections; in others it will

159

necessitate publishing identification keys and providing basic training in identification and taxonomy (e.g., through workshops). Museums must also encourage other organizations to prepare well documented voucher specimens and to deposit them in appropriate institutions (Appendix D).

Museums must also serve the general public. Much of this service should be with co-operating organizations, as most other organizations do not have public education as part of their mandate. The Royal B.C. Museum is committed to integrating collections development with research, and research with education. We believe that no research is complete without its results being published and thereby made available to the scientific community at large. Research is also incomplete until its results are made available to a wider audience that includes the general public.

Museums must educate the public in more ways than just publishing the results of scholarly research. They must be socially responsible and responsive, as reflected in research, collecting and education. Museums should actively inform and educate the general public about socially relevant and topical issues that pertain to museum mandates. For biodiversity, these issues include conservation, habitat loss, exploitation of natural resources and numerous related topics.

Museum curators offer expertise that is declining in university curricula and in university faculties, including comparative biology, taxonomy, systematics, collecting techniques and standards, and identification (Stuessy and Thomson 1981; Nelson 1987; Scudder 1987; Danks 1988; ASC 1989; Crowe et al. 1989; Kosztarab and Schaefer, 1990; see papers by Foster, Lertzman, McPhail and Marshall in this volume). This decline is serious for both the practice of systematics and the cultural transmission of values about biological diversity to the general public. Museums should take a lead to strengthen existing training and to offer new training through formal university courses, extension and non-credit courses, and in other ways.

It is difficult to explain to the tax-paying public why museum research on biological diversity is important and merits their support (see Lederman 1984). Arguments can be advanced on economic grounds or in terms of the catastrophic environmental consequences we can expect if society does not improve its understanding and appreciation of biological diversity. Such arguments, however, rest on only a few values. Many (most?) important benefits of such knowledge are indirect, diffuse and delayed. They affect the quality of human existence, not an artificial standard of living (Lederman 1984). Maser (1988:150-151) summarizes some reasons why humanity should preserve biological diversity in ecologically intact ecosystems such as coastal old-growth forests in the Pacific Northwest:

First, old-growth forests are our link to the past, to the historical forest. The historical view tells us what the present is built on, and the present in turn tells us what the future is projected on.... To lose the old-growth forests is to cast ourselves adrift in a sea of almost total uncertainty with respect to the sustainability of future forests. We must remember that knowledge is only in past tense, learning is only in present tense and prediction is only in future tense. To have sustainable forests, we need to be able to know, to learn and to predict. Without old-growth, we eliminate learning, limit our knowledge and greatly diminish our ability to predict.

Second, we did not design the forest, so we do not have a blueprint, parts catalogue or maintenance manual with which to understand and repair it.... Nor do we have a service department in which the necessary repairs can be made. Therefore, how can we afford to liquidate the old-growth that acts as a blueprint, parts catalogue, maintenance manual and service station — our only hope of understanding the sustainability of the redesigned, plantation forest?

Third, we are...playing 'genetic roulette' with forests of the future. What if our genetic simplifications run amuck, as they so often have around the world? Old-growth forests are thus imperative because they — and only they — contain the entire genetic code for living, healthy, adaptable forests.

Fourth, intact segments of the old-growth forest from which we can learn will allow us to make the necessary adjustments in both our thinking and our subsequent course of management to help assure the sustainability of the redesigned forest.

Humanity cannot and will not preserve biological diversity until we understand and appreciate it. Museums must be vigorous partners in efforts to improve this understanding. Museums' fundamental roles are to prepare and update the "parts catalogue" to which Maser refers — to study, document, name and organize information about plants and animals, and transmit that information to humanity — thereby enhancing human appreciation of organic diversity and enabling wise management decisions in the interests of future generations.

Notes

1. In B.C. we are surrounded by animals and plants that are unknown to science, have no names and hence cannot be identified. Usually, their ecology and distribution are mysteries as well (e.g., see Fig. 1A). For a summary of our shockingly poor knowledge of biological diversity on Vancouver Island, see Moore (1991); for an analysis and discussion of the sort of basic field and museum work needed to improve this knowledge, see Kosztarab and Schaefer (1990).
2. "Taxonomy" refers to the theory and practice of naming and classifying organisms. "Systematics" refers more broadly to the science concerned with biological diversity generally (Mayr 1969). For a good overview of the significance of systematics research see Danks (1988).
3. Subspecies are formally recognized and named geographic subdivisions of a species (Mawr, 1969). For example, the distinctive forms of Hairy Woodpecker, Northern Saw-whet Owl, Short-tailed Weasel and other species on the Queen Charlotte Islands have been afforded formal subspecies status (see Appendix C). In plants, there is also common formal recognition of "variety" and "form" (Davis and Heywood 1963).
4. Obviously, small collections can also be valuable for education and public programming. Most such needs, however, can be met with undocumented specimens or with specimens that are of minor research value. It is generally desirable for such museums to contribute or exchange valuable specimens for comparable specimens of less scientific value. At the other extreme, however, it is unwise to over-concentrate specimens in just a few museums, because of the consequences of major catastrophes like fires or earthquakes. On balance, for reasons of care, security and access, it is desirable to house scientifically valuable specimens in a network of major museums that have good conservation and curatorial support, and that are well known and readily accessible.
5. I am grateful to Stan Orchard for this example.

Acknowledgements

I would like to thank the following people who helped me in various ways while I prepared this manuscript: Ed Bousfield, Rob Cannings, Adolf Ceska, Christine Dionne, Mike Fenger, Jerry Jackson, Jacklyn Johnson, Frances Jones, Phil Lambert, Gerald Luxton, Dave

Nagorsen, Andrew Neimann, Bob Ogilvie, Stan Orchard, Alex Peden, Patrick von Aderkas and Liz Williams. I remain indebted to the organizations that fund my research and that of my graduate students: the Natural Sciences and Engineering Research Council (Individual Operating Grants to E.H. Miller, Biology Department, University of Victoria), the Royal British Columbia Museum and the Friends of the Royal B.C. Museum.

Literature Cited

Aderkas, P. von. 1986. Collection of ferns and fern allies in Nova Scotia. Proc. N.S. Inst. Sci. 37: 93-148.

Allen, G.A., and R.A. Cannings. 1984. Museum collections and life-history studies. *In* Museum collections: their roles and future in biological research, *edited by* E.H. Miller. B.C. Prov. Mus., Occ. Paper 25, Victoria, BC. pp. 169-194.

Anonymous. 1991. Old-growth stand of cedars destroyed. Times-Colonist, Victoria, BC., March 24 1991: A3.

Andersen, M.E., and A.E. Peden. 1988. The eelpout genus *Pachycara* (Teleostei: Zoarcidae) in the northeastern Pacific Ocean, with descriptions of two new species. Proc. Cal. Acad. Sci. 46: 83-94.

ASC (Association of Systematics Collections). 1989. Symposium on education of curators/systematists. Assoc. Syst. Coll. Newsletter 17: 61-67.

Baker, A.J. 1984. Museum collections and the study of geographic variation. *In* Museum collections: their roles and future in biological research, *edited by* E.H. Miller. B.C. Prov. Mus., Occ. Paper 25, Victoria, BC. pp. 55-77.

Barrowclough, G.F. 1984. Museum collections and molecular systematics. *In* Museum collections: their roles and future in biological research, *edited by* E.H. Miller. B.C. Prov. Mus., Occ. Paper 25, Victoria, BC. pp. 43-54.

Becker, P.R., B.J. Koster, S.A. Wise and R. Zeisler. 1990. Alaskan marine mammal tissue archival project. *In* Biological trace element research, *edited by* G.N. Schrauzer. Humana Press, Clifton, NJ. pp. 329-334.

Bousfield, E.L. 1982. The amphipod superfamily Talitroidea in the northeastern Pacific region. 1. Family Talitridae: systematics and distributional ecology. Nat. Mus. Can. Pub. Biol. Ocean. 11.

Bowen, W.D. (Editor). 1990. Population biology of sealworm (*Pseudoterranova decipiens*) in relation to its intermediate and seal hosts. Can. Bull. Fish. Aquatic Sci. 222.

Brayshaw, T.C. In press. Catkin bearing plants of British Columbia. 2nd rev. ed. Royal B.C. Mus., Occ. Pap. 18, Victoria, BC.

BSC (Biological Survey of Canada). 1982. Status and research needs of Canadian soil arthropods. Ent. Soc. Can. Bull. 14 (Suppl.).

Bunnell, F.L., and G.R. Williams. 1980. Subspecies and diversity — the spice of life or prophet of doom. *In* Threatened and endangered species and habitats in British Columbia and the Yukon, *edited by* R. Stace-Smith, L. Johns, and P. Joslin. B.C. Ministry of Environment, Victoria, BC. pp. 246-259.

Cannings, R.A. 1981. The larva of *Sympetrum madidum* (Hagen) (Odonata: Libellulidae). Pan-Pac. Ent. 57: 341-346.

Cannings, R.A. In press. New synonymy and description of *Rhadiurgus variabilis* (Zetterstedt) (Diptera: Asilidae) with notes on geographical variation. Can. Ent.

Cannings, R.A., and A.P. Harcombe (Editors). 1990. The vertebrates of British Columbia: scientific and English names. Royal B.C. Museum, Heritage Record 20, Victoria, BC.

Catzeflis, F.M. 1991. Animal tissue collections for molecular genetics and systematics. Trends Ecol. Evol. 6: 168.

Crowe, T.M., A.C. Kemp, R.A. Earle and W.S. Grant. 1989. Systematics is the most essential, but most neglected, biological science. S. Af. J. Sci. 85: 418-423.

Danks, H.V. 1988. Systematics in support of entomology. Ann. Rev. Entomol. 33: 271-296.

Danks, H.V., G.B. Wiggins and D.M. Rosenberg. 1987. Ecological collections and long-term monitoring. Bull. Ent. Soc. Can. 19: 16-18.

Daugherty, C.H., A. Cree, J.M. Hay and M.B. Thompson. 1990. Neglected taxonomy and continued extinctions of Tuatara (*Sphenodon*). Nature 347: 177-179.

Davis, P.H., and V.H. Heywood. 1963. Principles of angiosperm taxonomy. Oliver and Boyd, Edinburgh, Scotland.

Dessauer, H.C., and M.S. Hafner (Editors). 1984. Collections of frozen tissues: value, management, field and laboratory procedures, and directory of existing collections. Assoc. Syst. Coll., Lawrence, KS.

Diamond, J.M. 1990. Old dead rats are valuable. Nature 347: 334.

Douglas, G.W. 1982. The sunflower family (Asteraceae) of British Columbia. Vol. 1 — Senecioneae. B.C. Prov. Mus., Occ. Paper 23, Victoria, BC.

Edwards, R.Y. 1984. Research: a museum cornerstone. *In* Museum collections: their roles and future in biological research, *edited by* E.H. Miller. B.C. Prov. Mus., Occ. Paper 25, Victoria, BC. pp. 1-12.

Erwin, T.L. 1988. The tropical forest canopy: the heart of biological diversity. *In* Biodiversity, *edited by* E.O. Wilson. National Academy Press, Washington, DC. pp. 123-129.

Fjeldså, J. 1987. Museum collections of birds — relevance and strategies for the future. Acta Reg. Soc. Sci. Litt. Gothoburgensis, Zoologica 14: 223-228.

Gibbons, A. 1991. Systematics goes molecular. Science 251: 872-874.

Green, D.M., and R.W. Campbell. 1984. The amphibians of British Columbia. B.C. Prov. Mus., Handbook 45, Victoria, BC.

Holmgren, P.K., N.H. Holmgren and L.C. Barnett (Editors). 1990. Index herbariorum. Part I: the herbaria of the world. 8th ed. New York Botanical Garden, Bronx, NY.

Jackson, J.A. 1970. Character variation in the Hairy Woodpecker. Ph.D. Thesis. University of Kansas, Lawrence, KS.

Kim, K.C. 1987. Assessing and monitoring our biological diversity: a national biological survey. Proc. Penn. Acad. Sci. 61: 127-132.

Knox, P. 1988. Artist's work depicts endangered plant life. Globe and Mail, Toronto, Ontario, Sept. 15 1988: A1-A2.

Kosztarab, M., and C.W. Schaefer (Editors). 1990. Systematics of the North

American insects and arachnids: status and needs. Virginia Agric. Exp. Station and Virginia Polytech. Inst. and State Univ., Info. Ser. 90-1, Blacksburg, VA.

Lambert, P. 1990. A new combination and synonymy for two subspecies of *Cucumaria fisheri* Wells (Echinodermata: Holothuroidea). Proc. Biol. Soc. Wash. 103: 913-921.

Lederman, L.M. 1984. The value of fundamental science. Sci. Am. 251(5): 40-47.

Lee, W.L., B.M. Bell and J.F. Sutton. 1982. Guidelines for acquisition and management of biological specimens. Assoc. Syst. Coll., Lawrence, KS.

Li, W.-H., and D. Graur. 1991. Fundamentals of molecular evolution. Sinauer Associates, Sunderland, MA.

Macdonald, D. (Editor). 1984. The encyclopedia of mammals. Facts on File Publications, New York, NY.

Maser, C. 1988. The redesigned forest. R. & E. Miles, San Pedro, CA.

Maser, C. 1989. Forest primeval: the natural history of an ancient forest. Sierra Club Books, San Francisco, CA.

May, R.M. 1988. How many species are there on earth? Science 241: 1441-1449.

Mayr, E. 1969. Principles of systematic zoology. McGraw-Hill, New York, NY.

Mayr, E. 1990. A natural system of organisms. Science 348: 491.

Mee, M. 1988. In search of flowers of the Amazon forest. Nonesuch Expeditions. Antique Collectors' Club, Ithaca, NY.

Merwe, N.F. van der, J.A. Lee-Thorp, J.F. Thackeray, A. Hall-Martin, F.J. Kruger, H. Coetzee, R.H.V. Bell and M. Lindeque. 1990. Source-area determination of elephant ivory by isotopic analysis. Nature 346: 744-746.

Miller, E.H. (Editor). 1984a. Museum collections: their roles and future in biological research. B.C. Prov. Mus., Occ. Paper 25, Victoria, BC.

Miller, E.H. 1984b. Museum collections and the study of animal social behaviour. *In* Museum collections: their roles and future in biological research, *edited by* E.H. Miller. B.C. Prov. Mus., Occ. Paper 25, Victoria, BC. pp. 139-162.

Miller, E.H. 1988. Description of bird behavior for comparative purposes. *In* Current Ornithology. Vol. 5, *edited by* R.F. Johnston. Plenum Pub. Corp., New York, NY. pp. 347-394.

Miller, E.H. 1991. Communication in pinnipeds, with special reference to non-acoustic signalling. *In* Behaviour of pinnipeds, *edited by* D. Renouf. Chapman and Hall, London, England. pp. 128-235.

Moore, K. 1991. Profiles of the undeveloped watersheds on Vancouver Island. Unpubl. Report, Friends of Ecological Reserves, Victoria, BC.

Mound, L. 1991. Developments in London. Assoc. Syst. Coll., Newsletter 19(1): 9-10.

Myers, N. 1979. The sinking ark. A new look at the problem of disappearing species. Pergamon Press, Oxford, England.

Nagorsen, D.W. 1990a. The mammals of British Columbia: a taxonomic catalogue. Royal B.C. Mus., Memoir 4, Victoria, BC.

Nagorsen, D.W. 1990b. Mammals. *In* The vertebrates of British Columbia: scientific and English names, *edited by* R.A. Cannings and A.P. Harcombe. Royal B.C. Mus., Heritage Record 20, Victoria, BC. pp. 39-43.

Nagorsen, D.W., R.W. Campbell and G.R. Giannico. 1991. Winter food habits of Marten,

Martes americana, on the Queen Charlotte Islands. Can. Field-Nat. 105: 55-59.

Nelson, J.S. 1987. The next 25 years: vertebrate systematics. Can. J. Zool. 65: 779-785.

Nicholson, T.D. 1991. Preserving the Earth's biological diversity: the role of museums. Curator 34: 85-108.

O'Brien, S.J., and E. Mayr. 1991. Bureaucratic mischief: recognizing endangered species. Science 251: 1187-1188.

Orchard, S.A. 1990. Amphibians in B.C.: forestalling endangerment. BioLine (Assoc. Prof. Biologists of B.C.) 9(2): 22-24.

Ouellet, H.R. 1977. Biosystematics and ecology of *Picoides villosus* (L.) and *P. pubescens* (L.), (Aves: Picidae). Ph.D. Thesis. McGill University, Montreal, PQ.

Pääbo, S., R. Wayne and R. Thomas. 1992. On the use of museum collections for molecular genetic studies. Ancient DNA Newsletter 1: 4-6.

Paggi, L., G. Nascetti, R. Cianchi, P. Orecchia, S. Mattiucci, S. D'Amelio, B. Berland, J. Brattey, J.W. Smith and L. Bullini. 1991. Genetic evidence for three species within *Pseudoterranova decipiens* (Nematoda, Ascaridida, Ascaridoidea) in the North Atlantic and Norwegian and Barents Seas. Int. J. Parasitol. 21: 195-212.

Parker, T. 1990. A commentary on the Clean Water Act and taxonomy. Assoc. Syst. Coll., Newsletter 18(6): 81-82.

Peden, A.E. 1972. Redescription and distribution of the rare deep-sea eel *Xenomystax atrarius* in the eastern Pacific Ocean. J. Fish. Res. Bd. Can. 29: 1-12.

Peden, A.E. 1981. Recognition of *Leuroglossus schmidti* and *L. stilbius* (Bathylagidae, Pisces) as distinct species in the North Pacific Ocean. Can. J. Zool. 59: 2396-2398.

Peden, A.E. 1990. Marine fishes. *In* The vertebrates of British Columbia: scientific and English names, *edited by* R.A. Cannings and A.P. Harcombe. Royal B.C. Museum, Heritage Record 20, Victoria, BC. pp. 5-17.

Remsen, J.V., Jr., and R.M. Zink. 1988. Louisiana State University, Museum of Natural Science. Amer. Birds 42: 366-369.

Roberts, L. 1988. Hard choices ahead on biodiversity. Science 241: 1759-1761.

Schofield, W.B. 1989. Structure and affinities of the bryoflora of the Queen Charlotte Islands. *In* The outer shores. Proceedings of the Queen Charlotte Islands First Int. Symp., Univ. British Columbia, August, 1984, *edited by* G.G.E. Scudder and N. Gessler. Queen Charlotte Islands Museum Press, Skidegate, BC. pp. 109-119.

Scudder, G.G.E. 1987. The next 25 years: invertebrate systematics. Can. J. Zool. 65: 786-793.

Seutin, G., B.N. White and P.T. Boag. 1991. Preservation of avian blood and tissue samples for DNA analyses. Can. J. Zool. 69: 82-90.

Short, L.L., Jr. 1982. Woodpeckers of the world. Delaware Mus. Nat. Hist., Greenville, DE.

Stap, D. 1990. A parrot without a name. A.A. Knopf, New York, NY.

Stuessy, T.F. 1990. Plant taxonomy: the systematic evaluation of comparative data. Columbia Univ. Press, New York, NY.

Stuessy, T.F., and K.S. Thomson. 1981. Trends, priorities and needs in systematic biology. Assoc. Syst. Coll., Lawrence, KS.

Thomas, W.K., S. Pääbo, F.X. Villablanca and A.C. Wilson. 1990. Spatial and temporal continuity of kangaroo rat populations shown by sequencing mitochondrial DNA from museum specimens. J. Mol. Evol. 31: 101-112.

Vigilant, L., R. Pennington, H. Harpending, T.D. Kocher and A.C. Wilson. 1989. Mitochondrial DNA sequences in single hairs from a southern African population. Proc. Natl. Acad. Sci. USA 86: 9350-9354.

Vogel, J.C., B. Eglington and J.M. Auret. 1990. Isotope fingerprints in elephant bone and ivory. Nature 346: 747-749.

White, C.M. 1988. Peregrine *Falco peregrinus*. *In* Handbook of North American birds. Vol. 5, *edited by* R.S. Palmer. Yale Univ. Press, New Haven, CT. pp. 324-335.

White, C.M., and D.A. Boyce, Jr. 1988. An overview of Peregrine Falcon subspecies. *In* Peregrine Falcon populations: their management and recovery, *edited by* T.J. Cade, J.H. Enderson, C.G. Thelander, and C.M. White. The Peregrine Fund Inc., Boise, ID. pp. 789-810.

Wilson, E.O. (Editor). 1988a. Biodiversity. National Academy Press, Washington, DC.

Wilson, E.O. 1988b. The current state of biological diversity. *In* Biodiversity, *edited by* E.O. Wilson. National Academy Press, Washington, DC. pp. 3-18.

Appendix A:

Example of a Formal Published Description of a New Species: A New Fish Species From British Columbia

The following excerpts illustrate how a new species is described. They are from the description of a new species of eelpout (family Zoarcidae), a group of deep-sea marine fishes (Anderson and Peden, 1988). The description begins with a statement of the newly proposed scientific name plus a summary of all earlier published scientific names. The "holotype" or "type specimen" is a single specimen chosen to act as a fixed point for the name. The holotype of this species was deposited in the large, widely accessible collections of the National Museum of Natural History (Smithsonian Institution, Washington, D.C.), where it was given catalogue number 280120 (USNM is a formal abbreviation for this museum). Gender, size and details of the holotype are given, including the ship from which the specimen was taken, how the specimen was captured, etc. "Paratypes" are other representative specimens of the new species; details are as for the holotype (BCPM is the formal abbreviation for the B.C. Provincial Museum, which is now called the Royal B.C. Museum). Paratypes are designated to better characterize the species by including some of its natural variation. "Additional material" was also examined when the species' description was being prepared.

"Diagnosis" summarizes the new species' distinguishing features within the genus and "Description" is a detailed overall description. Finally, the distribution of specimens and the etymology of the new species' name are given. The distribution and evolution of the species and its relatives are also discussed in the original paper; that discussion is not included below.

For further information about formalities and protocols for describing new species, see Davis and Heywood (1963) and Mayr (1969).

Figure 9. Drawing of the holotype of *Pachycara lepinium* (from Fig. 6 of Andersen and Peden 1988).

"**Pachycara lepinium n. sp.**

Lycodes sp. Hubbs et al., 1979: 14
Lyenchelys "D." Pearcy et al., 1982: 387

HOLOTYPE.-USNM 280120 (male, 465 mm SL); British Columbia, W of Tasu Sound, Queen Charlotte Islands; 52°38.0'N, 132°05.8'W; trap, 2,744 m; TALAPUS set 8; 4 Feb. 1980.

PARATYPES.-British Columbia: BCPM 980-121 (1); same as holotype. BCPM 980-98 (1); W of Tasu Sound; trap, 2,889 m; EASTWARD HO set 6; Aug. 1979. NMC 86-0445 (1); W of Tasu Sound; trap, 2,744 m; EASTWARD HO set 5....

ADDITIONAL MATERIAL.-SIO 59-364 (1); off Cabo Colnet, Mexico; 31°00.5'N, 118°06.0'W; trap, 1,728 m; 27 Oct. 1959. SIO 59-366 (1); off Cabo Colnet, Mexico; 31°02.7'N, 116°59.3'W; trap, 2,140 m; 27-28 Oct. 1959.

DIAGNOSIS.-A species of *Pachycara* as described by Anderson ... distinguished by the following combination of characters: pelvic fins present, their length 11.5-17.3% HL; mediolateral branch of lateral line originating posterior to pectoral fin margin; scales present on nape, extending to interorbital region; pectoral fin length....

DESCRIPTION.-Counts and measurements presented in Table 1 were compiled from all known specimens, 221-597 mm SL (including one gravid female). The following description is based on 6 adult males, 8 adult females and 16 juveniles of both sexes. Head large, ovoid, wider in adults than juveniles. Body relatively short, deep, broader in cross section in adults than juveniles. Tail laterally compressed, more so posteriorly, tapering gradually....

DISTRIBUTION.-Off the Queen Charlotte Islands, British Columbia, south to off Guadalupe Island, Mexico, at depths of 1,728 to 2,970 metres over brown and green mud bottoms. Often taken in traps and trawls with *P. gymninium*, but this species not yet known from the Gulf of California.

ETYMOLOGY.-From the Greek λεπιζ (scale) and ινιον (nape) in reference to the species' scaly head."

Appendix B:

The Importance of Cross-referencing Original Sources of Information: The Tailed Frog (*Ascaphus truei*) in British Columbia

The following information is from S.A. Orchard (ms. in prep.). The excerpt shows the nature of documentation for each locality record shown on the species' distribution map. With this information, an investigator can re-examine all original sources including specimens in the museums indicated. "A" through "G" are codes for different museum collections.

<u>Ascaphus truei</u> TAILED FROG

Map Grid	Locality	A B C D E F G	Literature/correspondence
92 I/4	Lytton: 15 miles S	- - - - - - -	Carl 1943; Carl and Cowan 1945; Logier and Toner 1961; Mills 1948; Slipp and Carl 1943; Taylor 1979
92 K/1	Brittain River: 4 miles from mouth	- 1 - - - - -	
92 K/7	Quatam River: 9.5 km E of mouth	- - - - - 1 -	Schueler et al. 1980
92 K/10	Bute Inlet	- - - - - - -	Metter 1968
	Hovel Bay	1 - - - - - -	Carl 1955; Logier and Toner 1961
93 D/7	Snootli Creek	- - - - - 1 -	Fred Schueler (in litt.)
103 I/4	Agate Creek at Khyex River	- - - - - 3 -	Fred Schueler (in litt.)

Appendix C:

Names, Naming and Subspecies

Formal rules govern the naming of plants and animals. The rules are necessary for establishing conventions and maintaining order. For most purposes it is sufficient to use the two-part name, or "binomen", with the first word's lead letter capitalized and both words italicized or underlined (e.g., *Salmo gairdneri; Symphoricarpos occidentalis*). A binomen is unique for each species of animal or plant. The name of the person(s) who originally discovered and named the species and the year it was described also are often given: for example, *Salmo gairdneri* Richardson, 1888. More complicated conventions must be applied when a species name changes, when two species are merged, etc. (see Davis and Heywood 1963; Mayr 1969). A trend is growing to also formalize common names corresponding to scientific names. These typically have first letters capitalized too:

Rainbow Trout	*Salmo gairdneri*
Western Snowberry	*Symphoricarpos occidentalis*

In a recent publication, a standard five-letter code, useful for rapid field data and computer entry and for computer analysis, has been proposed for B.C. vertebrates:

Rainbow Trout	*Salmo gairdneri*	F-SAGA
Tailed Frog	*Ascaphus truei*	A-ASTR
Deer Mouse	*Peromyscus maniculatus*	M-PEMA

Here, F = fish, A = amphibian and M = mammal (Cannings and Harcombe 1990).

Despite the existence of universally accepted systems of animal and plant names, various other naming schemes are in use (especially for common names). The main causes of such inconsistency are that biologists disagree in applying rules of names and in naming or deciding how an organism should be classified. Examples are the River Otter, called either *Lontra canadensis* (e.g., Nagorsen 1990a, 1990b) or *Lutra canadensis* (e.g., Macdonald 1984), and the Rainbow Trout of North America (*Salmo gairdneri*), judged by some scientists to be the same species that occurs in Asia, called *Salmo mykiss* (Peden 1990). Where species are distinctively different from one another and familiar to non-specialists, formal systems of common names can be successful and should be encouraged. They are easier to remember, after all, and make more sense to most people so are more effective in most communication (see Mayr 1990). But whatever the naming system used, a reference for the system should always be given in technical reports, papers, books, etc.

Nearly universal agreement exists on the usefulness and naturalness of species names, but this is not so for subspecies (defined in footnote 3). Many workers believe that formal subspecies names are unnecessary and misleading, as they can give the impression that species are composed of a set of clearly defined geographic parts. Other workers believe that formal subspecies names are invaluable when applied to particularly distinctive parts of a species, especially for populations on islands (ecological "islands" also include lakes and ponds, mountain tops and other disjunct isolated areas where species occur). Formally recognized subspecies of such kinds are likely to reflect underlying biological reality, so merit special names (see O'Brien and Mayr 1991). At the very least, subspecies reflect genetic diversity within a species. The subspecies may therefore be the most useful level at which to describe and analyse biological diversity, at least for well known groups such as vertebrates, flowering plants or butterflies (Bunnell and Williams 1980). The Ministry of Environment has recognized this principle by listing both species and subspecies at risk on its Red and Blue lists (see Munro, this volume).

Subspecies are designated by an addition to the scientific name. Some examples are distinctive forms on the Queen Charlotte Islands such as Hairy Woodpecker (*Picoides villosus picoideus*) and Northern Saw-whet Owl (*Aegolius acadicus brooksi*).

Appendix D:

Voucher Specimens: Definition and Importance

In any environmental impact study or biological survey, organisms are collected, seen, photographed or otherwise documented, and then identified. Parts or all of the evidence upon which such identifications are made must be preserved. Unless materials are properly preserved, it is impossible to independently verify claims of an investigator, re-evaluate the species present in a sample using new taxonomic knowledge, or trace changes over time caused by environmental effects such as pollution or global warming (see Danks et al. 1987). Aside from these scientific reasons for preserving specimens, investigators have an overriding ethical responsibility to use fully the plants and animals that they collect.

"A voucher specimen is an organism or a sample thereof preserved to document data in an archival report" (Lee et al. 1982: 1).

Responsibilities of museums are to provide curatorial, legal and technical advice on how to collect, preserve, document and deposit voucher specimens, and to store and maintain the material, and make it accessible after it has been deposited. Responsibilities of investigators and collectors are to properly sample, preserve and document specimens and to ensure that biologically and statistically satisfactory samples of collections are placed in a repository promptly after a project is completed. As well, investigators should cross-reference voucher specimens in all reports and publications. Granting and sponsoring agencies have responsibilities too: ensuring that terms of employment or terms of reference of a project explicitly state that voucher material will be collected and deposited in an appropriate institution (which should be named), and building into the budget the costs of collecting, preserving and storing such material (e.g., alcohol, jars, storage cabinets, labour). Arrangements with repositories for training field investigators and for accepting voucher material must be made in the early planning stages of projects.

Finally, scientific journals have a responsibility to ensure that published papers refer to repositories where voucher specimens have been deposited and can be independently examined by other workers. *The Canadian Journal of Zoology* has such a policy, for example, and it is included in the journal's written guidelines for preparing manuscripts.

Diversity in Ecosystems
of
British Columbia

Terrestrial Diversity of British Columbia

Jim Pojar

B.C. Forest Service, Bag 5000, 3726 Alfred Avenue, Smithers, British Columbia, Canada V0J 2N0.

Abstract

Provincial terrestrial biodiversity is compared to that of surrounding jurisdictions and other provinces. Within a global context, endemism is low, but British Columbia has higher biodiversity than other similar jurisdictions with temperate climates. More than half of all the vascular plant species in Canada occur in B.C. More than 75% of Canada's byroflora is represented, as are 85% of all the species of lichens in northwestern North America. Provincial fauna is also diverse, with 70% of Canada's bird species and 80% of Canada's terrestrial mammal species represented here. True flies, beetles, wasps, bees and ants are the richest insect groups and all are well represented in B.C. Provincial ecosystem diversity is created by physical diversity, and by the variety of climates created by complex topography. Some centres of plant richness and rarity are the Strait of Georgia complex, the dry interior and the Peace River region. The Queen Charlotte Islands are floristically depauparate, but have some endemic taxa. Other notable biologically diverse areas of the province are seabird islands, fiordlands, the Rocky Mountain Trench, the Fraser Lowland, Brooks Peninsula, the Chilcotin Fraser Junction, Haines Triangle and numerous alpine areas.

Introduction

Some people may wonder, "What's a forester doing giving this talk?" Well, things change. Some people know that when the Bhagwan, Jack Ward Thomas, assumed his robes, I was his Model. Indeed, I am the very model of a modern professional forester. So we start with some questions and answers, keeping in mind that I am dealing with

In *Our Living Legacy: Proceedings of a Symposium on Biological Diversity*, edited by M.A. Fenger, E.H. Miller, J.A. Johnson and E.J.R. Williams, pp. 177-190. Victoria, B.C.: Royal British Columbia Museum.

terrestrial diversity only, and with species, ecosystem and landscape aspects of that diversity.

How Biologically Diverse is British Columbia?

British Columbia is very diverse, considering its northerly latitudes (48°14' to 60°N), the preponderance of rock, ice and snow in much of its landscape and the relatively recent glaciation it experienced. This province is definitely not Madagascar or even California in terms of its biological diversity. However, B.C. probably has greater biological diversity than Oregon and Washington combined and is certainly more diverse physically, ecologically and biologically than any other Canadian province or territory, or Alaska. Native vascular plants illustrate this point: about 2,500 species occur in B.C., 1,775 in Alberta and 1,600 in Alaska and the Yukon combined; indeed, the total for Canada is only about 4,153, well over half of which occur in B.C.

The seven largest families of vascular plants in B.C. are the Asteraceae, Poaceae, Cyperaceae, Brassicaceae, Fabaceae, Rosaceae and Scrophulariaceae. Globally, endemism in the flora is generally low, although there are several endemic taxa on the coast, especially on the islands (including Vancouver Island and the Queen Charlotte Islands). However, if one considers all of northwestern North America, from northern California to Alaska and the Yukon, then B.C. has much higher endemism of species, genera and even higher levels. The centres of plant richness and rarity in B.C. are the dry southern interior, the Peace River area and the Strait of Georgia complex of grassland, parkland, dry forest, rock outcrop, vernal meadow and pool/seepage vegetation, collectively called "saanich" (Pojar 1990).

Mosses, liverworts and hornworts also display great richness in B.C. About 700 species of mosses occur in B.C., yet only 900 occur in all of northwestern North America (Alaska, Yukon, B.C., Alberta, Oregon, Washington, Idaho and western Montana). B.C. has about 250 species of liverworts or hornworts, nearly all those that inhabit northwestern North America.

More than 75% of the total bryoflora of Canada is represented in B.C. Thirty genera and sixty species found in Canada are found *only* in B.C. Bryophytes are especially diverse and prominent in coastal forests. Endemism is not high but is concentrated on the Queen Charlotte Islands and adjacent island groups. Phytogeographically unusual species are concentrated on the hypermaritime outer coast; the semi-arid southern interior also has a few. Rare or endangered bryophytes are mainly those at their northern or southern range limits, or are restricted endemic species, or are rare throughout their range.

Knowledge of the province's lichens is sketchy (Egan 1987; Noble et al. 1987). The only large areas that have been intensively collected are the Queen Charlotte Islands and Strait of Georgia area. The most recent B.C. checklist added 50 species to the North American lichen flora of 3,400 species. B.C. has most (nearly 85%) of the 1200 species that occur in northwestern North America. The province is rich in corticolous, arboreal and terrestrial species, and has noteworthy epiphytic communities of both lichens and bryophytes, as one would expect of a cordilleran region fronting the ocean. There is excellent representation of arctic-alpine and Pacific maritime elements. Some wide-ranging genera of macrolichens are very well represented in B.C. For example, 15 of the 20 North American species of *Hypogymnia* occur here, as do 20 of 28 species of *Umbilicaria* and 23 of 30 species of *Cetraria*.

Fungi play fundamental roles in nutrient and energy dynamics, and our lack of knowledge about them is appalling. Much of the existing knowledge is limited to economically important species and on showy, tasty, dangerous, pathogenic or especially lurid species. Current estimates for B.C. are at least 10,000 species of fungi overall, including up to 3,000 mushrooms. We know so little of lower fungi, including phycomycetes and slime moulds, that meaningful estimates of their species richness are impossible. It would be surprising if 20% of the province's species are known. B.C. has more kinds of fungi than the Prairie Provinces or Maritimes, and about the same number as Ontario and Quebec, which are enriched by a component of southern, deciduous forest species (S. Redhead pers. comm.).

The rich fauna of B.C. parallels the flora. B.C. has more than 70% of all Canadian bird species, in terms of both the provincial total (about 450) and breeding species (Bunnell and Williams 1980). B.C. has about 300 species of breeding birds, nearly half of the total for the continental United States and more than any other province; 162 of these do not breed anywhere else in Canada. Among these are Marbled Murrelet, Ancient Murrelet, Rhinoceros Auklet, Tufted Puffin, Spotted Owl, Flammulated Owl, White-headed Woodpecker, White-throated Swift, Anna's Hummingbird, Gray Flycatcher, Canyon Wren and Hutton's Vireo. Species with a centre of distribution in B.C. include Harlequin Duck, Barrow's Goldeneye, Bald Eagle, Blue Grouse, Common Poorwill, Rufous Hummingbird, Lewis' Woodpecker, Sage Thrasher, Black-headed Grosbeak, Golden-crowned Sparrow, Cassin's Finch and American Dipper. B.C. has internationally significant populations of Barrow's Goldeneye, Blue Grouse, White-tailed Ptarmigan, Black Oystercatcher, Ancient Murrelet, Marbled Murrelet, Cassin's Auklet, Rhinoceros Auklet, Rufous Hummingbird, Northwestern Crow, Varied Thrush, Townsend's Warbler and Golden-crowned Sparrow, as well as raptors like Osprey, Bald Eagle and Peregrine Falcon. The province is a major migration corridor and staging area for species such as Brant, Sandhill Crane, Hammond's Flycatcher and Vaux's Swift. The centres of bird diversity occur in the southwest and the southern interior of B.C. The Fraser Lowland (especially the Fraser River delta) provides internationally significant, critical wintering habitat for many species.

B.C. has about 110 species of terrestrial mammals, nearly 80% of the total for all of Canada. In comparison, Ontario has 77 and the continental United States has 315. Of B.C.'s terrestrial mammal species, 25 (plus 12 marine species) occur nowhere else in Canada. B.C. contains more large (> 1 kg) mammal species than any other province or state in North America (Bunnell and Kremsater 1990). The southern Okanagan has 14 species of bats, the richest area for bats in Canada (Nagorsen and Brigham 1993). B.C.'s fauna is notable not only for high species diversity but also diversity within species (e.g., subspecies, peripheral populations, island races) and high abundance. Furthermore, some large, remote areas of the province still have intact large mammal predator-prey systems with all or various combinations of deer, Elk, Moose, Caribou, mountain sheep, Mountain Goat, Gray Wolf, Cougar, Grizzly Bear, Black Bear and Wolverine (Bunnell 1991). Historically, B.C. has lost the Queen Charlotte Islands ("Dawson") Caribou, White-tailed Jackrabbit, Bison (re-introduced) and the Sea Otter (re-introduced).

About 21 species of amphibians occur in B.C., 19 of which are native. Amphibians constitute much biomass in forest ecosystems in B.C., particularly in coastal old-growth forests (Burton and Likens 1975; Pough et al. 1987; Welsh and Lind 1988). Centres of richness in B.C. are the south coast and the dry southern interior. There are four species of amphibians designated "at risk" in the province: Pacific Giant Salamander, Tiger Salamander, Coeur d'Alene Salamander and Tailed Frog.

Five of B.C.'s fifteen native species of reptiles are designated "at risk": Sharp-tailed Snake, Western Rattlesnake, Gopher Snake, Short-horned Lizard and Night Snake (Ministry of Environment 1991). This does not include two species of sea turtles, both of which are seriously endangered. As one would expect, the warm dry southern Okanagan Valley is the centre of the province's reptile diversity.

Invertebrates constitute the vast majority of kinds of organisms and are very abundant, but they are usually small and difficult to study. Much of our knowledge of B.C.'s invertebrates is in the early phase of taxonomy (see Foster, this volume). The numbers and kinds that occur in B.C. are, therefore, extremely poorly known. No estimates are possible for rich and ecologically critical groups like nematodes, flatworms, annelids, molluscs and mites. Around 600 species of spiders

(Canadian total 1,400) and 35,000 species of insects (Canadian total 55,000) occur in B.C. Probably about 50% of insect species in Canada and B.C. are unknown to science, however, since at least 75% of living things are animals, and 75% of these are insects (one-quarter of all species are beetles!). Insects are clearly of fundamental importance to any terrestrial ecosystem and to B.C.'s biodiversity.

The richest insect orders in B.C. are Diptera (true flies), Coleoptera (beetles) and Hymenoptera (wasps, bees, ants). Flies and beetles probably dominate in coastal forests. Flies increase dramatically in importance with increasing latitude, whereas Hymenoptera tend to increase with aridity.

B.C.'s insect fauna is more diverse than that of Alberta and is probably also more diverse than that of Ontario and Quebec combined. The insect fauna is poorly known, but it is apparent that significant endemism (of both species and subspecies) occurs in those groups that are relatively well known (e.g., endemic subspecies of butterflies in the Peace River area and on the Gulf Islands). The insect fauna of coastal northwestern North America is highly distinctive and has a large endemic component even in areas that were glaciated (e.g., Queen Charlotte Islands and Olympic Peninsula / Vancouver Island; Kavanaugh 1988). Few other areas are known to have been heavily glaciated and to have developed such a distinctive insect fauna with so much endemism (Kavanaugh 1988, 1989). Along with the southwest coast (Strait of Georgia area), the southern Okanagan is the most diverse region in the province. Both areas also contain the largest numbers of threatened and rare insect species.

The Kingdom Protista comprises eukaryotic microorganisms and their close descendants, and includes various, mostly aquatic, flagellates (e.g., euglenophytes, ciliates) as well as organisms traditionally classed as algae and fungi with flagellated spores (Margulis and Guerrero 1991). This large, diverse and ubiquitous group is extremely poorly known taxonomically and no estimates of richness and diversity in B.C. are possible.

Ecosystems

B.C. is more ecologically diverse than any province or territory in Canada. The province has six of the ten forest regions of Canada (Rowe 1972). Though it is largely forested, it also has extensive grasslands, wetlands and alplands. B.C. has 14 biogeoclimatic zones (Pojar et al. 1987) and each zone has numerous habitats, dry to wet, forested or non-forested (Meidinger and Pojar 1991). There is much beta (between-habitat) diversity and very few monotonous landscapes in the province. Forests predominate, covering over 55% of the total area, and include deciduous, mixed-wood and even some broad-leaved evergreen (in the Strait of Georgia) forest types. In this ruggedly mountainous region with sharp climatic gradients, general diversity increases markedly from low to high elevations (tidewater to timberline in many cases) and from west to east, from the wet coast to the dry interior. B.C. boasts a great range of plant communities, from "saanich" to muskeg, from coastal temperate rain forest to Great Basin shrub-steppe and dry forest, from tidal and freshwater wetlands to grasslands and alpine tundra.

Physical and Climatic Diversity of B.C.

Physical Diversity

The province spans more than 11 degrees of latitude and 25 degrees of longitude and covers nearly one million square kilometres. Mountains are a prominent physical feature of B.C. The province has a coastal system of mountains, lowlands, islands and fiords, several parallel mountain ranges, extensive plateaus in the northern and southern interior, and a portion of the Great Plains (east of the Rockies in the northeast corner of the province).

B.C. has a complex bedrock geology (see Farley 1979: Map 13), which contributes to biodiversity through the interactions of climate, topography and parent materials. The interactions result in different erosional and soil-forming processes and consequent biota on or near different bedrocks. The province would be even more diverse than it is if the massive Coast Mountains were not so dominated by acid-resistant, intrusive rock.

Climate

B.C.'s climate is largely determined by the sea and the mountains. Frontal systems spawned over the Pacific Ocean move onto the coast and eastward across the province encountering successive mountain barriers that are roughly perpendicular to upper air flow. The mountain ranges control the overall distribution of precipitation and the balance between Pacific and continental air masses in the various regions of B.C. The interplay of climate, physiography and organisms is expressed in how the province can be divided into nine terrestrial ecoprovinces (areas of similar climate, topography and geological history; Demarchi et al. 1990), and further stratified altitudinally and vegetatively into 14 biogeoclimatic zones (large geographic areas with broadly consistent climate, patterns of vegetation and soil; Pojar et al. 1987).

Glacial History

Repeated continental glaciation has reduced the species diversity of B.C.'s biota. However, glaciation has contributed to ecosystem and landscape diversity through its extensive effects on surficial geology and landforms, and has also allowed for various invasions, migrations and persistences — sometimes with evolutionary divergence — in Pleistocene refuges (e.g., parts of the Queen Charlotte Islands, Brooks Peninsula).

Disturbance History

Disturbance and change are inevitable in natural systems, but vary in type, frequency and rate in different areas of the province. Besides the underlying climatic, topographic and edaphic variation, the main drivers of diversity are disturbances. Fire, insect outbreaks and windthrow have been the major large-scale natural disturbances in dry interior zones, whereas fluvial processes, mass movements, windthrow and (in places) fire have prevailed on the coast. Areas with more active disturbance regimes tend to have more complex succession and greater habitat heterogeneity and therefore more ecosystem and landscape diversity. On the other hand, areas that experience less frequent catastrophic disturbances can develop mature ecosystems with great within-habitat diversity, such as coastal old-growth forests with complex structure and composition (Pojar et al. 1990; B.C. Ministry of Forests 1992).

Regional Biodiversity in British Columbia

In this section, I outline those parts of the province that hold high biological diversity or distinctive biological diversity.

Strait of Georgia and the Dry Southern Interior

The Strait of Georgia and Dry Southern Interior regions bear the torch for biological diversity in B.C. in many respects: unusual, rare and threatened ecosystems and species, centres of richness, and nationally significant biota. Unfortunately but not unexpectedly, the bulk of B.C.'s human population also resides in these regions.

The saanich of the southwest coast and the dry forest/shrub-steppe of the southern interior are extensions of widespread southern ecosystems and many of the interesting taxa in B.C. represent peripheral populations. Nevertheless, the B.C. versions of these zones and their assemblages of species are distinctive. Our mix of habitats and species includes contributions from adjacent northern zones and lacks some of the typical southern elements. Furthermore, peripheral populations usually have distinctive dynamics.

Junction Area and the Western Chilcotin

In contrast to the previous two "marginal" regions, the Junction Area and Western Chilcotin are in the middle of the southern half of the province. The Junction (confluence of the Chilcotin and Fraser rivers) - Riske Creek area, with its badlands, grasslands, California subspecies of Bighorn Sheep, Long-billed Curlew, alkaline lakes, plateau ponds and parkland, is a B.C. specialty. So too is the expansive western Chilcotin, spread out on a high, dry, cold plateau in the lee of the most massive Coast Mountains. This area has some peculiar similarities (climatic and otherwise) with the dry, cold southwestern Yukon and even the White/Inyo mountains of California. It includes highly individual mountain ranges (Chilcotin, Itcha, Rainbow) and is the only place in Canada where Lodgepole Pine functions as a climax species and also is a consistent timberline tree. Superficially monotonous with extensive stands of even-aged Lodgepole Pine, at a larger scale the area is actually a complex mosaic of age classes of forests, myriad wetlands and alpine islands, and parts of it are the haunt of the woodland subspecies of Caribou.

Nass Basin

The Nass Basin in northwestern B.C. is an irregularly shaped area of low relief, mostly below 800 metres elevation, and encircled by mountains but in communication with the coast via the broad valley of the Nass River. The area has a coastal-transitional climate and is subjected to frequent outbreaks of cold, dense arctic air in the winter. The snowpack in the northern part of the basin (Meziadin Lake / Bell-Irving River) is immense and long-lasting, strikingly so for such relatively low elevations. The landscape is largely forested (primarily by Western Hemlock), is dotted with numerous lakes and wetlands, is ribbed with bedrock ridges and drumlins and is the domain of salmon, bears and black flies.

Queen Charlotte Islands

The Queen Charlotte Islands are a microcosm of the entire coast of B.C., with snowy peaks, rugged mountains, heavily forested slopes, big trees, boggy lowlands, archipelago-fiord landscapes, high- and low-energy beaches, salmon streams and estuaries. But the Charlottes are unique not only in their size (940,000 hectares in 150 islands) and remoteness (50-130 km off the mainland) but also in their biota, which is insular and depauperate but has several endemic taxa and numerous biogeographical oddities (Scudder and Gessler 1989).

Brooks Peninsula

The Brooks Peninsula, a rugged mountainous abutment of northwestern Vancouver Island to the edge of the continental shelf, is one of the truly wild and remote places of the B.C. coast. Since the mid 1970s several biological expeditions have discovered evidence of a Pleistocene glacial refuge on the peninsula. The most ambitious effort involved the Royal B.C. Museum in 1981 (Hebda and Haggarty 1993).

Seabird Islands

There are thousands of islands along the B.C. coast and some of them, like the remote and exposed Triangle Island and Kerouard Islands, are important nesting sites for colonial seabirds. Although much of their fauna is marine-oriented, these islands provide key terrestrial habitat for our internationally significant populations of seabirds and pinnipeds (Drent and Guiguet 1961; Rodway 1991).

Fiordland Hecate lowland

From the Juan de Fuca Strait and Puget Sound at the southern end of fiordland to Cross Sound and the head of Lynn Canal at the northern end, a distance of over 1400 km, "the mountainous western margin of North America meets the Pacific Ocean in a deeply indented, island-studded coast ranking with the grandest fiord-coasts of the world in the character and magnitude of its physical features" (Peacock 1935: 634). Although the northern third of this coastland has, through historical accident, come to be known as southeast Alaska, fiordland is a single bioregion despite the geopolitics. Not only is the scenery stupendous, but one of the coast's strongest ecological themes still survives intact in some forested valleys of this region. I refer to the pristine ecosystem that includes primary watersheds, estuaries, coastal old-growth forest featuring bottomland Sitka Spruce and Western Hemlock stands, avalanche tracks, salmon and Grizzly Bears.

Westward from the steep-sided transverse valleys and fiords of B.C.'s central and north coast lies a maze of islands, great and small, in a low-lying strip of moderate but rough relief called the Hecate Lowland. This elaborate section of coast contributes greatly to the 27,000 km total length of provincial coastline. The Hecate Lowland has been heavily glaciated, and the bedrock bones are often exposed in a peculiar peatland landscape, a sort of lost world of bonsai trees, multicoloured muskeg and a rich variety of wetland ecosystems and species (Banner et al. 1988).

Rocky Mountain Trench / East Kootenay

The Rocky Mountain Trench is a remarkable and uniquely British Columbian topographic feature that extends from the 49th parallel northwestward for over 1,400 km to the Liard River. From about Golden south, the southern Trench and surrounding mountains encompass great habitat diversity because of combinations of wet mountains and dry rainshadow valleys, complex topography and an assortment of vegetation including dry and wet forests, shrub-steppe, and extensive wetland and riparian communities. This habitat diversity is reflected in a rich flora and fauna that include significant populations of seven species of large ungulates plus their natural predators.

Fraser Lowland

The Fraser Lowland (or Lower Fraser Valley west from Laidlaw) is essentially a large delta and the low hills that have been enveloped by the delta. Most of the area is either heavily urbanized or has been converted to intensive agricultural or industrial use, but the Lowland is still a centre of richness for birds and small mammals. The Fraser River estuary is disproportionately significant to the biota of the province. The Fraser delta includes wetlands that make up the largest single unit of wetland habitat in B.C., and is the most important migratory and wintering area for waterbirds in the province (Demarchi et al. 1990). The Lowland is also home to several species of vertebrates at the northern limit of their ranges including North American Opossum, Trowbridge's Shrew, Townsend's Mole, Coast Mole and Pacific Giant Salamander.

Sub-boreal Spruce Zone

The Sub-boreal Spruce zone (SBS) is the montane forested zone that dominates the landscape of the central interior of B.C. (Meidinger and Pojar 1991). On the face of it, the SBS is an unremarkable landscape of forests and wetlands inhabited by mostly boreal species and not unlike the great boreal forest that sweeps across the continent from Alaska to Newfoundland. The SBS has, however, been enriched by some of the biotic elements of the montane Douglas-fir forests to the south. Furthermore, the north central interior of the province is affected by mild, moist, maritime air masses while also being somewhat protected from cold continental air from east of the Rocky Mountains. So the climate of the SBS is less harshly continental than that of the true boreal forest, with favourable consequences for forest productivity and species diversity. But perhaps the most distinctive feature of the zone, in a continental sense, is the abundance of large natural lakes (such as Babine, Stuart, Francois) that feed major undammed salmon rivers —

the Skeena and Fraser. Although not strictly a feature of terrestrial diversity, the aquatic and terrestrial systems are closely linked, and I know of no other such regional ecosystem in North America (the Great Lakes - St Lawrence system is no longer natural).

Stikine Canyon

The Grand Canyon of the Stikine River, including Telegraph Creek and vicinity, is the centre of a distinctive boreal area that is unusually warm and dry. While dry grassland and scrub are common they are usually found on steep colluvial slopes. The only known boreal locality of *Juniperus scopulorum* (Rocky Mountain Juniper) is in the Stikine Canyon. Extensive hybridization occurs between *Juniperus scopulorum* and *J. horizontalis*, and several uncommon plant communities and geographically unusual species are found in this area. There are similar faunal distributions related to the warm south-facing slopes, with interesting disjunctions that often link with occurrences to the north, in the Yukon or to the grasslands of the southern interior plateau (e.g., in the hemipteran insects; G.G.E Scudder pers. comm.).

Peace River Area

The Great Plains extend into the northeastern part of B.C. east of the Rocky Mountains. Centred on the Peace River is an area (part of the Alberta Plateau) of plateaus, plains and lowlands, generally of low relief but dissected by the Peace and its tributaries. Originally largely covered by boreal spruce forest, the area now is dominated by a mix of post-fire successional forest of Trembling Aspen, Lodgepole Pine and White Spruce as well as open shrub-steppe and cleared agricultural land — especially in the Peace Lowland. The terrain and vegetation are similar to and connect with more extensive areas in Alberta, where the landscape is commonly termed "aspen parkland". The B.C. portion of these Peace River parklands is a secondary centre of rarity for some taxonomic groups largely because of the occurrence of typically eastern species that just barely grow into B.C. here (e.g., vascular plants, insects, birds). Furthermore, the mix of habitats including the deciduous forest and shrub-steppe of the steep south-facing "breaks" along the major rivers, and the extensive and often productive wetlands, contribute to high diversity and abundance of waterbirds and large mammals (eight ungulate species plus their predators).

Haines Triangle

Shifting the focus to far northwestern B.C. reveals a little known yet magnificent corner of the province, the Haines Triangle. Here the Tatshenshini Basin, an area of wide valleys and moderate relief, is buttressed by the massive St Elias Mountains to the west and the northern end of the Coast Mountains to the east. The mountains support huge icefields, and even though the area is close to the ocean and the coastal influence does penetrate a short distance up some of the drainages, most of the area has a subarctic climate. The flora is not rich but includes such B.C. rarities as *Salix setchelliana, Spiraea stevenii, Swertia perennis, Primula stricta* and *Woodsia alpina*. Similarly, the vertebrate fauna includes significant populations of Thinhorn Sheep, Collared Pika, Smith's Longspur and Gyrfalcon, among others.

Alpine Tundra

The alpine zone occurs on high mountains throughout much of B.C., yet surprisingly few studies of the biology and ecology of what is one of the most interesting and characteristic environments of the province have been done. Our alpine biota, although highly specialized, is generally impoverished. Nevertheless, some groups of organisms (such as lichens) are relatively diverse in alpine tundra. Perhaps more noteworthy is the diversity of alpine ecosystems and landscapes, with complex mosaics of habitats that closely track differences in microenvironment, and a wonderful range of alpine landscapes.

The alpine zone extends beyond B.C. on the high mountains to the south, north and east, and obvious similarities and relationships occur between much of B.C.'s alpine tundra and that of the surrounding cordillera. A few alpine areas in the province, however, seem to be ecologically unique. Foremost among these is the relatively small hypermaritime alpine zone of the Queen Charlotte Islands, which boasts several endemic taxa and is unusually productive and lushly vegetated, even on extremely steep and rocky terrain. Other distinctively British Columbian alpine environments include the northern Rocky Mountains and the high country of the western Chilcotin.

Acknowledgements

Much of the technical information and many of the slides for the symposium presentation are from B.C. experts, including staff of the Royal B.C. Museum (Rob Cannings, Dave Nagorsen, Bob Ogilvie, Wayne Campbell), Fred Bunnell, Syd Cannings, Dick Cannings, Dennis Demarchi, Andy MacKinnon, Del Meidinger, Stan Orchard and Scott Redhead.

Literature Cited

Banner, A., R.J. Hebda, E.T. Oswald, J. Pojar, and R. Trowbridge (Editors). 1988. Wetlands of Pacific Canada. National Wetlands Working Group, Wetlands of Canada. Polyscience, Ottawa ON.

B.C. Ministry of Forests. 1992. An old growth strategy for British Columbia. B.C. Ministry of Forests, Victoria, BC.

Bunnell, F.L. 1991. Biodiversity: what, where, why, and how. *In* Proceedings, Wildlife Forestry Symposium, Prince George, BC., 7-8 March 1990, *edited by* A. Chambers. pp. 29-45.

Bunnell, F.L., and L.L Kremsater. 1990. Sustaining wildlife in managed forests. Northwest Env. J. 6: 243-269.

Bunnell, F.L., and R.G. Williams. 1980. Subspecies and diversity — the spice of life or prophet of doom. *In* Threatened and endangered species and habitats in British Columbia and the Yukon, *edited by* R. Stace-Smith, L. Johns, and P. Joslin. B.C. Ministry of Environment, Victoria, BC. pp. 246-259.

Burton, T.M., and G.E. Likens. 1975. Salamander populations and biomass in the Hubbard Brook Experimental Forest, New Hampshire. Copeia 3: 541-546.

Demarchi, D.A., R.D. Marsh, A.P. Harcombe, and E.C. Lea. 1990. The environment. *In* The birds of British Columbia, Vol. 1, by R.W. Campbell, N.K. Dawe, I. McT. Cowan, J.M. Cooper, G.W. Kaiser, and M.C.E. McNall. Royal B.C. Museum and Canadian Wildlife Service, Victoria, BC. pp. 55-144.

Drent, R.H., and C.J. Guiguet. 1961. A catalogue of British Columbia sea bird colonies. B.C. Provincial Museum, Occ. Pap. 14., Victoria, BC.

Egan, R.S. 1987. A fifth checklist of the lichen-forming, lichenicolous and allied fungi of the continental United States and Canada. Bryologist 90: 70-173.

Farley, A.L. 1979. Atlas of British Columbia. Univ. British Columbia Press, Vancouver, BC.

Hebda, R.J., and J.C. Haggarty (Editors). 1993. The Brooks Peninsula: an ice-age refuge in British Columbia. Royal B.C. Museum, Victoria, BC. (In press.)

Kavanaugh, D.H. 1988. The insect fauna of the Pacific Northwest coast of North America: present patterns and affinities and their origins. Mem. Ent. Soc. Can. 144: 125-149.

——————— 1989. The ground-beetle (Coleoptera: Carabidae) fauna of the Queen Charlotte Islands, its composition, affinities, and origins. *In* The outer shores. Proceedings of the Queen Charlotte Islands First Int. Symp., Univ. British Columbia, August, 1984, *edited by* G.G.E. Scudder and N. Gessler. Queen Charlotte Islands Museum Press, Skidegate, BC. pp. 131-146.

Margulis, L., and R. Guerrero. 1991. Kingdoms in turmoil. New Scientist 1761:46-49.

Meidinger, D., and J. Pojar (Editors). 1991. Ecosystems of British Columbia. B.C. Ministry of Forests, Research Branch Special Report Series 6, Victoria, BC.

Nagorsen, D.W., and R.M. Brigham. 1993. The bats of British Columbia. Lone Pine Publishing, Edmonton, and Royal B.C. Museum, Victoria, BC.

Noble, W.J., T. Ahti, G.F. Otto, and I.M. Brodo. 1987. A second checklist and bibliography of the lichens and allied fungi of British Columbia. Syllogeus (Nat. Mus. Can.) 61. Ottawa. ON.

Peacock, M.A. 1935. Fiord-land of British Columbia. Bull. Geol. Soc. Amer. 46: 633-696.

Pojar, J., K. Klinka, and D.V. Meidinger. 1987. Biogeoclimatic ecosystem classification in British Columbia. For. Ecol. Manage. 22: 119-154.

Pough, H.F., E.M. Smith, D.H. Rhodes, and A. Collazo. 1987. The abundance of salamanders in forest stands with different histories of disturbance. For. Ecol. and Manage. 20: 1-9.

Rodway, M.S. 1991. Status and conservation of breeding seabirds in British Columbia. *In* Seabird status and conservation: a supplement, *edited by* J.P. Croxall. Int. Council for Bird Preservation, Tech. Pub. 11, Cambridge, England. pp. 43-102.

Rowe, J.S. 1972. Forest regions of Canada. Can. Dep. Environment, Can. Forestry Service, Publ. No. 1300, Ottawa, ON.

Scudder, G.G.E., and N. Gessler (Editors). 1989. The outer shores. Proceedings of the Queen Charlotte Islands First Int. Symposium, Univ. British Columbia, August, 1984. Queen Charlotte Islands Museum Press, Skidegate, BC.

Welsh, H.H., Jr., and A.J. Lind. 1988. Old-growth forests and the distribution of the terrestrial herpetofauna. *In* Management of amphibians, reptiles, and small mammals in North America, *edited by* R.C. Szaro, K.E. Sieverson, and D.R. Patton. USDA Forest Service, Rocky Mtn Station, Gen. Tech. Rep. RM-166, Fort Collins, CO. pp. 439-458.

Biodiversity: The Marine Biota of British Columbia

Verena Tunnicliffe

Biology Department, University of Victoria,
Victoria, British Columbia, Canada V8W 2Y2.

Abstract

There has been no formal assessment of biological diversity in the marine waters of British Columbia, either in terms of the species that occur, or in terms of the possible changes that could result from human activities. Based on published information, I estimate species richness in these waters. Considering major groups within the algae, marine mammals, fishes and invertebrates, I estimate that about 5,000 documented species, or 3-4% of the world's marine species, occur in B.C. Interestingly, B.C. also has a similar proportion of the world's coastline, so the provincial marine fauna is neither particularly enriched nor depauperate compared with the world average. The complexity of B.C. coastal habitats imposes limitations on sampling. As a result, numerous species remain unrecorded. Variability in habitat promotes high regional diversity. An unusual habitat near these coastal waters is that of the hydrothermal vents on the Juan de Fuca and Explorer Ridges; here, an endemic fauna of species (and higher taxa) occurs that is largely new to science. A complete analysis of the responsiveness and vulnerability of B.C.'s marine habitats is required before the effects of species or gene-pool loss can be assessed.

Introduction

Assessing biological diversity of a region has three phases: (1) the record of taxa present, (2) the documentation of relationships among taxa and habitat, and (3) the documentation or prediction of what happens when certain taxa are eliminated. At present, the science of ecology can contribute good generalized models of interaction

In *Our Living Legacy: Proceedings of a Symposium on Biological Diversity*, edited by M.A. Fenger, E.H. Miller, J.A. Johnson and E.J.R. Williams, pp. 191-200. Victoria, B.C.: Royal British Columbia Museum.

processes, and studying extinction in the ancient record and the modern world provides many useful comparisons.

This paper will restrict itself to species that inhabit the waters of British Columbia. It addresses the diversity of habitat present and the gaps in our knowledge. Finally, I will mention some special habitats and inhabitants with which I have particular familiarity.

Methods

The information presented here is from literature compilations of species records. The organisms included are those that are obligate marine dwellers. Birds and transient terrestrial mammals are not examined. Scagel et al. (1989) provide a synopsis of the known marine plants of British Columbia. The discoveries of new species and new records of species distributions are not unusual in this group. My guess is that the list is almost complete but that further additions will be made. Sources for marine mammals are Watson (1981) and King (1983). Given the high visibility and extensive research on these animals, the species records are reliable. Fish records were compiled from Hart (1973). There have been several new species and new ranges published since then.

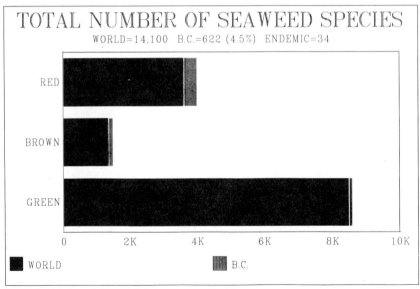

Figure 1. Bar diagram illustrating the numbers of species of algae in the world and in B.C. The bars represent: Rhodophyta (red), Phaeophyta (brown) and Chlorophyta (green).

192

The group that gives the greatest difficulty is the invertebrates, which encompasses some thirty phyla. My source is a checklist prepared by Austin (1985) that includes about 3,700 species. Unlike the algae and marine mammals that occur only in surface waters, and unlike fishes that tend to be relatively easy to obtain, the invertebrates are notoriously difficult to document. Their wide habitat range, small size and relatively cryptic behaviour mean that many remain undocumented. The relatively few scientists examining this group in B.C. add to their obscurity. Since the 1985 publication over 300 more species records have been documented (W.C. Austin pers. comm.).

Comparing B.C. fauna to the world fauna is aided by the synopsis of organisms compiled by Parker (1982) that was consulted for fishes and algae. Estimates were made as not all taxa are indicated as either marine or freshwater. Austin (1985) provides estimates of the total marine invertebrates of the world. Information on offshore hydrothermal vent communities is drawn from Tunnicliffe (1988, 1991).

Results

Algae

The plants treated here are the algae, and include benthic macroscopic multicellular forms. Two species of marine grasses are treated as well. There are three major divisions, or phyla, of the algae: Chlorophyta (green algae), Phaeophyta (brown algae) and Rhodophyta (red algae). Fig. 1 shows the approximate numbers of these algae in the world and in B.C. Of about 8,600 species of marine green algae in the world, 102 are found in B.C. Of about 4,000 species of red algae, 390 are found in B.C. Of about 1,500 brown, 130 are in B.C. Despite the small number of species, the last group has the greatest biomass. Overall, B.C. is home to about 4.5% of the world's species of marine algae. Most of these are known elsewhere, but Scagel et al. (1989) list 34 species with ranges only in B.C.

Marine Mammals

There are about 108 species of whales and pinnipeds in the world. Whales are widely distributed because of their nektonic and migratory habits. Eighteen species of whales have been recorded in B.C., although many very rarely (Fig. 2). Only five pinniped species occur regularly. While 20% of the world's marine mammals are seen here, none are found exclusively in B.C.

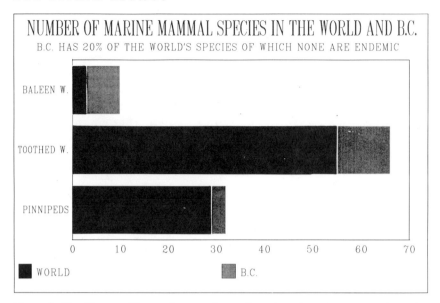

Figure 2. Bar diagram illustrating numbers of species of marine mammals in the world and in B.C. The bars represent: Mysticeti (baleen whales), Odontoceti (toothed whales), and seals, sea lions and walruses (pinnipeds).

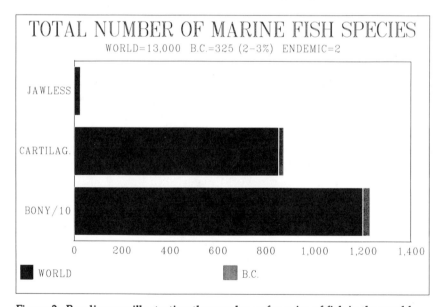

Figure 3. Bar diagram illustrating the numbers of species of fish in the world and in B.C. The bars represent: Cyclostomata (jawless), Chondrichthyes (cartilag.); and Osteichthyes (bony). Note that the bar for the bony fishes is one-tenth of its real height.

194

Fishes

Parker (1982) estimates that there are at least 19,000 fish species in the world; about 60% of these are marine (Cohen 1970). Using these figures, about 12,000 fish species are marine. Fig. 3 illustrates the distribution of the three major categories. The Osteichthyes include the most species. B.C. is home to four species of "jawless fishes" (Cyclostomata: lampreys and hagfish), twenty species of rays and skates and one ratfish (*Hydrolagus*). The remaining three hundred species are bony fishes. Several other species have also been described, but about 2-3% of the world's fish fauna is represented in B.C. Only two endemic species are present: a prickleback and a sculpin.

Marine Invertebrates

Representatives of 25 phyla of invertebrates are found in the ocean waters of B.C. (Fig. 4). For the sake of this study, I estimate that there are about 110,000 species of marine invertebrates in the world. B.C. is home to 3,771 species, about 3% of the world fauna. The largest number of species (1,129) are arthropods, followed by molluscs (826). Overall there are 293 endemic species (Fig. 5). Endemic species reflect much

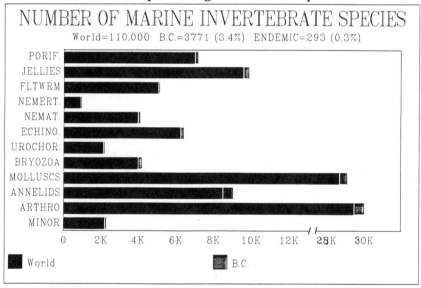

Figure 4. Bar diagram illustrating the number of marine invertebrate species in the world and in B.C. The bars represent: Porifera or sponges (Porif.); Coelenterata/Ctenophora, or anemones and jellyfish (jellies); flatworm groups (fltwrm); Nemertea or ribbon worms (Nemert.); Nematoda or round worms (Nemat.); Echinodermata (Echino.); Urochordata or sea squirts (Urochor.); Bryozoa or moss animals; molluscs; annelids; Arthropoda (Arthro.); and grouped minor phyla. Note the break in the horizontal axis.

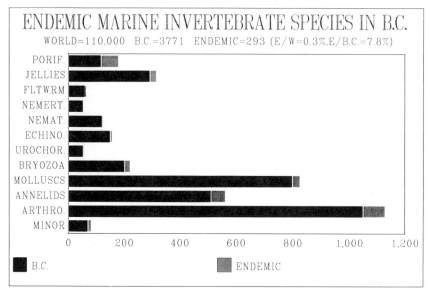

Figure 5. Bar diagram illustrating the distribution of endemic marine invertebrate species of B.C. in different taxonomic groups. For abbreviations, see Fig. 4.

work on certain groups such as sponges and oligochaetes. With greater collection effort, distribution ranges of many of these local species likely will be extended.

Discussion

Nearly 5,000 marine species are known to occur in B.C. I found no comparative information for other coasts. An interesting calculation concerns the available habitat. By using estimates of the total length of coastline of the global landmasses published in the Times Atlas of the Ocean (1983), B.C. has 2-4% of the world's coast. About 3-4% of the world's species are found here. I am surprised that these values are so similar, which probably reflects B.C.'s position midway along the polar/tropical diversity gradient and midway in the range of habitats from uniform to highly insular.

Few endemic species occur in B.C.: about 329 are documented.

B.C. falls within a larger biotic "province", the Oregonian, that extends from Point Conception in California to Dixon Entrance at the end of the Alaskan Panhandle. These biotic "provinces" are delineated on species similarities among localities; as such, the same species tend to be found throughout. It is not surprising, therefore, that B.C. shares most of its species at least with Washington, if not also with Oregon and northern California. Valentine (1966) shows that within the mollusc fauna about one-third of the species is shared between the Aleutian and Oregonian biotic provinces; about the same is true between the Oregonian and Californian provinces. Many species, therefore, are spread throughout the entire northwest Pacific.

The species lists I have compiled are variable in their completeness. A long-term effort by University of British Columbia botanists has resulted in a fine algal list that is probably highly representative of the local flora. Because algal habitats are limited to hard substrates within the photic zone (less than about 50 metres depth in coastal waters), they are accessible to collecting. Algae are not usually considered a group worth special attention or protection, but their economic potential poses some hazards. A major coastal community is that of the kelp forest. Here, high rates of primary production and high habitat heterogeneity foster a wide range of inhabitants. Among others, the Sea Otter is dependent on large sea urchin populations. Hawkes (1990) notes that harvest licences for B.C. kelp beds are being considered, and emphasizes the potential damage to algae and associated species.

The high visibility of marine mammals ensures complete documentation for this group although some, such as the Bering Sea Beaked Whale (*Mesoplodon stejnegeri*), are known only from a few beached or dead specimens throughout the northern Pacific. The geographic spread of the whales is so broad that stewardship by a single country is impossible. This feature has obvious benefits and disadvantages for protection.

Knowledge of the B.C. fishes is extensive because of the fisheries industry. Non-target species are frequently available for identification. The economic value of the group contributes to their vulnerability. Despite a large effort to assess their stocks, it becomes very difficult to manage and protect species that demand different treatment because of different life-history traits. Within the rockfish group (Scorpaenidae), for instance, age to maturity ranges from one to over 12 years (Phillips 1964). Among territorial species, recruitment to fishing grounds can be highly unpredictable.

The marine invertebrates of this province are poorly known, although some of the groups have received attention. The area of their potential habitat is too large to have been sampled adequately with present resources. In my work on the rock walls of fiords, it is common to find undescribed species, and most remain that way as few specialists exist. A biogeographic profile of the provincial invertebrate fauna has not been compiled but would give an interesting perspective on the effects of glaciation and ocean currents on recruitment to this coast.

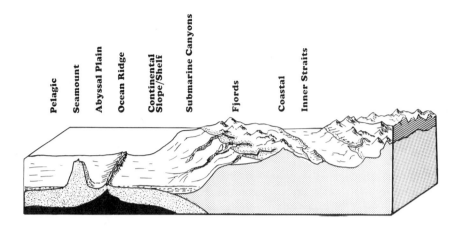

MARINE HABITATS OF BRITISH COLUMBIA

Figure 6. Representation of the different types of habitats that contribute to high habitat diversity in B.C. marine waters.

A major feature of the territorial waters around B.C. is the variable habitat. Fig. 6 shows the range available to organisms on this coast. The B.C. coast is up to 27,300 km long and includes 99 fiords over 10 km in length (Thomson 1981). The depth range of its waters is zero to 3,700 metres and the continental shelf is, typically, about 45 km wide. The variety of major habitat types sponsors a wider range of inhabitants. In addition, a good deal of variability exists within habitats. For example, rocky coasts on exposed shores demand different survival adaptations than those in sheltered fiords.

The variable habitat is one reason why so many species occur in these waters. But it is also why the number of species recorded are suspect: it is hard to get access to all the areas. The inner straits and the coastal, slope/shelf and pelagic realms are fairly well known. The other areas — including our many fiords — are poorly known as they are difficult to sample with standard techniques. Recently, we have worked on one of several offshore seamounts and found that dredging and towed camera work gave very few returns. A few dives with the PISCES IV submersible has revealed a rich biomass of species from several Pacific "provinces". Some have not been recorded in B.C. before.

The recent discovery of hydrothermal vents in B.C. waters illustrates that a major ecosystem can go unnoticed due to inaccessibility. The Juan de Fuca and Explorer Ridges lie off Vancouver Island and enter B.C. waters, giving us jurisdiction over two major ventfields: those on Endeavour Segment and on southern Explorer Ridge (see Tunnicliffe, 1991). The fauna is part of one that extends southward along Juan de Fuca Ridge and Gorda Ridge ending off Cape Mendocino (northern California). Of 53 species collected at these vents, 49 are known only at hydrothermal vents (Tunnicliffe 1988). In the summer of 1990, a new site in Canadian waters was examined, a sedimented vent site on Middle Valley Segment of northern Juan de Fuca Ridge, where we located at least eight more new species that are known nowhere else. The venting sediments here provide an unusual habitat. While it is unlikely that the vents on this ridge will be exploited for their mineral potential, we have found that the dredging, drilling and submersible activities are quite intrusive.

This vent fauna is a particularly valuable one for the lessons it can teach. Not only are the species new to science, but for the most part, the genera, families and even orders are also new. Newman (1985) suggested that the unusual degree of differentiation of hot-vent fauna from the rest of the marine world reflects the length of time that they have been in isolation. Systematic and palaeontologic comparisons of this fauna have illustrated its descent from Mesozoic ancestors (Tunnicliffe 1992). It appears to have evolved in isolation since that time and thus provides a view into ancient adaptations.

Summary

I have presented an overview of the marine species of B.C. Such information is necessary before we predict effects of human-caused changes on levels of biological diversity. B.C. appears to contain similar proportions of both the world's fauna and the world's coastline. The number of endemic species is not high because B.C. is within a large oceanic "province" that extends southward.

Species counts are useful in models of diversity behaviour and gradients. But documenting species is one thing, while knowing something about their position in the ecosystem is another. Eradicating local populations and gene pools may cause instabilities in community dynamics and species genomes. An integrated approach to the B.C. marine system would be timely, as the province derives much of its wealth from ocean resources, marine transport and associated tourism.

Literature Cited

Austin, W.C. 1985. An annotated checklist of marine invertebrates in the cold temperate Northeast Pacific. Three vols. Khoyatan Marine Laboratory, Cowichan Bay, BC.

Cohen, D. 1970. How many recent fishes are there? Proc. Calif. Acad. Sci. 37: 341-346.

Hart, J.L. 1973. Pacific fishes of Canada. Fisheries Research Board of Canada, Bull. 190, Ottawa, ON.

Hawkes, M.W. 1990. Benthic marine algal flora (seaweeds) of B.C.: diversity and conservation status. BioLine (Publication of Assoc. Prof. Biologists of B.C.) 9(2): 18-21.

King, J.E. 1983. Seals of the world. Second rev. ed. Cornell Univ. Press, Ithaca, NY.

Newman, W.A. 1985. The abyssal hydrothermal vent invertebrate fauna: a glimpse of antiquity? Bull. Biol. Soc. Wash. 6: 231-242.

Parker, S. (Editor). 1982. Synopsis and classification of living organisms. McGraw-Hill Inc., New York, NY.

Phillips, J.B. 1964. Life history studies on ten species of rockfish (genus *Sebastodes*). Calif. Dept. Fish. and Game Fish., Bull. 126.

Scagel, R.F., P.W. Gabrielson, D.J. Garbary, L. Golden, M.W. Hawkes, S.C. Lindstrom, J.C. Oliveira and T.B. Widdowson. 1989. A synopsis of the benthic marine algae of British Columbia, Southeast Alaska, Washington and Oregon. Univ. British Columbia, Dept. Botany, Phycological Contribution No. 3, Vancouver, BC.

Thomson, R.E. 1981. Oceanography of the British Columbia coast. Can. Spec. Publ. Fish. Aquat. Sci. 56.

Times Atlas of the Oceans. 1983. Van Nostrand Reinhold Co. Inc., New York, NY.

Tunnicliffe, V. 1988. Biogeography and evolution of hydrothermal-vent fauna in the eastern Pacific Ocean. Proc. R. Soc. Lond. B233: 347-366.

Tunnicliffe, V. 1991. The biology of hydrothermal vents: ecology and evolution. Annu. Rev. Oceanogr. Mar. Biol. 29: 319-407.

Tunnicliffe, V. 1992. The nature and origin of the modern hydrothermal vent fauna. Palaios 7:338-350.

Valentine, J.W. 1966. Numerical analysis of marine molluscan ranges on the extratropical northeastern Pacific shelf. Limnol. Oceanogr. 11: 198-211.

Watson, K. 1981. Whales of the world. Nelson Publ., Scarborough, England.

The Nature and State of Biodiversity in the Freshwater Fishes of British Columbia

J.D. McPhail

Fish Museum, Department of Zoology, University of British Columbia, Vancouver, British Columbia, Canada V6T 1Z4.

Abstract

The freshwater fish fauna of B.C. is composed mostly of post-glacial immigrants. These species colonized the province from Columbia, Great Plains and Bering glacial refuges. Many of them now have restricted distributions in B.C. but appear to be in no danger elsewhere. There is, however, a "made in B.C." component to our fauna. The pairs of benthic and limnetic sticklebacks that are indigenous to the central Strait of Georgia region are used as examples of how rapidly diversity can evolve if complete geographic isolation and novel ecological opportunities are available. I argue that we have a particular responsibility to document and protect this "made in B.C." diversity. I also argue that the greatest threat to the conservation of our native fish fauna is the "adipose fixation" of both governmental and private conservation agencies. This attitude leads to the exclusion of fishes other than trout and salmon from fisheries management plans, environmental impact assessments and conservation programs.

Introduction

My task is to report on the nature and state of biodiversity in freshwater environments. I will deal first with the nature of freshwater biodiversity and second with the state of that diversity in British Columbia. I will limit the discussion to freshwater fish for two reasons: they are the group I know best and they have a remarkable propensity

In *Our Living Legacy: Proceedings of a Symposium on Biological Diversity*, edited by M.A. Fenger, E.H. Miller, J.A. Johnson and E.J.R. Williams, pp. 201-209. Victoria, B.C.: Royal British Columbia Museum.

to form geographically isolated populations. The importance to biodiversity of this aspect of their biology will become apparent later.

The Nature of Freshwater Fish Biodiversity in B.C.

When asked to give this paper, the first thing I did was to draw up a list of freshwater fishes that are of special conservation concern in B.C. This exercise provided two unexpected insights. First, such lists are highly idiosyncratic and as much a reflection of the interests of the list-maker as the importance of species on the list. And second, if we stick to formally recognized taxa, we really don't have much of a freshwater fish fauna. What we have are the rag-tag ends of the Columbia fauna with a few Great Plains species in the Peace River region and a few Bering species in the extreme northern part of the province. In B.C., these species are not so much rare or threatened as they are restricted in their distributions. Their B.C. populations are peripheral to the main distributions, but the species themselves are in no danger over most of their geographic ranges. These peripheral species dominate lists of rare or threatened B.C. freshwater fishes and, unfortunately, distract attention and effort away from the problems of documenting and preserving our indigenous diversity.

Biodiversity is the result of evolutionary processes. We tend to think of evolution as slow, with species gradually changing over time, eventually giving rise to new species. To some extent this is true, but evolution can also proceed rapidly, with new species formed in a relatively short time. In theory, the circumstances conducive to the rapid formation of new species (speciation) are fairly well known; in practice, examples are uncommon and finding a single species in the process of splitting into two is exceedingly rare.

Two important elements required for rapid speciation are complete geographic isolation and novel ecological opportunities (Mayr 1982; Carson 1987). Our recent glacial history provides an abundance of both. Most of B.C. was covered in ice about 15,000 years ago and, as a consequence, the pre-glacial freshwater fauna was either wiped out or pushed into unglaciated refuges beyond the boundaries of the province. The one possible exception is the central coast region. This area may have contained an ice-free refuge that included parts of the east coast of Graham Island, much of what is now Queen Charlotte Sound and parts of northern Vancouver Island (Warner et al. 1982). This supposition is supported by solid botanical evidence and a growing body of evidence from freshwater fishes (McPhail and Lindsey 1986; Ogilvie 1989; McPhail 1993a). Because of this recent glacial history, the B.C. fish fauna consists mostly of immigrants from unglaciated regions outside the province (Fig. 1). Consequently, most B.C. fishes are recent colonists and, at first glance, insufficient time appears to have elapsed for significant evolution within B.C.; however, this is not the case. Many B.C. fish populations differ markedly in their ecology and morphology from populations of the same species in unglaciated areas. Why? The

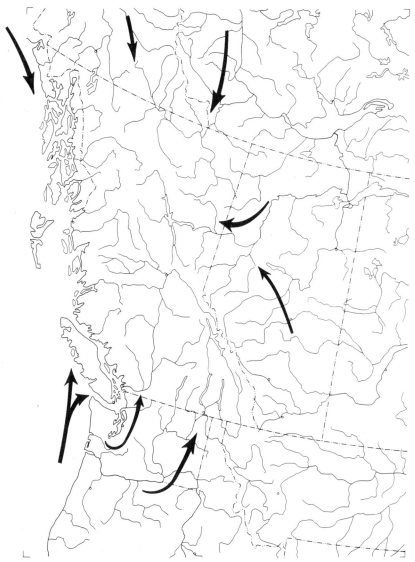

Figure 1. Map of B.C. showing the major post-glacial dispersal routes into the province.

answer lies in the two preconditions for rapid evolutionary change: geographic isolation and novel ecological conditions.

Among vertebrates, freshwater fishes are notoriously poor dispersers. This predisposes them to geographic isolation. They need freshwater connections to transfer between river systems and such connections are rare and ephemeral. Early in deglaciation, however, there was massive disruption of drainage patterns associated with the retreating ice-sheets (e.g., for a while, the entire upper Fraser drained into the Columbia system through a series of connections with the Okanagan and Similkameen rivers; Fulton 1969). As deglaciation proceeded, connections between drainages were cut and the present drainage patterns were established. After the disruption of glaciation, these new drainage patterns were stable and rarely altered. For species unable to disperse through the sea, this drainage stability resulted in complete geographic isolation.

The novel ecological opportunities encountered by many species in B.C. are also a direct consequence of glacial history. This is because only a fraction of the species that survived glaciation in the ice-free refuges were able to disperse into newly deglaciated areas before drainage connections were cut. As a consequence, the species that managed to use these ephemeral dispersal routes were eventually cut off from the main body of their species. Once isolated, the dispersing species were no longer able to exchange genes with more central populations and no longer subject to ecological constraints imposed by more diverse faunas found in unglaciated regions. Thus, isolated from the homogenizing effects of gene flow and released from interactions with many of their normal competitors and predators, these peripheral populations were able to respond rapidly to new selective regimes.

Since time precludes listing all the cases of rapid divergence in our fish fauna, I will discuss one example, using it to make a general argument for the existence of a "made in B.C." biodiversity that we have a special obligation to preserve.

The Remarkable Sticklebacks of the Strait of Georgia

The Threespine Stickleback (*Gasterosteus aculeatus*) is a small fish found in the coastal waters of B.C. and in similar environments throughout the northern hemisphere. Locally, it is especially abundant in the marine, brackish and fresh waters of the Strait of Georgia region. During the last (Fraser) glaciation this area was completely covered in ice. The ice sheets depressed the underlying land, and when the ice retreated, many low areas became temporarily submerged beneath the sea. Eventually the land rebounded and by about 10,000 years ago the present land forms were established (Clague 1981).

The distribution of freshwater sticklebacks in the Strait of Georgia region closely conforms to the limits of this postglacial submergence. This and the presence of freshwater sticklebacks on most islands suggest that sticklebacks colonized the region through the sea. Indeed, McPhail (1993a) argues that *G. aculeatus* is a marine species and that the freshwater populations represent peripheral isolates. Certainly most lake-dwelling populations are now completely isolated from the sea and it is in these isolated lake populations that we see evidence of rapid genetic divergence. Relative to marine populations, lake populations have diverged biochemically as well as in their morphology, reproductive behaviour and feeding ecology (Withler and McPhail 1985; McPhail 1977, 1993a). Most of these divergent populations occur by themselves, but two divergent populations coexist in six small lakes on islands in the central Strait of Georgia. Indeed, these populations do not even exchange genes and they use different food resources: one form (the "limnetics") feeds on plankton and the other form (the "benthics") feeds on bottom-dwelling invertebrates (Bentzen and McPhail 1984; McPhail 1984, 1993a; Ridgway and McPhail 1984). This is an extraordinary phenomenon! The Threespine Stickleback is one of the most intensively studied fish in world (Bell and Foster 1993), yet nothing like this has been recorded from elsewhere in its vast geographic range. The capacities to coexist without gene exchange (reproductive isolation) and to exploit different food sources (resource partitioning) are the essential characteristics of biological species (Mayr 1963). Apparently these benthic-limnetic species pairs have evolved within the last 11,000 to 12,000 years, a remarkably short time for speciation. Even more remarkable, however, is that the six lakes are on three islands (four on Texada Island, one on Lasqueti Island, and one in the Nanoose area of Vancouver Island; Fig. 2). This three-island distribution suggests that these pairs of benthic and limnetic sticklebacks may have evolved independently on each of the islands (McPhail 1993a). If so, they are an extraordinary example of parallel evolution! Also, detailed studies on two of the lakes (Enos and Paxton) disclose that, although reproductive isolation and resource partitioning are well developed in these benthic-limnetic pairs, they are not perfect (McPhail 1984, 1992, 1993b). These key attributes of species appear to be still evolving and it is this dynamic aspect of the speciation process that makes these pairs a unique resource for the study of evolution.

Although I use sticklebacks as an example, they are not unique. For salt-tolerant species, the coast of B.C. was a major dispersal route (Fig. 1). Anadromous species like lampreys, smelts, trout and salmon have repeatedly and independently colonized lakes and streams and taken up a totally freshwater existence. Characteristically these freshwater populations have diverged from their migratory ancestors in both ecology and morphology. In the southeastern part of the province, the isolation in headwater situations of fragments of the Columbia fauna has resulted in their significant divergence from relatives immediately to the south. In the Peace and Liard river systems, several Great Plains minnows (Cyprinidae) have hybridized so completely that one of the putative parent species can no longer be found in the area, but the other parental species (and what appear to be self-perpetuating populations of hybrids) continues to coexist.

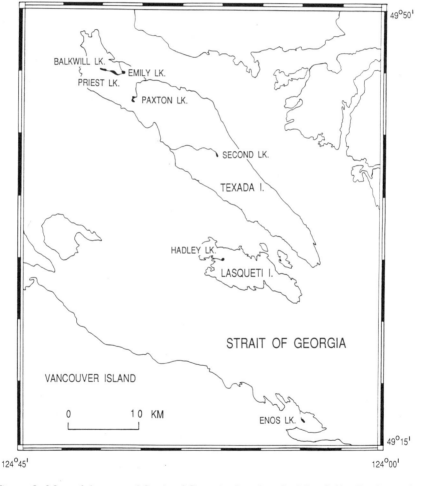

Figure 2. Map of the central Strait of Georgia showing the island distributions of three pairs of benthic and limnetic sticklebacks.

The State of Freshwater Fish Biodiversity in B.C.

All of the above divergences are unique to B.C. and share the common features of a recent origin and a high degree of geographic isolation. They also are all poorly known, and most have not received formal taxonomic names — not because they are unimportant, but because the resources and personnel necessary to document these complex situations are lacking. This underlines a major feature of fish biodiversity in the province: most of our "made in B.C." diversity is at the population level and involves widespread but taxonomically difficult groups. Although these animals lack scientific names, they are of considerable scientific value and an important part of our natural heritage. Before this "made in B.C." biodiversity can be protected it must be documented, which is not a job for amateurs. The nature and complexity of this recently evolved diversity require the skills of professional systematists, particularly those with strong backgrounds in evolutionary ecology and zoogeography.

Ironically, the most serious threat to our native B.C. fishes comes from governmental fisheries agencies. Fisheries biologists are notorious for their "adipose fixation" (a reference to the adipose fin that is a diagnostic feature of trout, salmon and related game fish species): they view native non-salmonid species as "coarse" or "trash" fish. At best, they ignore most of the native fauna; at worst, they actively seek to destroy it. A favourite euphemism for this destruction is "lake rehabilitation". Translated, this means lakes are poisoned. Usually the purpose of the poisoning is to kill the natural fish community and replace it with a monoculture of Rainbow Trout (*Salmo gairdneri*). In the past, so many lakes were "rehabilitated" in southern B.C. that in many areas it is now hard to find a lowland lake that contains its original fish community.

Fortunately for the native fauna, large lakes are too expensive to "rehabilitate", and lakes closely associated with main rivers are too difficult to isolate from "re-infection". Thus, although badly abused, most native species are in no danger of extinction and in B.C. the practice of lake "rehabilitation" now is largely a thing of the past. Yet the mind-set among fisheries biologists who so enthusiastically embraced the wholesale destruction of native species is very much alive in B.C. Many fisheries biologists still view their conservation mandate as applying only to trout and salmon. Curiously, this "adipose fixation" is shared by private conservation organizations: the presence of salmon or trout is used regularly to argue for the preservation of watersheds, but one rarely hears a whisper of concern about other fishes. Apparently even the most committed conservationists see no inconsistency in protecting all plants and animals except fish in national and provincial parks. This attitude leads to the exclusion of fish, other than trout and salmon, from management plans, environmental impact assessments and conservation programs. It is the greatest long-term threat to our native fish fauna.

Acknowledgements

Bob Carveth drew the illustrations and helped in many other ways. Without his assistance this work would never have been completed. Most of our knowledge of the distribution and taxonomy of freshwater fishes in B.C. is due to the energy and inspiration of one man, Dr C.C. Lindsey. It is a debt that can be acknowledged but never repaid.

Literature Cited

Bell, M.A., and S.A. Foster. (Editors). 1993. The evolutionary biology of the threespine stickleback. Oxford Univ. Press, London, England. (In press.)

Bentzen, P., and J.D. McPhail. 1984. Ecology and evolution of sympatric sticklebacks (*Gasterosteus*): specialization for alternative trophic niches in the Enos Lake species pair. Can. J. Zool. 52: 2280-2286.

Carson, H.L. 1987. The genetic system, the deme, and the origin of the species. Annu. Rev. Genetics 21: 405-423.

Clague, J.J. 1981. Late Quaternary geology and geochronology of British Columbia. Part 2: Summary and discussion of radiocarbon-dated Quaternary history. Geol. Surv. Canada, Paper 80-34.

Fulton, R.J. 1969. Glacial lake history, southern interior plateau, British Columbia. Geol. Surv. Canada, Paper 69-37.

Mayr, E. 1963. Animal species and evolution. Harvard Univ. Press, Cambridge, MA.

Mayr, E. 1982. Processes of speciation. *In* Mechanisms of speciation, *edited by* C. Barigozzi. Alan R. Liss Inc., New York, NY. pp. 1-19.

McPhail, J.D. 1977. Inherited interpopulation differences in size at first reproduction in threespine stickleback *Gasterosteus aculeatus* L. Heredity 38: 53-60.

McPhail, J.D. 1984. Ecology and evolution of sympatric sticklebacks (*Gasterosteus*): morphological and genetic evidence for a species pair in Enos Lake, British Columbia. Can. J. Zool. 62: 1402-1408.

McPhail, J.D. 1992. Ecology and evolution of sympatric sticklebacks (*Gasterosteus*): evidence for a species-pair in Paxton Lake, Texada Island, British Columbia. Can. J. Zool. 70: 361-369.

McPhail, J.D. 1993a. Speciation and the evolution of reproductive isolation in the sticklebacks (*Gasterosteus*) of southwestern British Columbia. *In* The evolutionary biology of the threespine stickleback, *edited by* M. A. Bell and S. A. Foster. Oxford Univ. Press, London, England. (In press.)

McPhail, J.D. 1993b. Ecology and evolution of sympatric sticklebacks (*Gasterosteus*): origin of the species pairs. Can. J. Zool. (In press.)

McPhail, J.D., and C.C. Lindsey. 1986. Zoogeography of the freshwater fishes of Cascadia. *In* The zoogeography of North American freshwater fishes, *edited by* C. Hocutt and E. O. Wiley. John Wiley and Sons, New York, NY. pp. 615-637.

Ogilvie, R.T. 1989. Disjunct vascular flora of northwestern Vancouver Island in relation to Queen Charlotte Islands' endemism and Pacific coast refugia. *In* The outer shores, *edited by* G.G.E. Scudder and N.A. Gessler. Queen Charlotte Islands Museum Press, Skidegate, BC. pp. 125-127.

Ridgway, M.S., and J.D. McPhail. 1984. Ecology and evolution of sympatric sticklebacks (*Gasterosteus*): mate choice and reproductive isolation in the Enos Lake species pair. Can. J. Zool. 62: 1813-1818.

Warner, B.G., R.W. Mathewes and J.J. Clague. 1982. Ice-free conditions on the Queen Charlotte Islands, British Columbia, at the height of late Wisconsin glaciation. Science 218: 675-677.

Withler, R.E., and J.D. McPhail. 1985. Genetic variability in freshwater and anadromous sticklebacks (*Gasterosteus aculeatus*) of southern British Columbia. Can. J. Zool. 63: 528-533.

Diversity at Risk

Designation of Endangered Species, Subspecies and Populations by COSEWIC

W.T. Munro

B.C. Ministry of Environment, Land and Parks,
Wildlife Branch, 780 Blanshard St., Victoria,
British Columbia, Canada V8V 1X4.

Abstract

The mandate of the Committee on the Status of Endangered Wildlife in Canada is to determine the status of species, subspecies and populations at risk at the national level in Canada. The Wildlife Branch of the Ministry of Environment is mandated to determine the status of species in British Columbia. The processes and criteria used by each organization to determine status and the results of those determinations are described.

Introduction

The Committee on the Status of Endangered Wildlife in Canada (COSEWIC) was formed in 1977 as a result of a recommendation of the Federal-Provincial Wildlife Conference. It arose from the need for a single, official, scientifically sound national listing of wild species, subspecies and separate populations at risk in Canada. Its mandate is to determine the status of these elements at the national level. All native vascular plants and vertebrate animals are included.

COSEWIC consists of one representative from the wildlife agency of each province or territory, one from each of four federal agencies (Canadian Wildlife Service, National Museums of Canada, Department of Fisheries and Oceans, Parks Canada), and one from each of three nationally based private conservation agencies (Canadian Nature Federation, Canadian Wildlife Federation, World Wildlife Fund

In *Our Living Legacy: Proceedings of a Symposium on Biological Diversity*, edited by M.A. Fenger, E.H. Miller, J.A. Johnson and E.J.R. Williams, pp. 213-227. Victoria, B.C.: Royal British Columbia Museum.

Canada). The Committee elects its own chairperson for a two-year term. In addition to the above representatives, COSEWIC includes subcommittee chairpersons responsible for work on five biological groups (birds, terrestrial mammals, fish and marine mammals, amphibians and reptiles, plants).

Subcommittees are the organizational units of COSEWIC responsible for obtaining status reports, reviewing their quality and presenting them. All status determinations are based on these reports. The subcommittee chairpersons, chosen for their scientific knowledge, are expert advisors to COSEWIC in their fields. They are proposed by the COSEWIC chairperson and confirmed by all members at the next annual meeting. Subcommittee chairpersons choose and recruit sub-committee members who are experts in the field of the subcommittee.

The purpose of subcommittees is to obtain and present status reports at general meetings. Status reports may be provided by COSEWIC member jurisdictions or individuals, or may be obtained by COSEWIC through contract. Subcommittees ensure high scientific quality, propose an appropriate national status and present status reports to the membership for formal assignment of status.

Status

Status reports are the instruments of status determination. Each report is an up-to-date description of distribution, abundance and population trends. COSEWIC meets annually in April to declare official status for all species (subspecies, populations) for which status reports have been prepared and circulated to members.

COSEWIC recognizes five risk categories: vulnerable, threatened, endangered, extirpated and extinct. Definitions of these categories are in Appendix A. Sometimes a status report is reviewed and the species (subspecies, population) is determined to be not at risk. In such cases the committee classifies it as "report accepted - no designation required".

Since its inception COSEWIC has evaluated 276 status reports. For 64 of them it found no designation was required. Designations based on the remaining 212 reports are summarized in Table 1.

Designations are based on the best information available at the time of study, so are subject to change. They may be reviewed whenever new information is presented. It is possible to re-designate a species (subspecies, population) into a more critical category should its prospects for continued survival deteriorate or to a less critical category when risks diminish. Examples of criteria used for determining status are in Appendix B.

Table 1.		Classification of species by COSEWIC			
		Number of species:			
	Vulnerable	Threatened	Endangered	Extirpated	Extinct
Mammals:					
• terrestrial (29)	15	5	5	3	1
• marine (16)	4	3	6	2	1
Birds (38)	17	8	9	1	3
Reptiles & amphibians (12)	6	2	4	0	0
Fish (53)	34	10	3	2	4
Vascular plants (64)	21	22	19	2	0
Totals (212)	97	50	46	10	9

The findings of COSEWIC are published in the form of detailed status reports that are sold through the Canadian Nature Federation in Ottawa. Following each annual meeting a news release is issued listing new designations. Up-to-date lists of all species (subspecies, populations) evaluated and designations assigned are available annually from the COSEWIC Secretariat. Designations of endangered, threatened and vulnerable forms in B.C. are summarized in Appendix C.

It should be emphasized that COSEWIC is concerned only with drawing official attention to the possible loss of wild species (subspecies, populations) that have been historically maintained in Canada. It is not mandated to take any action to alter the fortunes of species beyond establishing status and publishing the information on which status is based. There are no legal consequences or requirements following the declaration of status. The purpose of status designation by COSEWIC is to provide a national scientific consensus that may be used by jurisdictions in exercising their mandates.

In 1980, the first vertebrate species were designated as "endangered" under the B.C. Wildlife Act. The criteria used to determine the species to be included have been outlined by Munro and Low (1980).

The Wildlife Branch does not consider for designation anything (native or non-native) that has been introduced into B.C. or neighbouring regions, or those that are common elsewhere but whose range barely extends into B.C. The former category is termed "exotic" and the latter "peripheral".

The reasons for not including exotics are: (1) they have been introduced by humans and are not part of our wildlife heritage, (2) they may compete with native species, (3) they may not be adapted to conditions found in B.C. and (4) efforts to maintain viable populations of some of them can be more profitably directed to native forms.

The reasons for not including peripheral forms are: (1) they are at the edge of their range and often increase or decrease markedly as a result of climatic trends, habitat availability and inherent biological adaptability; (2) they are common elsewhere and will expand into B.C. if conditions become appropriate; and (3) efforts required to maintain them as viable can be more profitably directed to forms that are truly threatened or endangered. We will not ignore peripheral species (subspecies, populations): we will protect them and their habitat and let them undergo natural fluctuations.

All designations are applied in a provincial context and not on a local basis. Designations are made on a provincial basis because that is the geographic and political boundary within which we operate. If one chose small enough local areas, everything could likely be considered to be endangered somewhere in B.C. Thus a species (subspecies, population) will not be designated as threatened if it is considered threatened in one area but exists in viable numbers elsewhere in B.C. Forms not designated as endangered or threatened will continue to be of management concern and will be considered under general management and protection programs.

Species or subspecies that may be designated as threatened or endangered by other administrations elsewhere in Canada or the world will be designated as such in B.C. only when it is warranted by their status in B.C. For example, while the Bald Eagle is rare and officially classed as an endangered species in Ontario, it is abundant and not considered endangered in B.C. Status elsewhere, however, is one factor weighed when considering status in B.C.

Originally, subspecies were excluded from designation, except where there was significant public concern for their separate management. Upon review we found the arguments put forward for their inclusion sufficiently strong to change our policy. Subspecies and populations are now included for designation as threatened or endangered.

Recently, the Wildlife Branch reviewed and modified the criteria for designating species as threatened or endangered. The revised criteria and designations are presented in Appendix D.

Literature cited

Munro, W.T., and D.J. Low. 1980. Designation of threatened and endangered species. *In* Threatened and endangered species and habitats in British Columbia and the Yukon *edited by* R. Stace-Smith, L. Johns and P. Joslin. B.C. Ministry of Environment, Lands and Parks, Victoria, BC. pp. 65-74.

Appendix A:
Definitions of Conservation Status used by COSEWIC

Form:	"Species" means any species, subspecies or geographically separate population.
Vulnerable Species:	Any indigenous species of fauna or flora that is particularly at risk because of low or declining numbers, occurrence at the fringe of its range or in restricted areas, or for some other reason, but is not threatened.
Threatened Species:	Any indigenous species of fauna or flora that is likely to become endangered in Canada if the factors affecting its vulnerability do not become reversed.
Endangered Species:	Any indigenous species of fauna or flora that is threatened with imminent extinction or extirpation throughout all or a significant portion of its Canadian range.
Extirpated Species:	Any indigenous species of fauna or flora no longer existing in the wild in Canada but occurring elsewhere.
Extinct Species:	Any species of fauna or flora formerly in Canada but no longer existing anywhere.

Appendix B:
Examples of criteria used by COSEWIC to change status

Change Between:	Downlisting:	Uplisting
Extirpated and Endangered	a) Species, subspecies or population successfully reproducing at F_2 generation in free-ranging or fenced conditions. b) Survival of offspring to reproductive age is greater than mortality over that time period. c) Habitat essential to species' survival is adequately protected in foreseeable future (30 years) through management or preservation. d) Population is stable or increasing but numbers still very small. e) Backup source of breeding-age individuals is available in Canada (captive, cultivated or wild individuals from another country).	a) No record for 3-5 years of individuals or populations in suitable habitat in spite of investigation.
Endangered and Threatened	a) At least three populations separated spatially by some considerable distance. b) All sites adequately protected and in regulated areas. c) Most or all populations showing significant increase with natural expansion of range occurring, or all existing habitat at site is occupied. d) Habitat, in addition to that currently occupied, is available for continued range expansion. e) Threat still evident in part or most of species' former range in Canada. f) Captive or cultivated stock for wild release maintained if needed.	a) Very few native populations exist (perhaps three or fewer remaining, protected or not). b) Populations declining drastically due to habitat loss, excessive harvest, natural catastrophes, environmental stresses or other factors caused by humans.

Appendix B (cont.)

Change between:	Downlisting	Uplisting:
Threatened and Vulnerable	a) Populations recovered or increased to point where extinction or extirpation unlikely if currently available; habitat is preserved or managed. b) All or most wild, free-ranging or fenced populations increasing or stable with no evidence of a decrease for last 3-5 years. c) Populations sufficiently low so species is uncommon within its range or confined to small geographic area; otherwise the species is removed from the list.	a) Several populations have decreased drastically in numbers; perhaps only 1/2 of those existing 5-10 years ago remain. b) Rapid decline in numbers caused by environmental factors, pollutants or other detrimental factors caused by humans. c) Factors causing decline still evident.
Vulnerable and No Designation Required	a) Significant increases in known population(s) found or established that substantially increase the number previously known in area of its documented range. b) Population currently occupies most suitable habitat remaining in former geographical range. c) Harvestable portion could be removed for subsistence, commercial or recreational use if such demands occur without seriously reducing population size or affecting potential for sustained growth.	

Appendix C:
British Columbian forms designated by COSEWIC*

Endangered

Southern Maidenhair Fern
Leatherback Turtle
Anatum Peregrine Falcon
Spotted Owl

Salish Sucker
Sea Otter
Vancouver Island Marmot

Vulnerable

Macoun's Meadowfoam
Pacific Giant Salamander
Caspian Tern
Common Barn Owl
Cooper's Hawk

Flammulated Owl
Great Gray Owl
Peale's Peregrine Falcon
Trumpeter Swan

Threatened

Giant Helleborine
Mosquito Fern
Western Blue Flag
Burrowing Owl
Ferruginous Hawk
Loggerhead Shrike
Tundra Peregrine Falcon
Marbled Murrelet
Enos Lake Stickleback
Shorthead Sculpin
North Pacific Humpback Whale
Wood Bison
Charlotte Unarmored Stickleback
Giant Stickleback

Green Sturgeon
White Sturgeon
Lake Lamprey
Pacific Sardine
Speckled Dace
Umatilla Dace
Fringed Myotis
Keen's Long-eared Bat
Pallid Bat
Queen Charlotte Islands Ermine
Spotted Bat
Wolverine-Western Population
Western Woodland Caribou

* Names follow conventions used by COSEWIC (Eds).

Appendix D: The Red and Blue lists (as of March 1993)

Provincial Red List of higher vertebrate species and subspecies that are candidates for legal designation as endangered or threatened:

Amphibians
Ambystomatidae
Ambystoma tigrinum	Tiger Salamander
Dicamptodon tenebrosus	Pacific Giant Salamander

Plethodontidae
Plethodon idahoensis	Coeur d'Alene Salamander
Rana pipiens	Leopard Frog

Reptiles
Iguanidae
Phrynosoma douglassii	Short-horned Lizard

Colubridae
Contia tenuis	Sharp-tailed Snake
Hypsiglena torquata	Night Snake
Pituophis melanoleucus catenifer	Gopher Snake (ssp. *catenifer*)

Birds
Gaviiformes
Aechmophorus occidentalis	Western Grebe

Pelecaniformes
Pelecanus erythrorhynchos	American White Pelican
Phalacrocorax pelagicus pelagicus	Pelagic Cormorant (ssp. *pelagicus*)
Phalacrocorax penicillatus	Brandt's Cormorant

Falconiformes
Accipiter gentilis laingi	Northern Goshawk (ssp. *laingi*)
Buteo regalis	Ferruginous Hawk
Falco mexicanus	Prairie Falcon
Falco peregrinus anatum	Peregrine Falcon (ssp. *anatum*)

Galliformes
Centrocercus urophasianus	Sage Grouse

Charadriiformes
Bartramia longicauda	Upland Sandpiper
Fratercula corniculata	Horned Puffin
Sterna forsteri	Forster's Tern
Uria aalge	Common Murre
Uria lomvia	Thick-billed Murre

Cuculiformes
Coccyzus americanus	Yellow-billed Cuckoo

Strigiformes
    Athene cunicularia   Burrowing Owl
    Strix occidentalis   Spotted Owl
Piciformes
    Picoides albolarvatus   White-headed Woodpecker
    Sphyrapicus thyroideus nataliae   Williamson's Sapsucker (ssp. *nataliae*)
Passeriformes
    Ammodramus caudacutus   Sharp-tailed Sparrow
    Ammodramus savannarum   Grasshopper Sparrow
    Anthus spragueii   Spague's Pipit
    Dendroica castanea   Bay-breasted Warbler
    Dendroica tigrina   Cape May Warbler
    Eremophila alpestris strigata   Horned Lark (ssp. *strigata*)
    Icteria virens   Yellow-breasted Chat
    Oporornis agilis   Connecticut Warbler
    Oreoscoptes montanus   Sage Thrasher
    Pooecetes gramineus affinis   Vesper Sparrow (ssp. *affinis*)
    Progne subis   Purple Martin
    Spizella breweri breweri   Brewer's Sparrow (ssp. *breweri*)

Mammals
Insectivora
    Scapanus townsendii   Townsend's Mole
    Sorex bendirii   Pacific Water Shrew
    Sorex palustris brooksi   Water Shrew (ssp. *brooksi*)
    Sorex tundrensis   Tundra Shrew
Chiroptera
    Antrozous pallidus   Pallid Bat
    Lasiurus blossevilli   Southern Red Bat
    Myotis keenii   Keen's Long-eared Myotis
    Myotis septentrionalis   Northern Long-eared Myotis
Lagomorpha
    Lepus americanus washingtonii   Snowshoe Hare (ssp. *washingtonii*)
    Lepus townsendii   White-tailed Jackrabbit
Rodentia
    Aplodontia rufa rufa   Mountain Beaver (ssp. *rufa*)
    Clethrionomys gapperi   Southern Red-backed Vole
      occidentalis   (ssp. *occidentalis*)
    Marmota vancouverensis   Vancouver Island Marmot
    Microtus townsendii cowani   Townsend's Vole (ssp. *cowani*)
    Synaptomys borealis artemisiae   Nothern Bog Lemming
                            (ssp. *artemisiae*)
    Tamias minimus selkirki   Least Chipmunk (ssp. *selkirki*)

Tamias ruficaudus ruficaudus	Red-tailed Chipmunk (ssp. *ruficaudus*)
Tamias ruficaudus simulans	Red-tailed Chipmunk (ssp. *simulans*)
Thomomys talpoides segregatus	Northern Pocket Gopher (ssp. *segregatus*)

Carnivora
Enhydra lutris	Sea Otter
Gulo gulo vancouverensis	Wolverine (ssp. *vancouverensis*)
Mustela erminea haidarum	Ermine (ssp. *haidarum*)
Mustela frenata altifrontalis	Long-tailed Weasel (ssp. *altifrontalis*)

Artiodactyla
| *Bison bison athabascae* | Bison (Wood Bison, ssp. *athabascae*) |
| *Ovis dalli dalli* | Thinhorn Sheep (Dall's Sheep, ssp. *dalli*) |

Provincial Blue List of higher vertebrate species and subspecies considered to be vulnerable or sensitive:

Amphibians
Ascaphidae
| *Ascaphus truei* | Tailed Frog |

Pelobatidae
| *Scaphiopus intermontanus* | Great Basin Spadefoot Toad |

Reptiles
Emydidae
| *Chrysemys picta* | Painted Turtle |

Boidae
| *Charina bottae* | Rubber Boa |

Colubridae
| *Coluber mormon* | Western Yellow-bellied Racer |
| *Pituophis melanoleucus deserticola* | Gopher Snake (ssp. *deserticola*) |

Viperidae
| *Crotalus viridis* | Western Rattlesnake |

Birds
Pelecaniformes
| *Phalacrocorax auritus* | Double-crested Cormorant |

Ciconiiformes
Ardea herodias	Great Blue Heron
Botaurus lentiginosus	American Bittern
Butorides striatus	Green-backed Heron

Anseriformes
Clangula hyemalis — Oldsquaw
Cygnus buccinator — Trumpeter Swan
Melanitta perspicillata — Surf Scoter
Falconiformes
Buteo swainsoni — Swainson's Hawk
Cathartes aura — Turkey Vulture
Falco peregrinus pealei — Peregrine Falcon (ssp. *pealei*)
Falco rusticolus — Gyrfalcon
Haliaeetus leucocephalus — Bald Eagle
Galliformes
Lagopus leucurus saxatilis — White-tailed Ptarmigan (ssp. *saxatilis*)
Tympanuchus phasianellus columbianus — Sharp-tailed Grouse (ssp. *columbianus*)
Gruiformes
Grus canadensis — Sandhill Crane
Charadriiformes
Brachyramphus marmoratus — Marbled Murrelet
Fratercula cirrhata — Tufted Puffin
Heteroscelus incanus — Wandering Tattler
Larus californicus — California Gull
Limnodromus griseus — Short-billed Dowitcher
Limosa haemastica — Hudsonian Godwit
Numenius americanus — Long-billed Curlew
Phalaropus lobatus — Red-necked Phalarope
Pluvialis dominica — Lesser Golden-plover
Ptychoramphus aleuticus — Cassin's Auklet
Recurvirostra americana — American Avocet
Sterna caspia — Caspian Tern
Synthliboramphus antiquus — Ancient Murrelet
Strigiformes
Aegolius acadicus brooksi — Northern Saw-whet Owl (ssp. *brooksi*)
Asio flammeus — Short-eared Owl
Glaucidium gnoma swarthi — Northern Pygmy-owl (ssp. *swarthi*)
Otus flammeolus — Flammulated Owl
Otus kennicottii kennicottii — Western Screech-owl (ssp. *kennicottii*)
Otus kennicottii macfarlanei — Western Screech-owl (ssp. *macfarlanei*)
Tyto alba — Common Barn-owl
Apodiformes
Aeronautes saxatalis — White-throated Swift
Archilochus alexandri — Black-chinned Hummingbird

Piciformes
- *Melanerpes lewis* — Lewis' Woodpecker
- *Picoides villosus picoideus* — Hairy Woodpecker (ssp. *picoideus*)
- *Sphyrapicus thyroideus thyroideus* — Williamson's Sapsucker (ssp. *thyroideus*)

Passeriformes
- *Calcarius pictus* — Smith's Longspur
- *Catherpes mexicanus* — Canyon Wren
- *Chondestes grammacus* — Lark Sparrow
- *Cyanocitta stelleri carlottae* — Steller's Jay (ssp. *carlottae*)
- *Dendroica palmarum* — Palm Warbler
- *Dendroica virens* — Black-throated Green Warbler
- *Dolichonyx oryzivorus* — Bobolink
- *Empidonax flaviventris* — Yellow-bellied Flycatcher
- *Empidonax wrightii* — Gray Flycatcher
- *Pinicola enucleator carlottae* — Pine Grosbeak (ssp. *carlottae*)
- *Vireo huttoni* — Hutton's Vireo
- *Vireo philadelphicus* — Philadelphia Vireo
- *Wilsonia canadensis* — Canada Warbler

Mammals

Insectivora
- *Sorex arcticus* — Black-backed Shrew
- *Sorex trowbridgii* — Trowbridge's Shrew

Chiroptera
- *Euderma maculatum* — Spotted Bat
- *Myotis ciliolabrum* — Western Small-footed Myotis
- *Myotis thysanodes* — Fringed Myotis
- *Plecotus townsendii* — Townsend's Big-eared Bat

Lagomorpha
- *Sylvilagus nuttallii* — Nuttall's Cottontail

Rodentia
- *Aplodontia rufa rainieri* — Mountain Beaver (ssp. *rainieri*)
- *Clethrionomys gapperi galei* — Southern Red-backed Vole (ssp. *galei*)
- *Perognathus parvus* — Great Basin Pocket Mouse
- *Reithrodontomys megalotis* — Western Harvest Mouse
- *Spermophilus saturatus* — Cascade Mantled Ground Squirrel
- *Synaptomys borealis borealis* — Northern Bog Lemming (ssp. *borealis*)
- *Tamias minimus oreocetes* — Least Chipmunk (ssp. *oreocetes*)
- *Zapus hudsonius alascensis* — Meadow Jumping Mouse (ssp. *alascensis*)

Carnivora
    Gulo gulo luscus — Wolverine (ssp. *luscus*)
    Martes pennanti — Fisher
    Mustela erminea anguinae — Ermine (ssp. *anguinae*)
    Taxidea taxus — Badger
    Ursus americanus emmonsii — Black Bear (ssp. *emmonsii*)
    Ursus arctos — Grizzly Bear
Artiodactyla
    Bison bison bison — Bison (Plains Bison, ssp. *bison*)
    Cervus elaphus roosevelti — Elk (Roosevelt Elk, ssp. *roosevelti*)
    Ovis canadensis californiana — Bighorn Sheep (California Bighorn Sheep, ssp. *californiana*)
    Ovis canadensis canadensis — Bighorn Sheep (Rocky Mountain Bighorn Sheep, ssp. *canadensis*)
    Ovis dalli stonei — Thinhorn Sheep (Stone Sheep, ssp. *stonei*)
    Rangifer tarandus — Caribou (southeastern populations)

Editors' note: The full reference for these lists, with explanations of criteria for inclusion, can be found in:

B.C. Wildlife Branch. 1993. Birds, mammals, reptiles and amphibians at risk in British Columbia: the 1993 Red and Blue lists. B.C. Ministry of Environment, Lands and Parks, Victoria, B.C.

Biodiversity at Risk: Soil Microflora

C.P. Chanway

Department of Forest Sciences, University of British Columbia, Vancouver, British Columbia, Canada V6T 1Z4.

Abstract

The soil microflora comprises thousands of fungal and bacterial species that are integral components of all terrestrial ecosystems. Many of these are soil-dwelling saprophytic fungi that reach population sizes of up to one million spores and hyphal fragments per gram of soil. Soil bacteria live in close association with saprophytic fungi, often performing tasks that are similar or complementary to those performed by fungi. Up to 80% of bacterial species in natural communities have not been described. A single gram of soil may contain billions of bacterial cells. Together, bacteria and fungi in the top 20 cm of fertile soil may reach five tonnes per hectare. It is no wonder that these organisms exert a major influence on the soil, mainly by facilitating decomposition of organic matter and by influencing soil structure.

Mutualistic soil fungi and bacteria are a particularly important part of the soil microflora. Their importance to ecosystem functioning is underscored by the fact that most plant species, including all conifers growing in the wild, are mycorrhizal. Nitrogen-fixing species provide a critical link in the successional processes that are characteristic of our forests. In addition, several species of mutualistic bacteria may proliferate around, on, or more rarely, in root or mycorrhizal tissue without forming any special root structure. These can fix nitrogen, stimulate tree growth or suppress pathogen populations through their activity.

Certain plant species appear to be able to form stable associations above ground due to their common ability to associate with similar plant-beneficial micro-organisms below ground. Therefore, the role of below-ground diversity in the microbial community may be central to establishing and maintaining healthy and resilient forests. At present, we know little about the ecology of soil micro-organisms and critical gaps exist in our knowledge. Examples of plantation failure caused by the apparent loss of important soil micro-organisms after harvesting demonstrate the danger of our ignorance.

In *Our Living Legacy: Proceedings of a Symposium on Biological Diversity*, edited by M.A. Fenger, E.H. Miller, J.A. Johnson and E.J.R. Williams, pp. 229-238. Victoria, B.C.: Royal British Columbia Museum.

Introduction

Microfloral species in forest soils may outnumber the plant species we see above the ground. This review describes the major categories of soil microflora (i.e., bacteria, fungi, algae, viruses) and outlines their importance to the healthy functioning of forest ecosystems, with emphasis on mycorrhizal fungi. In addition, some critical gaps in our knowledge of soil microflora are identified, with particular reference to current forestry practices in British Columbia.

What We Do Know

More than 100,000 species of fungi are known to science and probably about 200,000 have yet to be discovered and described by scientists. Tens of thousands of the known species spend part or all of their life cycles in the soil. These soil fungi constitute a diverse group and include familiar organisms such as yeasts and toadstool-forming fungi. Soil fungi are extremely important agents of decomposition and are also instrumental in maintaining good structure in forest soils. In addition, 5,000 species of fungal plant pathogens and an equal number of symbiotic fungal species are known. The population of soil fungi is substantial: a single gram of soil may contain up to a million fungal propagules (i.e., spores, resting stages, hyphal fragments) and may be ramified by metres of strands or hyphae.

Living in close association with the soil fungi are the soil bacteria. These micro-organisms comprise thousands of species, including actinomycetes (filamentous bacteria), but differences among strains of a species are tremendous and effectively increase functional diversity several-fold. Soil bacteria have critical roles in the forest ecosystem as decomposers, nutrient cyclers and nitrogen fixers. In addition, many strains may stimulate or inhibit plant growth or cause disease. Numerically, soil bacteria are often more predominant than soil fungi, though the reverse may be true in acidic forest soils. Bacterial populations may reach a billion cells per gram of soil, especially near roots (Rouatt and Katznelson 1961).

The cyanobacteria, or blue-green algae, are also common microbial inhabitants of forest soils. At least 50 soil-dwelling species capable of fixing nitrogen asymbiotically (i.e., in the absence of a plant host) are known. Some of these species may contribute up to 80 kg of nitrogen each year to a hectare of land (Postgate 1982). Forest ecosystems also support many species of lichen, which is a symbiosis between a fungus and a green or blue-green alga. If the symbiosis involves blue-green algae, a significant amount of nitrogen may also be fixed biologically and added to forest soils. Finally, viruses are important components of soil ecosystems because: they parasitize bacteria and fungi and hence control microbial populations, and they may facilitate the transfer of DNA from one cell to another, thereby increasing genetic diversity of bacteria and fungi.

The "rhizosphere" is the zone of soil influenced by living plant roots. The rhizosphere microflora is a particularly important subset of the soil microflora in plant-dominated ecosystems. Because a substantial amount of the chemical materials that plants produce through photosynthesis is exuded from their roots (up to 40%; Whipps and Lynch 1986), the rhizosphere is rich in nutrients and typically supports a population of microflora 10-50 times as great as that in other soil (Richards 1987). Rhizosphere microflora includes fungi and bacteria that are both beneficial and deleterious to plants, as well as micro-organisms that have no apparent effect on plant growth.

The beneficial component of the microflora includes symbiotic and asymbiotic (free-living) bacteria and fungi. Asymbiotic microbes may proliferate around, on or even within root tissues and stimulate growth of the host plant in ways that we do not understand. Asymbiotic fungi that stimulate plant growth have been studied little, but the plant-growth-promoting rhizobacteria (PGPR) are better understood (Kloepper et al. 1989). Treatment of seeds or roots with PGPR can enhance growth of seedlings or plants by 20-30% or more. We do not know how these micro-organisms stimulate plant growth, but mechanisms that involve plant hormones, mobilization of nutrients or suppression of pathogens are probably involved.

Nitrogen is a key component of all biological systems because it is the structural cornerstone of proteins, which are fundamental to life as we know it. The earth's atmosphere contains about 78% nitrogen in an inert and nutritionally useless form. The capability of living organisms to convert atmospheric nitrogen into a usable organic form is referred to as biological nitrogen fixation (BNF). BNF occurs only in certain kinds of bacteria, some of which have been mentioned. About 60% of the world's usable nitrogen supply originates from BNF, while the rest comes from natural oxidative processes such as lightning and forest fires or human activities such as air pollution and fertilizer manufacture (Postgate 1982).

Most biologically fixed nitrogen originates from symbiotic bacteria found in the root systems of certain plants (e.g., legumes) in specialized structures called nodules. Because so much atmospheric nitrogen may be fixed when in symbiosis with an appropriate legume host (up to 300 kg per hectare per year), the best known symbiotic bacteria belong to the legume-root-nodule bacteria in the genera *Rhizobium* and *Bradyrhizobium*. Many non-leguminous plants, however, also form nitrogen-fixing root-nodules after infection with filamentous bacteria in the genus *Frankia*. This so-called actinorrhizal symbiosis, exemplified by nodule formation in alder (*Alnus*), is analogous to the *Rhizobium*-legume system but appears to be less evolutionarily advanced and less effective in fixing nitrogen. The first manipulations of *Frankia* in pure culture occurred less than 15 years ago so we know comparatively little about this remarkable symbiosis. Further research on its evolutionary status and its effectiveness in fixing nitrogen will likely yield important new information.

Mycorrhizal fungi are symbiotic root-infecting fungi that have been broadly categorized into one of two groups. One group, the ectomycorrhizae, is commonly associated with forest trees including those in the families Pinaceae (pine), Betulaceae (birch) and Fagaceae (beech). Infection of plant roots results in the formation of a recognizable mycorrhiza (literal meaning "fungus root") characterized by a dense fungal sheath surrounding the surface of the infected root and proliferation of the fungus between the outer cells of the root to form the Hartig Net. Little or no actual penetration of root cells occurs.

The second group is the endomycorrhizae. These are commonly associated with almost all taxonomic groups of plants including the economically important legumes and grasses. Most mycorrhizal fungi in our coniferous forest species are ectomycorrhizal, but a notable exception is the Western Red-cedar, which has only endomycorrhizal fungi. These fungi typically do not form an extensive sheath or a Hartig Net but do penetrate cortical cells of roots. Vesicular-arbuscular mycorrhizae constitute the most abundant type of endomycorrhiza in nature and are so named because of the specialized organs they form (vesicles and arbuscules) after penetration of cortical cells. Several other intermediate types of mycorrhizae have characteristics of both ecto- and endomycorrhizae (Harley and Smith 1983).

Actual or potential benefits to the host plant owing to its mycorrhizal condition are numerous. They include: lengthened root life; increased efficiency of ion uptake from the soil; increased volume of soil that can be explored for nutrients; mobilizing of "immobile" nutrients such as phosphorus; supporting populations of beneficial bacteria; tolerance to drought, toxic metals and pathogens; enhanced soil aggregation; and "short circuiting" normal nutrient cycling processes. Certain of these benefits may be realized only if the appropriate fungus is present when a particular environmental stress occurs.

From a purely functional point of view, then, the diversity of soil microflora as a group is substantial. It includes micro-organisms that fix atmospheric nitrogen, produce plant hormones, help sequester nutrients and water for plants, help detoxify compounds that are toxic to plants and suppress plant pathogens. When diversity among species and genetic strains is superimposed on this functional diversity, the overall variation among soil microflora is immense.

What We Don't Know

Perhaps the most startling revelation regarding soil microflora, considering its crucial ecological roles, is that we are aware of only a small fraction of existing species. Current methods of detection depend on our ability to guess what nutritional and environmental conditions these micro-organisms require, so that we can culture them in the laboratory and characterize them morphologically and physiologically. Recent results from molecular studies, however, indicate that up to 80% of the micro-organisms in natural communities have not been cultured (Ward et al. 1990). In all likelihood, a large proportion of species diversity below the ground in our forest ecosystems is unknown. Furthermore, we have not even sufficiently catalogued the microbial diversity that we can detect with current technology.

While we have achieved a basic understanding of some of the principles governing the behaviour of soil micro-organisms, we still have much to learn about their ecology. The remainder of this discussion addresses this point.

In nature, the mycorrhizal state is obligatory for forest-tree species and symbiotic fungi cannot complete their life cycles in the absence of their host (Harley and Smith 1983; Marx and Cordell 1989). Individual tree seedlings can have mycorrhizal associations with several different fungal species simultaneously and these different fungi may be physiologically specialized. Some species may be more efficient at extracting water from a dry environment, for example, while others may be more valuable in gathering certain limiting nutrients.

By entering into symbiosis with a number of different fungi at once, seedlings can substantially broaden their ability to survive in a changing environment. Therefore, diversity of mycorrhizal fungi serves to buffer the deleterious effects of environmental fluctuations or stresses (Perry et al. 1989). A corollary is that the maintenance of mycorrhizal diversity below the ground is necessary to ensure biodiversity in the above-ground vegetative components of an ecosystem.

The most dramatic demonstrations of the roles that soil microflora play are from studies of artificially manipulated or disturbed ecosystems. Several studies have demonstrated that soil transfer from forests has been necessary to facilitate plantation establishment on afforested sites (Mikola 1970). The identity of the key agent(s) through soil transfer has not always been known, but mycorrhizal fungi are usually implicated (Perry et al. 1987). For example, one clearcut in southern Oregon was planted with Douglas-fir four times without success. Adding small amounts of soil from the rooting zone of an established Douglas-fir plantation to planting holes increased seedling survival by about 50% and markedly improved growth rate (Amaranthus and Perry 1989). Furthermore, only seedlings that received soil transfers were alive three years after planting. No effects were detected when pasteurized soil was used.

A similar situation has been described when harsh environments such as mine spoils or highly eroded sites are planted with seedlings of forest-tree species. Without the appropriate fungal partner, seedling growth is poor or nonexistent (Marx 1975, 1980; Valdes 1986). In a recent review, Perry et al. (1989) provide several further examples of changing species composition and degradation of ecosystems caused by eliminating important rhizosphere micro-organisms (e.g., mycorrhizal fungi and associated nitrogen-fixing bacteria) in response to human disturbance. Tranquillini (1979) concluded that the range of conifer species at high elevations depends on ectomycorrhizal fungi that are adapted to cool temperatures. It appears that, without the appropriate mycorrhizal partners, forest seedlings will not grow in certain environments. These studies illustrate the importance of the appropriate rhizosphere micro-organisms in determining the range of plant species that will grow and in stabilizing above-ground floral diversity.

It would, therefore, be prudent to strive to maintain biological diversity of mutualistic mycorrhizal fungi when managing an ecosystem. Some (all?) of the mutualists can survive on dead or dying matter for a while, but at least certain groups of them decline rapidly in the absence of their host (Amaranthus 1990). One way to maintain a stable and diverse below-ground mutualist community is to retain suitable species of host plants above the ground to act as refuges for rhizosphere micro-organisms. In forestry, where host trees are necessarily removed by harvest, maintaining alternate host plant species may be extremely important.

For example, some ericaceous species such as Hairy Manzanita and bearberry form mycorrhizae with fungi that are also compatible with Douglas-fir and other conifers (Zak 1976). Thus ericaceous plants may act as alternate hosts and hence as sources of mycorrhizal fungi for regenerating conifer seedlings (Acsai and Largent 1983). Amaranthus and Perry (1989) demonstrated this potential by planting Douglas-fir

234

seedlings into three areas previously dominated by different vegetation types before clearing grasses, oak or Whiteleaf Manzanita (*Arctostaphylos viscida*). After two years, Douglas-fir seedlings on the manzanita site had 50% greater basal- area growth than seedlings on the grass site. Seedlings on the oak site were smallest. The growth response on the manzanita site was correlated with the kind of mycorrhizae in the root system but not to the number of infected root tips. When seedlings were inoculated with soil from a nearby stand of the ericaceous species *Arbutus menziesii* (Arbutus) their growth on the manzanita site was more than 300% of that on the grass site. Again, pasteurized soil had no effect. Inoculation of seedlings on the grass or oak site with Arbutus soil did not affect basal area or height.

This specific response of Douglas-fir seedlings to the biotic component of soil previously occupied by manzanita and Arbutus suggests that Douglas-fir, manzanita and Arbutus constitute a species "guild" defined by their association with common mycorrhizal fungi. If such guilds are typical of the ecosystems that we manage for fibre production, it would be useful to be able to recognize them so that alternate hosts could be left intact during harvesting, thereby maintaining a diverse source of inoculum for regenerating conifers. This strategy is currently being tested in central B.C. with bearberry as a source of mycorrhizal inoculum for planted conifer seedlings (G. Hunt pers. comm.).

Many of our current forest harvesting and management practices have the potential to have negative impacts on the structure and population size of important communities of soil micro-organisms. Clearcutting results in the removal of host plants and, in some cases, large nurse logs and woody debris that may serve as moist reservoirs of mutualistic fungi and bacteria. Soil compaction restricts the movement of air and water in soil and hence reduces the population size and diversity of soil micro-organisms. Soil microclimate may also change when above-ground vegetation is removed, possibly to the detriment of native microbial populations. Slash burning has obvious potential negative effects including incineration of microbial propagules and reduction or removal of organic layers where many soil mutualists reside.

Several studies have investigated the impact of clearcutting and slash burning on mutualistic fungi, but no clear trend has emerged. Some authors have found a significant reduction in fungal populations after these activities have occurred (Perry and Rose 1983; Parke et al. 1984; Amaranthus et al. 1987). Others have found no changes in population size (Parke et al. 1983) or have even noted increased population size (Pilz and Perry 1984). Mycorrhizal species diversity was

235

reduced in clearcuts in the latter study, however. The discrepant findings noted may be due in part to how long sites were left before regeneration. Young clearcuts appear to sustain fewer deleterious effects than old ones. There may be a window shortly after harvesting, possibly briefer than one year, within which little damage to microbial mutualists occurs (Harvey et al. 1980). The longer the period between harvesting and planting, however, the more difficult it may be for conifers to regenerate (Amaranthus 1990). If slash burning is not too intense or too frequent, it may have only a minor effect on mycorrhizal fungi. We have little information on the effects on soil mutualists of intensive silvicultural management such as the use of herbicides and fertilizers, or monoculture plantations.

We are severely limited in our knowledge but can reach several conclusions regarding biodiversity of soil microflora. Below-ground biodiversity is substantial and may be crucially important to above-ground biodiversity and ecosystem stability. This contention is supported by the fact that the presence or absence of appropriate below-ground mutualists can determine survival and growth of individual seedlings. Because of technological restrictions we probably know only a small fraction of the soil microflora in our ecosystems. Ecologically it appears that host plants may form "guilds" depending on their association with common below-ground mutualists. Species of plants that are not commercially harvested may be extremely important in this regard, but currently we have no way of recognizing species "guilds" in the field. With our poor knowledge we cannot predict these relationships or the effects that our forest-management practices have on soil microflora, so a major research effort to elucidate the role of soil micro-organisms in long-term sustainable forestry is required.

Literature Cited

Acsai, J., and D.L. Largent. 1983. Fungi associated with *Arbutus menziesii* and *Arctostaphylos uva-ursi* in central and northern California. Mycologia 75: 544-547.

Amaranthus, M. P. 1990. Rethinking the ecology and management of temperate forests: the living soil. *In* Proceedings, Forests — Wild and Managed: Differences and Consequences, *edited by* A.F. Pearson and D.A. Challenger. Univ. British Columbia, Vancouver, BC. pp. 55-65.

Amaranthus, M.P., and D.A. Perry. 1989. Interaction effects of vegetation type and Pacific madrone soil inocula on survival, growth, and mycorrhiza formation of Douglas-fir. Can. J. For. Res. 19: 550-556.

Amaranthus, M.P., D.A. Perry and S. Borchers. 1987. Reductions in native mycorrhizae reduce growth of Douglas-fir seedlings. *In* Proceedings, 7th

North American Conference on Mycorrhizae, *edited by* D.M. Syvia, L.L. Hung, and J.H. Graham. Gainesville, FL. p. 80.

Harley, J.L., and S.E. Smith. 1983. Mycorrhizal symbiosis. Academic Press, London, England.

Harvey, A.E., M.F. Jurgensen, and M.J. Larsen. 1980. Clearcut harvesting and ectomycorrhizae: survival of activity on residual roots and influence on a bordering forest stand in western Montana. Can. J. For. Res. 10: 300-303.

Kloepper, J.M., R. Lifshitz, and R.M. Zablotowicz. 1989. Free-living bacterial inocula for enhancing crop productivity. Trends in Biotechnol. 7: 39-44.

Marx, D.H. 1975. Mycorrhiza and establishment of trees on strip-mined land. Ohio J. Sci. 75: 288-297.

Marx, D.H. 1980. Ectomycorrhiza fungus inoculations: a tool for improving forestation practices. *In* Tropical mycorrhiza research, *edited by* P. Mikola. Oxford Univ. Press, Oxford, England. pp. 13-71.

Marx, D.H., and C.E. Cordell. 1989. The use of specific ectomycorrhizas to improve artificial forestation practices. *In* The biotechnology of fungi for improving plant growth, *edited by* J.M. Whipps and R.D. Lumsden. Cambridge Univ. Press, Cambridge, England. pp. 1-25.

Mikola, P. 1970. Mycorrhizal inoculation in afforestation. Int. Rev. For. Res. 3: 123-196.

Parke, J.L., R.G. Linderman, and J.M. Trappe. 1984. Inoculum potential of ectomycorrhizal fungi in forest soils of southwest Oregon and northern California. Forest Sci. 30: 300-304.

Parke, J.L., R.G. Linderman, and J.M. Trappe. 1983. Effect of root zone temperature on ectomycorrhiza and vesicular-arbuscular mycorrhiza formation in disturbed and undisturbed forest soils of southwest Oregon. Can. J. For. Res. 13: 657-665.

Perry, D.A., and S.L. Rose. 1983. Soil biology and forest productivity: opportunities and constraints. *In* International Union of Forest Research Organisations, Symposium on Forest Site and Continuous Productivity, *edited by* R. Ballard and S.P. Gessel. USDA Forest Service, General Tech. Rep., Portland, OR. pp. 229-238.

Perry, D.A., R. Molina, and M.P. Amaranthus. 1987. Mycorrhizae, mycorrhizospheres, and reforestation: current knowledge and research needs. Can. J. For. Res. 17: 929-940.

Perry, D.A., M.P. Amaranthus, J.G. Borchers, S.L. Borchers, and R.E. Brainerd. 1989. Bootstrapping in ecosystems. Bioscience 39: 230-237.

Pilz, D.P., and D.A. Perry. 1983. Impact of clearcutting and slash burning on ectomycorrhizal associations of Douglas-fir seedlings. Can. J. For. Res. 14: 94-100.

Postgate, J.R. 1982. The fundamentals of nitrogen fixation. Cambridge Univ. Press, Cambridge, England.

Richards, B.N. 1987. The microbiology of terrestrial ecosystems. John Wiley and Sons, New York, NY.

Rouatt, J.W., and H. Katznelson. 1961. A study of bacteria on the root surface and in the rhizosphere soil of crop plants. J. Appl. Bact. 24: 164-171.

Tranquillini, W. 1979. Physiological ecology of the alpine timberline. Springer-Verlag, New York, NY.

Valdes, M. 1986. Survival and growth of pines with specific ectomycorrhizae after 3 years on a highly eroded site. Can. J. Bot. 64: 885-888.

Ward, D.M., R. Weller, and M.M. Bateson. 1990. 16S rRNA sequences reveal numerous uncultured micro-organisms in a natural community. Nature 345: 63-65.

Whipps, J.M., and J.M. Lynch. 1986. The influence of the rhizosphere on crop productivity. Adv. Microb. Ecol. 9: 187-244.

Zak, B. 1976. Pure culture synthesis of bearberry mycorrhizae. Can. J. Bot. 54: 1297-1305.

Sustainable Forestry and Soil Fauna Diversity

Valin G. Marshall

Forestry Canada, Pacific Forestry Centre, 506 West Burnside Road, Victoria, British Columbia, Canada V8Z 1M5.

Abstract

Animal groups that are abundant in the soil are the protozoans, nematodes, gastropod mollusks, arthropods and annelid worms. The abundance and number of species in these groups in British Columbia's soils can be estimated only roughly. Aside from protozoans, the numerically dominant groups are nematodes and arthropods, which may reach densities of over a million individuals and hundreds of species per square metre of surface soil. Most soil animals occur within 30 cm of the surface. This fauna, in concert with the microflora, affects soil processes in several important ways. They break up litter, bring about mineralization, serve as a reservoir of nutrients, influence microbiostasis and alter the physical characteristics of soils. The soil fauna is also important in controlling pests and pathogens. Human activity increasingly threatens soil faunal diversity, such as reduction through many agricultural and forestry practices including pesticides, fertilizers and fire. The fauna is also adversely affected by removal of organic matter, acid deposition, forest harvesting and soil compaction. The ability to identify species is essential for effective use of available knowledge and to minimize adverse effects of human activity. Unfortunately, many species in most groups are undescribed or cannot be identified because of inadequate literature. We know little about the ecological functions of soil fauna. Our extremely poor knowledge of species richness, species composition and ecological functions of soil fauna should caution us to be very conservative in our management practices. Basic taxonomic and ecological research on B.C.'s soil fauna are desperately needed.

In *Our Living Legacy: Proceedings of a Symposium on Biological Diversity*, edited by M.A. Fenger, E.H. Miller, J.A. Johnson and E.J.R. Williams, pp. 239-248. Victoria, B.C.: Royal British Columbia Museum.

Introduction

Small soil animals and soil microflora are the major components of soil biota. The world beneath our feet contains nearly all of the animal phyla that are not exclusively marine (Kevan 1965). The numerically dominant soil animal groups are protozoans, nematodes, gastropod mollusks, arthropods and annelid worms. A few vertebrates, such as certain species of amphibians and small mammals, may also be considered as members of the soil fauna. The soils of British Columbia remain relatively unaltered by human activity and contain a richness of species whose identity and numbers can only be roughly estimated because of the paucity of information.

This paper discusses the density and diversity of major groups of soil fauna in four forest ecosystems of B.C., the role of these organisms in ecosystem functioning, the impact of human activities on the fauna, and research needs for taxonomy and habitat preservation. The discussion is focused on forest soils, but the general conclusions also apply to agricultural soils.

Numerical Abundance and Species Diversity

Numerical Abundance

Soil animals even occur in embryonic soils colonized by lichens and mosses, where drought-resistant forms of many groups become established (Sayre and Brunson 1971). As such soils develop, the numbers and diversity of species increase and they reach astronomical figures in mature forest soils. In forest soils, most animals live near the surface, 0-30 cm deep (Fig. 1). The diagram shows springtails and mites only, but the other members of the soil fauna follow a similar pattern (Persson et al. 1980). The soil near the surface is more suitable for fauna as better conditions of food supply, aeration, pore space, temperature and moisture prevail. This surface concentration also makes the fauna vulnerable to many human activities.

The densities of soil fauna in mature forest sites are generally high (Table 1). Values for the various taxa from the few sites in B.C. are within ranges for other temperate forests (Peterson and Luxton 1982). No data are available for Protozoa, which probably number in the millions per square metre (Lousier and Bamforth 1990). Over two million metazoans per square metre live in highly productive forest soils on Vancouver Island, even if the unicellular protozoans are excluded; of these, about half are nematodes. Mites and springtails predominate among the arthropods. Other groups are numerically less abundant, but

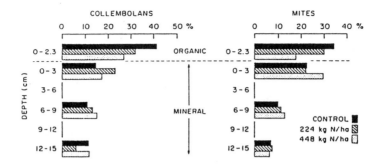

Figure 1. Vertical distribution of springtails (Collembolans) and mites in soil in a Douglas-fir plantation near Shawnigan Lake, Vancouver Island. Note the high numbers of individuals near the surface. (From Marshall 1974. Canadian Journal of Soil Science.)

numerically unimportant groups such as annelids may have high biomass. The total soil faunal biomass in a productive temperate forest may reach 8 grams per square metre dry mass, or about 80 kilograms per hectare (Petersen and Luxton 1982). This biomass is still greater when its turnover rate is considered. For example, the mean annual standing crop of testacean protozoans in an Alberta deciduous woodland was 0.72 grams per square metre wet mass, but the total annual production was 206 grams per square metre (Lousier and Bamforth 1990). Since this fauna feeds on decaying organic matter or is itself fed upon by other soil animals, the potential for nutrient cycling by these tiny creatures is tremendous.

Species Diversity

Various definitions have been given for "biological diversity" (Wilcove 1988; Probst and Crow 1991). The data in Table 1 show the richness and variety of soil animals in forest ecosystems within four different biogeoclimatic zones in B.C. The sites described may be regarded as "ecosystem associations".

The number of soil animal species is not known accurately, since no soil ecosystem in B.C. has been intensively studied for all faunal groups. The richest taxonomic group of metazoans is almost certainly the Nematoda or Arthropoda, based on preliminary observations and what is known from soils elsewhere. We may anticipate totals of about 300 genera and 550 species in a single association. The Arthropoda, which are best known, will constitute about half of the species, as in the American Pacific Northwest (Moldenke and Lattin 1990b). Among the

Table 1. Mean densities (number per square metre) of soil faunal taxa in four forest ecosystems in B.C.[a]

Taxon	No. of individuals				No. of genera				No. of species				B.C. species
	CDF	IDF	SAF	CWH	CDF	IDF	SAF	CWH	CDF	IDF	SAF	CWH	
PROTOZOA	(1000)[b]	x	x	x	(100)[b]	x	x	x	(150)[b]	x	x	x	250
ROTIFERA	13	x	x	14	(4)	x	x	x	(5)	x	x	x	50
NEMATODA	565	856	x	1684	11	x	x	x	(75)[c]	x	x	x	500
TARDIGRADA	6	9	x	23	(5)	x	x	x	(20)[d]	x	x	x	80
ARACHNIDA													
Acari	206	329	72	179	97	48	40	x	131	56	63	x	2000
Araneae	0.05	0.03	x	0.4	(15)	x	x	x	(40)[e]	x	x	x	200
Pseudoscorpionida	0.2	0.2	x	2	1	x	x	x	2	x	x	x	15
CRUSTACEA													
Copepoda	0.2	0	x	41	1	x	x	x	1	x	x	x	5
Isopoda	2	x	x	1	(3)	1	x	x	(5)	x	x	x	20
MYRIAPODA													
Chilopoda	0.03	0.2	x	0.2	1	x	x	x	(4)[f]	x	x	x	30
Diplopoda	0.05	0.03	x	0.3	1	1	x	x	(4)[f]	x	x	x	25
Pauropoda	2.8	x	x	0.3	4	1	x	x	6	1	x	x	20
Symphyla	2	x	x	2	1	x	x	x	1	x	x	x	5
HEXAPODA													
Collembola	19	87	7	115	31	10	12	x	53	10	40	x	200
Diplura	0.03	0.03	x	0.03	1	x	x	x	(1)	x	x	x	3
Insecta (other)	3	1	x	2	(17)	x	x	x	(30)	x	x	x	3000
Protura	0.3	x	x	0.6	1	x	x	x	3	x	x	x	25
GASTROPODA	0.4	x	x	0.006	(5)	x	x	x	(10)	x	x	x	100
ANNELIDA													
Enchytraeidae	6	0.2	x	0.6	7	x	x	x	9	x	x	x	100
Lumbricidae	(0.1)	x	x	0	1	x	x	x	1	x	x	x	20
Megascolecidae	0	x	x	0.1	0	0	0	2	0	0	0	2	5
Totals	1826	1282	79	2065	307	60+	52+	x	551	66+	103+	x	6643

a CDF = Coastal Douglas-fir near Shawnigan Lake, Vancouver Island (Marshall 1974 and unpubl. data).
   IDF = Interior Douglas-fir, Palmer Forsyth Road, near Kamloops (Marshall 1979 and unpubl. data).
   SAF = Subalpine forest near Sparwood (Lawrence 1986).
   CWH = Coastal Western Hemlock (CH phase) near Port McNeill (Battigelli 1990).

x = No data available.
() = Estimates based on number of species
     known from individual habitats in B.C.
b = J.D. Lousier (pers. comm.).

c = For a beech forest (Yeates 1979).
d = From Kathman and Cross (1991).
e = C. Dondale (pers. comm.); McIver et al. (1990).
f = From Behan-Pelletier (1991).

arthropods there are about 100 species in a single order of mites, the Oribatida. Therefore, the 131 species estimated for B.C. for all orders of soil mites may be very low. Similarly, estimates of 551 species for an ecosystem association may be low, because one square metre of soil may contain a thousand or more species of soil animals (Swift et al. 1979; Stanton and Lattin 1989). The soils of B.C. may harbour well over 6,600 species of soil animals, with over 2,000 species of mites and 3,000 species of insects (mainly immature stages).

The Roles of Soil Fauna in Ecosystem Functioning

Soil fauna and microflora affect the biological, chemical and physical character of soils in several important ways (Dindal 1990a). The fauna shreds litter to produce fragments that greatly increase substrate surface area, thereby facilitating access by microflora. This sets into motion a succession of flora and fauna, each advancing the process of decomposition and mineralization, thereby releasing nutrients for plant growth. The animals themselves are equipped with enzymes that enhance the decomposition of many simple biological substances and they release waste products with high concentrations of nutrients. The faunal biomass holds a significant reservoir of nutrients that is released upon death. The fauna disperses fungal spores directly and has an additional indirect effect through interaction of fauna and flora, especially through microbiostasis (nongermination or cell division of germinable propagules on natural soil). The soil fauna's contribution to biodiversity is enormous. When prevalent, earthworms exert a tremendous impact on mineralization, aeration, water-holding capacity and bulk density of soils.

Some soil fauna may spread plant-borne pathogens (Kevan 1965; Edwards et al. 1988), while others are active in the natural control of plant diseases. A local example of plant pathogen control is the ability of termites to suppress root rot. Research shows the potential for similar biological control in other forest systems. Another example involves interplanting Black Walnut (*Juglans nigra*) with Autumn Olive (*Elaeagnus umbellata*), which greatly reduces the diseases caused by *Gnomonia leptostyla* and *Mycosphaereela juglandis* (Kessler 1990). This reduction was brought about primarily through more active grazing by soil mites and collembolans on the fungal perithecia within infected fallen walnut leaves. A final example involves lumbricid earthworms. A high earthworm population (especially *Lumbricus terrestris*) in winter can consume the perithecia on fallen leaves that have apple scab (*Venturia inaequalis*), and can prevent the carry-over of the disease (Edwards 1981).

For other examples see Finnegan (1977), McNeil et al. (1978), Edwards (1981) and Lee (1985).

Effects of Human Activities on Soil Fauna

Human activity increasingly threatens soil faunal diversity. Both numerical abundance and diversity are being reduced by many agricultural and forestry practices. These include chemicals (especially insecticides and some fertilizers), fire, forest harvesting, erosion, acid deposition, flooding and soil compaction (Vlug and Borden 1973; Marshall 1977, 1979, 1980; Dindal 1980; Moldenke and Latin 1990a). In addition, the diets of soil fauna are being changed by practices of monoculture. Conversely, agricultural and silvicultural practices that encourage structurally more diverse plant communities enhance beneficial soil-invertebrate communities (House and Stinner 1987; Franklin et al. 1989).

Habitat alterations brought about by agricultural and forestry practices could reduce diversity: abundant species could disappear rapidly and common species may increase at the expense of rarer or more sensitive ones (Schindler 1989). Changes have already been documented with certain earthworm species (Lee 1985). Many ecologists support the theory that diversity promotes stability. Odum (1972) goes even further, stating that variety may be a necessity and not just the spice of life. Although this theory is still controversial, Franklin et al. (1989) noted that diversity buffers against the impact of disturbances by maintaining redundancies, which could be important when ecosystems are placed under new stresses such as global warming. An increase of 4-5°C in mean temperature — perhaps with a slight decrease in summer temperatures — is predicted for the Pacific Northwest during the next 50 years (Franklin et al. 1989). Such a change could profoundly alter the complex interactions among higher vegetation, soil microflora and soil fauna, both through direct impact on individual organisms and indirectly through increased impact of pathogens and pests.

Taxonomic Needs

Effective use of available knowledge to minimize future adverse effects on the soil fauna depends on our ability to recognize individual species. Despite this need, the soil fauna of Canada has not received the attention it deserves (Behan-Pelletier 1991). The recent publication by

Dindal (1990b) is a major advance in this direction, but it also points out the many gaps in our taxonomic knowledge. Much work remains to be done. Family and generic keys have now been published for most groups, but there are few keys to species. Even when they are available, such keys are often inadequate because of the large number of undescribed species in most soils. Marshall et al. (1982) estimated that fewer than half of the soil arthropod species in Canada have been described. Even when adults have been described, problems still exist. This is especially critical for soil insects, because most are immature stages. The small size of many soil animals contributes to difficulties with recognition and description, but numerous species of both microscopic and macroscopic fauna are still undescribed. This is exemplified by the recent discoveries of new species of *Tachidius*, a harpacticoid crustacean, which was found in soil well away from water bodies (Marshall, unpublished data), and native megascolecid earthworms in the genera *Arctiostrotus* and *Toutellus* in wet forests on Vancouver Island (Fender and McKey-Fender 1990). These previously undescribed megascolecid species form the last small remnant populations of an indigenous earthworm fauna in Canada, and they deserve special attention. We do not yet know the exact number of species present or their ecology; such information is urgently needed to assess the impact of harvesting old-growth forests.

Taxonomic problems are further complicated by a lack of specialists for many groups of soil fauna (Marshall et al. 1982). Even when taxonomists are available they are overwhelmed by the great array of species. Urgent "economic" concerns usually take precedence over basic taxonomic work. The following quotation concerning nematodes applies equally to many other groups: "Our present research program is concerned primarily with plant parasitic nematodes and, therefore, our identification service and technical support must be concentrated in this area. Regretfully, it has not been possible with our present staff of three nematologists to develop expertise required for specific identifications of the many and diverse groups of saprophagous nematodes that are of particular interest to you" (R.H. Mulvey and R.V. Anderson pers. comm.). When all groups including nematodes are considered, the situation has worsened since 1974 (Behan-Pelletier 1991).

Conclusions

We know little about the numbers or kinds of animals that live in B.C. soils or about their ecological roles. We know that the soil fauna contributes enormously to decomposition and nutrient release and to the natural control of pathogens and pests. Hill (1989) has remarked that if we eliminate any of these species we will inherit their jobs, and in many cases, we are totally ignorant of exactly what they do. Clearly we must strive to preserve these little allies that are essential for maintaining the productivity of our farms and forests.

Recommendations for preserving biodiversity have been made by Behan-Pelletier (1991) and Probst and Crow (1991). Most of these recommendations apply to the soil fauna. For B.C., urgent action should include cataloguing of all species through intensive surveys and inventories in all major habitats represented in the province, monitoring problem species, preserving habitats that are sufficient to maintain species at risk of extinction, establishing long-term research sites and increasing public awareness.

Literature Cited

Battigelli, J.P. 1990. Soil fauna of the Cedar-hemlock and hemlock-Amabalis Fir forest phases on northern Vancouver Island. Unpublished Contract Report. Pacific Forestry Centre, Victoria, BC.

Behan-Pelletier, V.M. 1991. Diversity of soil arthropods in Canada: systematic and ecological problems. In Symposium Proceedings, Systematics and Entomology. Ent. Soc. Can. and Ent. Soc. Alberta, Banff, Alberta, 1990.

Dindal, D.L. (Editor). 1980. Soil biology as related to land use practices. Proceedings, 7th International Soil Zoology Colloquium of the International Society of Soil Science. Office of Pesticide and Toxic Substances, Environmental Protection Agency, Washington, DC.

———— 1990a. Introduction. In Soil biology guide, edited by D.L. Dindal. John Wiley and Sons, New York, NY. pp. 1-14.

———— (Editor). 1990b. Soil biology guide. John Wiley and Sons, New York, NY.

Edwards, C.A. 1981. Earthworms, soil fertility and plant growth. In Proceedings, Workshop on the Role of Earthworms in the Stabilization of Organic Residues, Volume 1, edited by M. Appelhof. Beech Leaf Press, Kalamazoo, MI. pp. 61-85.

Edwards, C.A., B.R. Stinner, D. Stinner, and S. Rabatin (Editors). 1988. Biological interactions in soil. Elsevier, Amsterdam, The Netherlands.

Fender, W.M., and D. McKey-Fender. 1990. Oligochaeta: Megascolecidae and other earthworms from western North America. In Soil biology guide, edited by D.L. Dindal. John Wiley and Sons, New York, NY. pp. 357-378.

Finnegan, R.J. 1977. Establishment of a predacious red wood ant, Formica obscripes (Hymenoptera: Formicidae) from Manitoba to eastern Canada. Can. Entomol. 109: 1145-1148.

Franklin, J.F., D.A. Perry, T.D. Schowalter, M.E. Harmon, A. McKee, and T.A. Spies. 1989. Importance of ecological diversity in maintaining long-term site productivity. In Maintaining the long-term productivity of Pacific

Northwest forest ecosystems, *edited by* D.A. Perry, R. Meurisse, B. Thomas, R. Miller, J. Boyle, J. Means, C.R. Perry, and R.F. Powers. Timber Press, Portland, OR. pp. 82-97.

Hill, S. 1989. The world under our feet. Federation of Ontario Naturalists (Winter 1989): 15-19.

House, G.J., and B.R. Stinner. 1987. Influence of soil arthropods on nutrient cycling in no-tillage agroecosystems. Misc. Pub. Entomol. Soc. Am. 65: 44-52.

Kathman, R.D., and S.F. Cross. 1991. Ecological distribution of moss-dwelling tardigrades on Vancouver Island, British Columbia, Canada. Can. J. Zool. 69: 122-129.

Kessler, K.J., Jr. 1990. Destruction of *Gnomonia leptostyla* perithecia on *Juglans nigra* leaves by microarthropods associated with *Elaeagnus umbellata* litter. Mycologia 82: 387-390.

Kevan, D.K. McE. 1965. The soil fauna — its nature and biology. *In* Ecology of soil-borne plant pathogens: prelude to biological control, *edited by* K.F. Baker and W.C. Snyder. University of California Press, Berkeley, CA. pp. 33-51.

Lawrence, J.M. 1986. Soil fauna colonization of high elevation coal mine spoils in the Canadian Rockies. M.Sc. thesis, University of Victoria, Victoria, BC.

Lee, K.E. 1985. Earthworms: their ecology and relationships with soils and land use. Academic Press, Sydney, Australia.

Lousier, J.D., and S.S. Bamforth. 1990. Soil protozoa. *In* Soil biology guide, *edited by* D.L. Dindal. John Wiley and Sons, New York, NY. pp. 97-136.

Marshall, V.G. 1974. Seasonal and vertical distribution of soil fauna in a thinned and urea fertilized Douglas-fir forest. Can. J. Soil Sci. 54: 491-500.

——— 1977. Effects of manures and fertilizers on soil fauna. A review. Commonwealth Bureau of Soils, Special Publication No. 3.

——— 1979. Effects of the insecticide diflubenzuron on soil mites of a dry forest zone in British Columbia. Recent Advances in Acarology 1: 129-134.

——— 1980. Soil fauna. *In* Operational field trials against the Douglas-fir tussock moth with chemical and biological insecticides, *edited by* R.F. Shepherd. Can. For. Serv., BC-X-201. pp. 16, 18.

Marshall, V.G., D.K. McE. Kevan, J.V. Mathews, Jr., and A.D. Tomlin. 1982. Status and research needs of Canadian soil fauna. Bull. Ent. Soc. Can., Suppl. 14.

McIver, J.D., A.R. Moldenke, and G.L. Parsons. 1990. Litter spiders as bio-indicators of recovery after clearcutting in a western coniferous forest. Northwest Env. J. 6: 410-411.

McNeil, J.N., J. Delisle, and R.J. Finnegan. 1978. Seasonal predatory activity of the introduced red wood ant, *Formica lugubris* (Hymenoptera: Formicidae) at Valcartier, Quebec, in 1976. Can. Ent. 110: 85-90.

Moldenke, A.R., and J.D. Lattin. 1990a. Dispersal characteristics of old-growth soil arthropods: the potential for loss of diversity and biological function. Northwest Env. J. 6: 408-409.

——— 1990b. Density and diversity of soil arthropods as "biological probes" of complex soil phenomena. Northwest Env. J. 6: 409-410.

Odum, E.P. 1972. Ecosystem theory in relation to man. *In* Ecosystem structure and function. Proceedings, 31st Annual Biology Colloquium, Corvallis, OR, 1970. Oregon State University Press, Corvallis, OR. pp. 11-24.

Persson, T., E. Bååth, M. Clarholm, H. Lundkvist, B.E. Söderström, and B.

Sohlebius. 1980. Trophic structure, biomass dynamics and carbon metabolism of soil organisms in a Scots pine forest. *In* Structure and function of northern coniferous forests — an ecosystem study, *edited by* T. Persson. Ecological Bulletin 32: 419-459.

Petersen, H., and M. Luxton. 1982. A comparative analysis of soil fauna populations and their role in decomposition. Oikos 39: 288-388.

Probst, J.R., and T.R. Crow. 1991. Integrating biological diversity and resource management: an approach to productive, sustainable ecosystems. J. For. 89: 12-17.

Sayre, R.M., and L.K. Brunson. 1971. Microfauna of moss habitats. American Biology Teacher 33: 100-105.

Schindler, D.W. 1989. Biotic impoverishment at home and abroad. BioScience 39: 426.

Stanton, N.L., and J.D. Lattin. 1989. In defence of species. BioScience 39: 67.

Swift, M.J., O. Heal, and J. Anderson. 1979. Decomposition in terrestrial ecosystems. Studies in ecology 5. Blackwell Sci. Pub., Oxford, England.

Vlug, H., and J.H. Borden. 1973. Soil Acari and Collembola populations affected by logging and slash burning in a coastal British Columbia coniferous forest. Environ. Entomol. 2: 1016-1023.

Wilcove, D.S. 1988. National Forests: policies for the future. Volume 2: Protecting biological diversity. The Wilderness Society, Washington, DC.

Yeates, G.W. 1979. Soil nematodes in terrestrial ecosystems. J. Nematol. 11: 213-229.

Biodiversity Inventory in the South Okanagan

W.L. Harper
E.C. Lea
R.E. Maxwell

Wildlife Branch, B.C. Ministry of Environment , Lands and Parks, 780 Blanshard St., Victoria, British Columbia, Canada V8V 1X4.

Abstract

The British Columbia Ministry of Environment is conducting biophysical habitat inventory at 1:20,000 scale, to classify the landscape according to its ability to support rare, threatened or endangered species in the South Okanagan. The semi-arid habitats being mapped in the South Okanagan support one of the most diverse, rare and unique assemblages of plant and animal species in British Columbia and Canada. Some species of vertebrates, invertebrates and plants occur nowhere else in B.C., and some occur nowhere else in Canada. Unfortunately, threats to this biological diversity are as many and varied as the species they adversely affect. Habitat losses caused by urban, recreational and agricultural development account for most of these threats. Cattle grazing has also reduced the capacity of these dry, sensitive habitats to support many plants and animals. Only 9% of the landscape that supports this biological diversity has been left in a relatively natural state and the Ministry of Environment considers that many species are at risk. Of the 22 species of vertebrates being considered by government for designation as endangered or threatened in B.C., 12 occur in the South Okanagan. Based on permanent physical features in the landscape, biophysical habitats are classified and mapped for the whole range of plant successional stages. These habitat units are relatively homogeneous in soils, topography, bedrock geology, vegetation and animal use. We map the present successional stage or range condition class for each habitat unit and describe all potential successional stages in a legend. From these habitat maps we will generate interpretive maps that rate habitat suitability and capability for 54 species: 32 birds, 12 mammals, 8 reptiles and 2 amphibians. Other interpretive maps will classify and rate the landscape for plant and invertebrate

In *Our Living Legacy: Proceedings of a Symposium on Biological Diversity*, edited by M.A. Fenger, E.H. Miller, J.A. Johnson and E.J.R. Williams, pp. 249-262. Victoria, B.C.: Royal British Columbia Museum.

diversity. This mapping will help integrate planning with other resource agencies by identifying areas of high priority for habitat protection, acquisition, enhancement and species re-introduction programs.

Introduction

The southern end of the Okanagan Valley in Canada represents the northernmost extension of the Western Great Basin of North America. Low annual precipitation, hot summers and mild winters create a variety of semi-arid habitats for plants and animals. Though relatively extensive in the United States, the semi-arid habitats in the South Okanagan are one of the three most endangered biomes in Canada. These habitats also provide one of the most diverse, rare and unique assemblages of plant and animal species in British Columbia and Canada. Many species in the South Okanagan occur at the edge of their range. These small peripheral populations, with their ecologically marginal existence, often play a key role in maintaining genetic diversity within a species (Scudder 1989). The study area for biodiversity habitat inventory in this area includes valley bottom habitats from Okanagan Mountain Park to the Canada-United States border (Fig. 1).

In 1964, the Interpretation Section of the B.C. Parks Branch suggested that there was a last chance to preserve some of this area for the future. Sixteen years later Scudder (1980: 49-55) stated that "now must be the *very last chance*" to preserve some of this area. It is now 1991 and the Ministry of Environment, the Habitat Conservation Fund and The Nature Trust of B.C., having taken up the challenge, are mounting a concerted and co-ordinated effort to preserve a small portion of the rich biological diversity found in the South Okanagan.

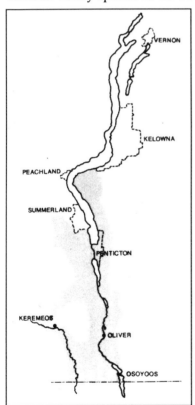

Figure 1. The South Okanagan biophysical mapping study area.

Vertebrate Diversity

B.C. has the greatest diversity of birds and mammals of any province in Canada (Bunnell and Williams 1980). The South Okanagan is a focal point for a significant part of this diversity. Not only does the South Okanagan have a wide range of animal species, but many of them are found nowhere else in the province and some occur nowhere else in Canada.

Of the 448 species of birds in B.C., 303 have been recorded in the South Okanagan (Cannings et al. 1987). The South Okanagan has the greatest abundance and diversity of bats in Canada. Of the 20 bat species that occur in Canada, 14 reside in the South Okanagan, which is the only place in Canada inhabited by the Spotted Bat and Pallid Bat. The hot, dry South Okanagan also supports the highest diversity of reptiles in B.C.

Since rare species are more at risk of extirpation than common species, efforts to maintain biological diversity usually concentrate on protecting the rarer species in an ecosystem. The South Okanagan Critical Areas Program (SOCAP) of The Nature Trust of B.C. has identified 54 vertebrate species of concern in the South Okanagan (Hlady 1990). Most of these are species that the Wildlife Branch considers "at risk", thus they are on the 1990 Red and Blue lists for B.C. (see update in Munro, this volume; B.C. Environment 1990). In the South Okanagan, Red and Blue-listed species of vertebrates addressed in this study include 12 mammals, 20 birds, 4 reptiles and 1 amphibian:

Mammals

Badger	*Taxidea taxus*
Fringed Myotis	*Myotis thysanodes*
Great Basin Pocket Mouse	*Perognathus parvus*
Nuttall's Cottontail	*Sylvilagus nuttallii*
Pallid Bat	*Antrozous pallidus*
Southern Red Bat	*Lasiurus blossevilli*
Spotted Bat	*Euderma maculatum*
Townsend's Big-eared Bat	*Plecotus townsendii*
Western Harvest Mouse	*Reithrodontomys megalotis*
Western Small-footed Myotis	*Myotis ciliolabrum*
White-tailed Jackrabbit	*Lepus townsendii*

Birds

American White Pelican	*Pelecanus erythrorhynchos*
Bobolink	*Dolichonyx oryzivorus*
Brewer's Sparrow	*Spizella breweri*
Burrowing Owl	*Athene cunicularia*

Birds (cont.)

Canyon Wren	*Catherpes mexicanus*
Common Poorwill	*Phalaenoptilus nuttallii*
Flammulated Owl	*Otus flammeolus*
Grasshopper Sparrow	*Ammodramus savannarum*
Gray Flycatcher	*Empidonax wrightii*
Great Blue Heron	*Ardea herodias*
Lewis' Woodpecker	*Melanerpes lewis*
Long-billed Curlew	*Numenius americanus*
Peregrine Falcon	*Falco peregrinus*
Prairie Falcon	*Falco mexicanus*
Sage Thrasher	*Oreoscoptes montanus*
Western Bluebird	*Sialia mexicana*
White-headed Woodpecker	*Picoides albolarvatus*
White-throated Swift	*Aeronautes saxatalis*
Williamson's Sapsucker	*Sphyrapicus thyroideus*
Yellow-breasted Chat	*Icteria virens*

Reptiles

Gopher Snake	*Pituophis melanoleucus*
Night Snake	*Hypsiglena torquata*
Short-horned Lizard	*Phrynosoma douglassii*
Western Rattlesnake	*Crotalus viridis*

Amphibians

Tiger Salamander	*Ambystoma tigrinum*

Other vertebrate species addressed in the South Okanagan study that are not on the 1990 Red and Blue lists for B.C. include 3 mammals, 12 birds, 4 reptiles and 1 amphibian:

Mammals

Bighorn Sheep	*Ovis canadensis*
Mule Deer	*Odocoileus hemionus*
White-tailed Deer	*Odocoileus virginianus*

Birds

Eared Grebe	*Podiceps nigricollis*
Ferruginous Hawk	*Buteo regalis*
Golden Eagle	*Aquila chrysaetos*
Lark Sparrow	*Chondestes grammacus*
Long-eared Owl	*Asio otus*
Osprey	*Pandion haliaetus*
Sage Grouse	*Centrocercus urophasianus*
Sandhill Crane	*Grus canadensis*

Sharp-tailed Grouse	*Tympanuchus phasianellus*
Swainson's Hawk	*Buteo swainsoni*
Turkey Vulture	*Cathartes aura*
Yellow-headed Blackbird	*Xanthocephalus xanthocephalus*

Reptiles

Painted Turtle	*Chrysemys picta*
Rubber Boa	*Charina bottae*
Western Skink	*Eumeces skiltonianus*
Western Yellow-bellied Racer	*Coluber mormon*

Amphibians

Great Basin Spadefoot Toad	*Scaphiopus intermontanus*

There is a natural tendency for those concerned with maintaining biological diversity to concentrate on rare, threatened and endangered species. But biological diversity also depends on the more common species in an ecosystem. In the South Okanagan, this includes vertebrate forms such as the California Myotis (*Myotis californicus*), Snowshoe Hare (*Lepus americanus*), Deer Mouse (*Peromyscus maniculatus*), Mule Deer (*Odocoileus hemionus*), American Kestrel (*Falco sparverius*), Northern Flicker (*Colaptes auratus*), Rock Wren (*Salpinctes obsoletus*), Vesper Sparrow (*Pooecetes gramineus*), Western Meadowlark (*Sturnella neglecta*) and Pacific Treefrog (*Hyla regilla*) (see Nicholson et al. 1991). Often preservation of habitat for rare species will also protect more common species.

Invertebrate diversity in the South Okanagan is unquestionably greater than vertebrate diversity, but it is poorly documented. We do know that at least 125 invertebrate species found in the South Okanagan occur nowhere else in B.C. and 37 of these occur nowhere else in Canada (Cannings 1990).

The South Okanagan, as part of the Columbia Basin, also contains populations of three species of fish that are considered to be vulnerable: the Leopard Dace (*Rhinichthys falcatus*), Umatilla Dace (*Rhinichthys umatilla*) and Mottled Sculpin (*Cottus bairdi*) (Peden 1990).

Plant Diversity

Ten years ago Pojar (1980) considered the dry forest savanna ecosystems of Ponderosa Pine and bunchgrass as threatened, and riparian and lakeshore forests as very threatened. Currently, 33 rare ecosystem associations are rated as Priority 1 by SOCAP, including plant communities of sagebrush (*Artemisia tridentata* and *A. tripartita*),

Antelope-brush (*Purshia tridentata*), Ponderosa Pine (*Pinus ponderosa*), Black Cottonwood (*Populus balsamifera* ssp. *trichocarpa*) and Water Birch (*Betula occidentalis*). A number of wetland and dry lakebed ecosystems are also considered to be rare in the South Okanagan. The following is a list of the 67 rare, threatened or endangered plant species that occur in the South Okanagan study area (after Pavlick 1990 and Douglas 1991):

Shrub-Steppe

Nettle-leaved Giant-hyssop	*Agastache urticifolia*
Long-leaved Aster	*Aster ascendens*
Threadstalk Milk-vetch	*Astragalus filipes*
Dalles Milk-vetch	*Astragalus sclerocarpus*
Narrow-leaved Brickellia	*Brickellia oblongifolia* var. *oblongifolia*
Andean Evening-primrose	*Camissonia andina*
Palish Paintbrush	*Castilleja pallescens*
Obscure Cryptantha	*Cryptantha ambigua*
Cockscomb Cryptantha	*Cryptantha celosioides*
Fendler's Cryptantha	*Cryptantha fendleri*
Sierra Cryptantha	*Cryptantha nubigena*
Scarlet Gaura	*Gaura coccinea*
Hairstem Groundsmoke	*Gayophytum ramosissimum*
Shy Gilia	*Gilia sinuata*
Okanagan Stickseed	*Hackelia ciliata*
Whited's Halimolobos	*Halimolobos whitedii*
Rabbitbrush Goldenweed	*Haplopappus bloomeri*
Columbia Goldenweed	*Haplopappus carthamoides*
Small-flowered Ipomopsis	*Ipomopsis minutiflora*
Brandegee's Lomatium	*Lomatium brandegei*
Silvery Lupine	*Lupinus argenteus* var. *argenteus*
Velvet Lupine	*Lupinus leucophyllus*
Wyeth's Lupine	*Lupinus wyethii*
Great Basin Nemophila	*Nemophila breviflora*
Flat-topped Broomrape	*Orobanche corymbosa* ssp. *mutabilis*
Winged Combseed	*Pectocarya penicillata*
Showy Phlox	*Phlox speciosa*
Drummond's Campion	*Silene drummondii*
Munroe's Globe-mallow	*Sphaeralcea munroana*
Thick-leaved Thelypody	*Thelypodium laciniatum*
Many-flowered Thelypody	*Thelypodium milleflorum*

Disturbed Sites in Shrub-Steppe

Tufted Lovegrass	*Eragrostis pectinacea*

Dry Rocky Sites
Nuttall's Rockcress	*Arabis nuttallii*
Cushion Fleabane	*Erigeron poliospermus*
Richardson's Penstemon	*Penstemon richardsonii* var *richardsonii*
Okanagan Fameflower	*Talinum sediforme*

Dry Coniferous Forests
Lyall's Mariposa Lily	*Calochortus lyallii*
Strict Buckwheat	*Eriogonum strictum*
Northern Flaxflower	*Linanthus septentrionalis*

Meadows, Lakeshores, Streambanks, Vernal Pools
Swamp Onion	*Allium validum*
Tall Beggarticks	*Bidens vulgata*
Bearded Sedge	*Carex comosa*
Atkinson's Coreopsis	*Coreopsis atkinsoniana*
Awned Cyperus	*Cyperus aristatus*
Red-rooted Cyperus	*Cyperus erythrorhizos*
Purple Spike-rush	*Eleocharis atropurpurea*
Giant Helleborine	*Epipactis gigantea*
Small-flowered Hemicarpha	*Hemicarpha micrantha*
Torrey's Rush	*Juncus torreyi*
False-pimpernel	*Lindernia anagallidea*
Hairy Pepperwort	*Marsilea vestita*
Dwarf Montia	*Montia dichotoma*
Bristly Mousetail	*Myosurus aristatus*
Bushy Cinqefoil	*Potentilla paradoxa*
Mistassini Primrose	*Primula mistassinica*
Toothcup Meadow-foam	*Rotala ramosior*
Peach-leaf Willow	*Salix amygdaloides*
River Bulrush	*Scirpus fluviatilis*
Hairgrass Dropseed	*Sporobolus airoides*
Blue Vervain	*Verbena hastata*

Moist Saline Sites
Scarlet Ammannia	*Ammannia coccinea*
Annual Paintbrush	*Castilleja exilis*
Western Centaury	*Centaurium exaltatum*
Hutchinsia	*Hutchinsia procumbens*

High-Elevation Montane and Subalpine on Mt Kobau
Two-spiked Moonwort	*Botrychium paradoxum*
Blackened Sedge	*Carex epapillosa*
Kellogg's Knotweed	*Polygonum polygaloides* ssp. *kelloggii*

A Diversity of Threats

Throughout the world, environmental trends are strongly negative. Habitat loss, habitat fragmentation and global warming are affecting wildlife species everywhere. Conditions are no different in the South Okanagan, but because of the great diversity of species and limited geographic area in question, negative trends can be even more acute. In the South Okanagan intense competition occurs between humans and other species for limited space and resources. So far humans have been very successful: the population has increased and will continue to increase. Unfortunately, much of this growth is at the expense of other species.

Since Europeans first settled the Okanagan, four vertebrate species have been extirpated. The first was the Sage Grouse, followed soon after by the Sharp-tailed Grouse, both probably as a result of over-hunting. The next was the Burrowing Owl (last breeding record in 1970) and the most recent was the White-tailed Jackrabbit (last record in 1980). These last two probably disappeared as a result of over-hunting, habitat loss and habitat fragmentation. A fifth species, the Short-horned Lizard, may also be extirpated, since it has not been positively identified in B.C. since 1898. Population trends for many other species in the Okanagan are not good (e.g., Tiger Salamander, Sage Thrasher and Peregrine Falcon). Attempts to reintroduce extirpated species can be successful but are often very costly, as demonstrated in the Burrowing Owl recovery program in the South Okanagan. McNeely (1988) also points out the tremendous costs involved in replacing ecosystems, compared to the opportunity costs of protecting them. Local extirpation is not necessarily permanent, but it is expensive to recover from.

Threats to biological diversity are as many and varied as the species being threatened. Chief among them are urban and agricultural development and peoples' attitudes towards the species that are affected. While there are always occasional bursts of development activity, it is usually a gradual series of small impacts that adds up inexorably over time to major losses of habitat and populations. The pressure to expand municipal boundaries as the population grows and the community perception that nothing important exists in the "vacant" dry lands surrounding a city are major impediments to preventing and mitigating these losses. Increasing human populations also create habitat losses through increased demands for firewood (e.g., felling of snags), building materials (e.g., rock from talus slopes) and recreation (e.g., all-terrain vehicles, golf courses).

Agricultural development can be classified as intensive and non-intensive. Intensive agriculture completely changes the suitability of the land to support certain plants and animals. These lands should be considered as committed to agriculture, since any realistic attempt to rehabilitate them to natural conditions would be very expensive and would probably take generations. Examples include cultivation for forage crops and tree fruits. Insecticides and mowing may also directly kill many native species, so agriculturally developed areas are often deadly traps to any wildlife that are attracted to them.

Non-intensive agriculture includes livestock grazing and range improvement. Of the two, grazing may have less effect on a habitat's permanent ability to support a particular species, but it may result in a large depression in current suitability if the range condition or successional stage does not support native wildlife species. Aspen copses and areas around wetlands are particularly vulnerable because cattle concentrate there. Livestock trampling vegetation and amphibian breeding sites in wetlands is a particular problem; steeper terrain is not as severely affected.

Range improvement can affect native species since agronomic species, such as Crested Wheatgrass (*Agropyron pectiniforme*), are not preferred by most native wildlife and plantings of them greatly simplify an ecosystem. Range burning to remove sagebrush affects species that require the shrubs for cover or food. Some species, such as Sage Thrasher and Brewer's Sparrow, depend on sagebrush shrubs for nest sites. Control efforts using herbicides on Diffuse Knapweed (*Centaurea diffusa*) and other weeds also affect native non-target plant and animal species. Thus herbicides may simplify and destabilize an ecosystem, thereby reducing biological diversity.

Flood control and river channelling have impacts on both aquatic and riparian ecosystems. Extensive engineering projects to control water courses have greatly reduced available wetland and riparian habitats. Wetland areas, for example, are now reduced to 15% of their original area (Sarell 1990). Stocking lakes with fish species that prey or compete with amphibian larvae has reduced some native amphibian populations. Lake rehabilitations involve poisoning a lake with rotenone to remove species of fish that are undesirable for the sports fishery. This practice, used more in the past that at present, also reduced or extirpated native amphibian populations in the Okanagan.

Public attitudes can directly and indirectly affect many of the species that account for the rich biological diversity in the South Okanagan. Some species are killed on sight by many people (e.g., rattlesnakes, scorpions, spiders). Other species are widely judged to be unappealing or unattractive (e.g., snakes, bats, toads, lizards, insects, lower plants), so receive less attention and sympathy than more aesthetically pleasing species like deer, owls and showy wild flowers.

Many exotic species have been introduced to the South Okanagan, including Brown Bullhead (*Ictalurus nebulosus*), Largemouth Bass (*Micropterus salmoides*), Ring-necked Pheasant (*Phasianus colchicus*), California Quail (*Callipepla californica*), Gray Partridge (*Perdix perdix*), Chukar (*Alectoris chukar*), European Starling (*Sturnus vulgaris*), House Sparrow (*Passer domesticus*), Diffuse Knapweed (*Centaurea diffusa*), Purple Loosestrife (*Lythrum salicaria*) and Eurasian Water-milfoil (*Myriophyllum spicatum*). These can have deleterious effects on native animal and plant species when niche-overlap is sufficient to cause competition or when people manage for or against the exotic species.

The net result of all these impacts in the South Okanagan has been a tremendous alteration to natural ecosystems. An analysis by the Canadian Wildlife Service revealed that natural upland habitat remained in only 9% of the South Okanagan (Redpath 1990). The critical habitats that most endangered species depend on are now extremely limited. Of 22 species of vertebrates being considered for designation as endangered or threatened (1990 Provincial Red List), 12 occur in the South Okanagan. The South Okanagan has the dubious distinction of having more vertebrate species at risk than any other region in B.C.

Biophysical Habitat Inventory

Because habitat alteration by humans is the major cause of population declines of rare and endangered vertebrates (Cowan 1980), preserving biological diversity depends on establishing reserves where ecosystems can remain untouched or can be carefully controlled. The keys are planning and sound decision-making based on the best available information. The South Okanagan Conservation Strategy provides a plan for assembling the large amount of information needed to make sound resource-use decisions (Hlady, this volume). Ecosystem mapping is one very important part of this decision support system. Several groups and individuals are currently involved in species inventories, status reports, land evaluation and acquisition, public education, research, habitat protection and habitat management in the South Okanagan. Biophysical habitat mapping can be used to support all of these endeavours since it stratifies the landscape, based on permanent topographic and soil features, into ecosystem units of importance to various plant and animal species. Biophysical mapping provides a solid framework from which to plan and evaluate these conservation measures (Demarchi and Lea 1989).

Biophysical mapping is a multi-disciplinary, hierarchical classification system where ecosystems are defined and classified based on land-form boundaries that remain identifiable at all stages of secondary succession (Demarchi et al. 1990). At the highest level of the hierarchy are broad regional ecological units based on climate, physiography and broad plant and animal distributions, called Ecoregions (Demarchi 1988). The next level is Biogeoclimatic Ecosystem Classification as mapped by the Ministry of Forests (Pojar et al. 1987). Habitat Classes make up the lowest level of the classification hierarchy. These are relatively homogeneous units with respect to bedrock geology, terrain, land form, surficial materials, soils, climate, successional trends of vegetation and animal use that are separated into successional stages for mapping (Demarchi and Lea 1989).

Habitat Classes are divided into successional stages and described in a legend. The example in Table 1 for the Antelope-brush - Needle-and-Thread Grass Coarse Textured Soil Habitat Class, shows four different successional stages based on range condition. When identified on a map through air-photo interpretation and field work, these Habitat Classes are called Habitat Units (Fig. 2). The example in Figure 2 shows three different Habitat Classes: AN - Antelope-brush - Needle-and-Thread Grass; CLw - Cliff, warm aspect; and WS - Wheatgrass Sage. In the first year of field work, we identified over 100 habitat classes in the South Okanagan (Lea et al. 1991). In addition to stratification by Ecoregion, Biogeoclimatic Subzone and Habitat Class, each mapped Habitat Unit is rated into one of three nominal categories of tree or

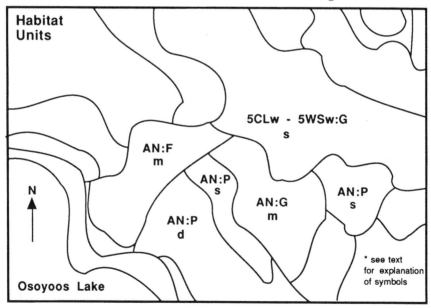

Figure 2. Example of biophysical habitat unit map.

Table 1. Example of a habitat class legend (after Lea et al. 1991).

A N	Antelope-brush - needle-and-thread grass coarse-textured soils	Biogeoclimatic Unit BGxh1*

Description: Occurs on coarse-textured soils of glaciofluvial materials, often with a capping of aeolian sands; occurs on flat to gently sloping areas; rapidly drained, droughty sites.

Map Symbol	Stage	Successional Trends	
		Dominants	**Associates**
AN:P	Poor	shrub/grassland of: antelope-brush cheatgrass needle-and-thread grass sand dropseed	big sagebrush rabbit-brush red three-awn Sandberg's bluegrass snow buckwheat prickly-pear cactus long-leaved phlox
AN:F	Fair	shrub/grassland of: antelope-brush needle-and-thread grass cheatgrass sand dropseed prickly-pear cactus	big sagebrush rabbit-brush red three-awn Sandberg's bluegrass snow buckwheat long-leaved phlox bluebunch wheatgrass
AN:G	Good	shrub/grassland of: antelope-brush needle-and-thread grass bluebunch wheatgrass	big sagebrush rabbit-brush red three-awn sand dropseed Sandberg's bluegrass snow buckwheat prickly-pear cactus
AN:E	Excellent	shrub/grassland of: antelope-brush bluebunch wheatgrass needle-and-thread grass	big sagebrush rabbit-brush red three-awn sand dropseed Sandberg's bluegrass

Comments: litter layers are important for maintaining soil moisture and for humus accumulation; cryptogams are valuable for preventing soil erosion

Shrub Density Classes: s - sparse, m - moderate, d - dense.

*BGxh1 = Very Dry Hot Bunchgrass Zone, Okanagan variant (Nicholson et al. 1991).

shrub density (see Table 1). The Habitat Unit polygons and data attached to each polygon are then digitized into a Geographic Information System (GIS), using PAMAP 2.2.

Habitat Unit mapping provides an ecological framework for assessing the land for its current suitability and potential capability to support the species of animals and plants that are considered to be a conservation priority in the South Okanagan (hereafter, we distinguish current "Suitability" from future "Capability"). To interpret all these habitat units for the 54 vertebrate species of concern, we will use algorithms or models to generate maps of habitat Suitability and Capability. We have chosen this method for a number of reasons, not the least of which is the size of the task. More importantly, since we know these models reflect only our current, imperfect knowledge of the habitat requirements of these species, the output maps should be easy to update as more and better information becomes available. Another advantage of using species algorithms to generate interpretive maps is that all the assumptions for generating the interpretive maps are explicit.

A habitat Suitability algorithm consists of ratings for each habitat class (High, Moderate, Low or Nil) and adjustments for polygon attributes such as shrub density and adjacency to other habitat requirements. An example for the Pallid Bat (Table 2) shows the rating for the three Habitat Classes shown in Figure 2 and adjustments for shrub density and distance from suitable roosting and breeding sites. Because Pallid Bats forage for large insects on the ground, and because insects require vegetation as forage, we have assumed that the Suitability for foraging increases as range condition improves in the AN habitat class. Also, the warm aspect cliff habitat is rated as High for breeding, roosting and hibernation. Applying this algorithm would generate an interpretive map of habitat Suitability based on the current range condition of the plant communities shown in Figure 3.

One of the great values of biophysical mapping is in delineating potential habitat at the successional stage that has optimum Capability.

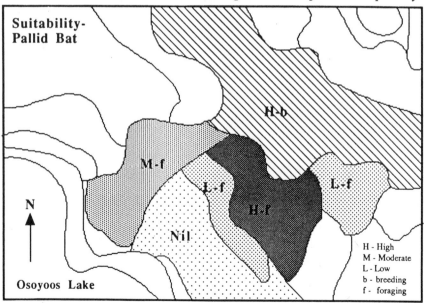

Figure 3. Example of habitat suitability map for Pallid Bat.

Table 2. Example of habitat suitability algorithm for Pallid Bat in three
habitat classes.

Ratings for Habitat Classes:

Habitat Class*	N/A	Poor	Fair	Good	Excellent	Habitat Use
		Range Condition (Successional Stage)				
AN	-	L	M	H	H	foraging
CLw	H	-	-	-	-	breeding, roosting
WSw	-	L	M	H	H	foraging

* - AN   = Antelope-brush needle-and-thread grass, coarse textured soils.
CLw = Cliff, warm aspect.
WSw = Bluebunch wheatgrass - selaginella shallow soil, warm aspect.

Adjustments for polygon attributes:

Assumption 1. Density of shrubs affect the suitability of individual polygons as foraging
habitat.

a. If shrub density is sparse or moderate, then no adjustment.

b. If shrub density is dense, then High and Moderate polygons are stepped down to
Low, and Low polygons are stepped down to Nil.

Assumption 2. Proximity to breeding and roosting sites affect the suitability of individual
polygons as foraging habitat.

a. If polygon is less than 5 km from any roosting site, then no adjustment.

b. If polygon is 5 to 10 km from a Moderate or High roosting site, then polygon rating
is stepped down one level (e.g. High becomes Moderate, Moderate becomes Low,
and Low becomes Nil.

c. If polygon is 5 to 10 km from a Low roosting site, then polygon is stepped down to
Nil.

d. If polygon is over 10 km from any roosting site, then polygon is stepped down to
Nil.

If range condition was allowed to go to Good or Excellent and shrub
density was reduced to sparse (perhaps through fire), then all the AN
habitats would be High even though the current Suitability may be
considerably lower. By comparing Suitability and Capability maps,
resource managers can find the best places to plan habitat
improvements. This mapping can also help integrate planning with
other resource agencies, identify areas of high priority for habitat

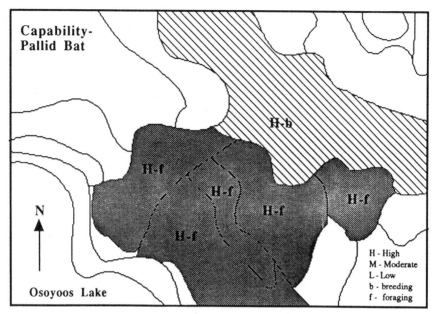

Capability-Pallid Bat

H-b

H-f

H-f

H-f

H-f

H-f

N

Osoyoos Lake

H - High
M - Moderate
L - Low
b - breeding
f - foraging

Figure 4. Example of habitat capability map for Pallid Bat.

protection and acquisition, identify areas of high priority for species re-introduction programs and help stratify the landscape for animal censuses. The GIS format also makes possible statistical area summaries, graphical display and modelling of land-use options.

These standard techniques of biophysical habitat mapping will be applied to the species of vertebrates of high conservation priority, indicated above. Interpretive maps for some species of insects and plants can also be generated when their habitat requirements are sufficiently well known. As more research is done, the base mapping can be interpreted for other species of concern and mapping for the 54 vertebrate species can be re-interpreted as new and better information becomes available.

Acknowledgements

Funding for this project is supplied by the Habitat Conservation Fund of B.C. Dennis Demarchi reviewed the draft manuscript. Ted Miller, Mike Fenger, Jacklyn Johnson and Liz Williams provided valuable editorial comments. We thank staff of the Royal B.C. Museum, the University of British Columbia and the Penticton Office of the Ministry of Environment for their continued support.

Literature Cited

B.C. Environment. 1990. Managing wildlife to 2001: a discussion paper. B.C. Ministry of Environment, Lands and Parks, Wildlife Branch, Victoria, BC.

Bunnell, F.L., and R.G. Williams. 1980. Subspecies and diversity — the spice of life or prophet of doom. *In* Threatened and endangered species and habitats in British Columbia and the Yukon, *edited by* R. Stace-Smith, L. Johns, and P. Joslin. B.C. Ministry of Environment, Victoria, BC. pp. 246-259.

Cannings, R.A., R.G. Cannings, and S. Cannings. 1987. Birds of the Okanagan Valley, British Columbia. Royal B.C. Museum, Victoria, BC.

Cannings, R.A. 1990. List of invertebrates of special interest. *In* South Okanagan conservation strategy 1990-1995, *edited by* D.A. Hlady. B.C. Ministry of Environment, Victoria, BC. Appendix 4.

Cowan, I. McT. 1980. The basis of endangerment. *In* Threatened and endangered species and habitats in British Columbia and the Yukon, *edited by* R. Stace-Smith, L. Johns, and P. Joslin. B.C. Ministry of Environment, Victoria, BC. pp. 3-20.

Demarchi, D.A. 1988. Ecoregions of British Columbia [MAP (1:2,000,000)]. B.C. Ministry of Environment, Victoria, BC.

Demarchi, D.A., D. Clark, and E.C. Lea. 1990. Ecological (biophysical) inventory, classification and mapping conducted by the British Columbia Ministry of Environment. *In* Symposium proceedings, wildlife forestry: a workshop on resource integration for wildlife and forestry managers, *edited by* A. Chambers. B.C. Ministry of Forests, FRDA Report 160, Prince George, BC. pp. 107-114.

Demarchi, D.A., and E.C. Lea. 1989. Biophysical habitat classification in British Columbia: an interdisciplinary approach to ecosystem evaluation. *In* Symposium proceedings, land classification based on vegetation: applications for resource management, *compiled by* D.E. Ferguson, P. Morgan, and F.D. Johnson. U.S. Dept. of Agric., Forest Service, Intermountain Research Station, Gen. Tech. Rep. INT-257, Ogden, UT. pp. 275-276.

Douglas, G.W. 1991. Rare, endangered and threatened native vascular plants of British Columbia. B.C. Ministry of Forests, Victoria, BC.

Hlady, D.A. 1990. South Okanagan conservation strategy 1990-1995. B.C. Ministry of Environment, Victoria, BC.

Lea, E.C., R.E. Maxwell, and W.L. Harper. 1991. Biophysical habitat units of the South Okanagan study area. Draft, Wildlife Working Report. B.C. Ministry of Environment, Victoria, BC.

Luttmerding, H.A., D.A. Demarchi, E.C. Lea, T. Vold, and D.V. Meidinger. 1990. Describing ecosystems in the field. B.C. Ministries of Environment and Forests, Victoria, BC. Second edition.

McNeely, J.A. 1988. Economics and biological diversity: developing and using economic incentives to conserve biological resources. Inter. Union for Cons. of Nature and Nat. Res., Gland, Switzerland.

Nicholson, A., E. Hamilton, W.L. Harper, and B.M. Wikeem. 1991. Bunchgrass Zone. *In* Ecosystems of British Columbia, *edited by* D. Meidinger and J. Pojar. B.C. Ministry of Forests, Victoria, BC. pp. 125-137.

Pavlick, L. 1990. Priority plant species. *In* South Okanagan conservation strategy, *edited by* D. Hlady. B.C. Ministry of Environment, Victoria, BC. Appendix 3.

Peden, A. 1990. Threatened fish diversity in fresh waters of B.C. Biological diversity and endangered biota: Part 1. Bioline (Assoc. Prof. Biologists of B.C.), 9(2): 25-31

Pojar, J. 1980. Threatened forest ecosystems of British Columbia. *In* Threatened and endangered species and habitats in British Columbia and the Yukon, *edited by* R. Stace-Smith, L. Johns, and P. Joslin. B.C. Ministry of Environment, Victoria, BC. pp. 28-39.

Pojar, J., K. Klinka, and D.V. Meidinger. 1987. Biogeoclimatic ecosystem classification in British Columbia. Forest Ecology and Management 22: 119-154.

Redpath, K. 1990. Identification of relatively undisturbed areas in the South Okanagan. Canadian Wildlife Service, Delta, BC.

Sarell, M. 1990. Survey of relatively natural wetlands in the South Okanagan. B.C. Ministry of Environment, Lands and Parks, Penticton, B.C.

Scudder, G.G.E. 1980. The Osoyoos-arid biotic area. *In* Threatened and endangered species and habitats in British Columbia and the Yukon, *edited by* R. Stace-Smith, L. Johns, and P. Joslin. B.C. Ministry of Environment, Victoria, BC. pp. 49-55.

Scudder, G.G.E. 1989. The adaptive significance of marginal populations: a general perspective. *In* Proceedings of the national workshop on effects of habitat alteration on salmonid stocks, *Edited by* C.D. Levings, L.B. Holtby, and M.A. Henderson. Can. Spec. Publ. Fish. Aquat. Sci. 105. pp. 180-185.

Strategies

Strategies for Protecting Biodiversity

Jim Walker

British Columbia Ministry of Environment, Lands and Parks, 810 Blanshard St., Victoria, British Columbia, Canada V8V 1X4.

Abstract

Traditional strategies for fish and wildlife agencies have been directed towards a few game species. The current wildlife strategy in British Columbia departs from this by classifying species "at risk" and those "not at risk". Existing mechanisms such as Parks Plan 90, the Forest Wilderness System Plan and the Old-growth Strategy can lead to an overall system to protect biological diversity. The term "biodiversity" needs to be carefully defined and its underlying values need to be made explicit — otherwise, effective conservation will be jeopardized. This province has international obligations to protect biodiversity. Success will be reflected in how well we maintain species in their natural ranges. Natural changes (e.g., through global warming) will profoundly alter what species occur in B.C. and where they exist. Preservation of biological diversity must consider dynamic, natural, ecological and evolutionary processes. There are enough current designations for land to protect biological diversity. Most emphasis for protection focuses on threats from poaching, logging, mining and development; there is less appreciation of threats by introduced species, diseases and hybridization. Restoring natural biological diversity by transplanting wildlife is becoming more difficult due to conflicts with agriculture, forestry, etc. Current land-conservation initiatives offer excellent opportunities to preserve biological diversity in B.C. An umbrella strategy that integrates them and gives direction to the protection of natural areas is needed.

In *Our Living Legacy: Proceedings of a Symposium on Biological Diversity*, edited by M.A. Fenger, E.H. Miller, J.A. Johnson and E.J.R. Williams, pp. 265-270. Victoria, B.C.: Royal British Columbia Museum.

Introduction

As the last participant in the session on strategies, I sense some advantages and some disadvantages. A disadvantage is that many points have already been covered; an advantage is that one can bring a perspective to what has already been presented. We have heard a great deal about ecological problems, the dynamics involved and the practical difficulties in preserving biological diversity. Perhaps the greatest contribution I can make is to outline what I think is possible in preserving and managing biological diversity, rather than what is desirable in an ideal world. So my paper will be a mixture of personal opinion, government policy, observation and experience.

Existing Strategies

What are the most important points in a strategy to protect biological diversity in British Columbia? Is there any sort of current strategy within the resource agencies of government that indicates some appreciation for the need to protect biological diversity per se? There are at least two positive indications.

Historically, the emphasis of most fish and wildlife agencies across North America has been to manage those species of animals that were most in demand by the public. This emphasis led to greatest effort being directed towards a few species, mainly game species. Over the past decade, priorities in wildlife management have changed, and now reflect concern for biological diversity. In "Managing Wildlife to 2001: A Discussion Paper", all species of higher vertebrates have been categorized as "at risk" or "not at risk" based on six criteria: (1) abundance, (2) distribution, (3) habitat integrity, (4) population trends, (5) reproductive potential and (6) national and international status. Species at risk are further subdivided into Red List (endangered and threatened) and Blue List (sensitive and vulnerable; i.e., showing significant downward trends in population or habitat). Species not at risk are subdivided into Yellow (requiring management emphasis due to public demand) and Green (requiring no management emphasis). (B.C. Ministry of Environment 1991; see Appendix D to Munro, this volume.)

This system is important both for what it implies about the thinking and direction of the Ministry of Environment and for its management usefulness. It reflects the thinking that we have to be concerned about everything, not just large, well known species that are hunted as game.

Mechanisms are being developed or are in place that provide the opportunity to preserve areas in order to protect biological diversity. Several strategies for protected areas are currently in place. I hope that documents and discussions such as Parks Plan 90, the Forest Wilderness System Plan and the Old-growth Strategy will lead to an improved system of areas of which the preservation of biological diversity will be a main goal (B.C Ministry of Forests 1989, 1990, 1992; B.C. Ministry of Parks 1990).

Other mechanisms exist within the B.C. government for protecting natural areas, such as ecological reserves and wildlife management areas. I believe that a comprehensive umbrella strategy is needed that will place these diverse mechanisms in context and bring an overall sense of direction to the establishment and protection of natural areas in B.C.

Other Components

It is critical to educate and inform the general public about biological diversity and its values. Many agencies and institutions, including the Ministry of Environment, share the responsibility for this. Such education is essential to gain understanding and support for legislation.

This province prides itself on being the most environmentally active of any in Canada. It is also probably one of the most environmentally misinformed. Sometimes one gets the impression that environmental concepts are embraced and championed not because they help to make sense of the natural world, but because they seem to be useful in achieving a preconceived agenda. We — you, I and the general public — must have a common understanding of what we mean by "biodiversity" and must appreciate the values that the concept embraces. Only then can we profitably debate applications, strategies, etc. Unless we do this, the term will come to be used as vaguely as "sustainable development" is now. Since there is no generally accepted definition of sustainable development, we have interpreted it to mean that which best suits our own ends. It has been explained as an affirmation of the forestry concept of sustained yield, as a rationale for wilderness by environmentalists and as support for unrestricted exploration by mining companies. "Biodiversity" is fast becoming the rallying cry for many environmental groups as well as the general public, yet the word's meanings and ramifications are largely undefined, not understood or misunderstood. Biodiversity must become the conceptual cornerstone in forestry, agriculture, fisheries, wildlife management and related areas. If it becomes just another "flag of convenience" to try to get watersheds reserved where parks and wilderness arguments have failed, then the concept will have a short life in this province.

The general public must also be made aware that the concepts and values of biodiversity are universal, not "homegrown" or specific to B.C.: we are not trying to complete a B.C. collection of ecological baseball cards — we are trying to preserve historic ranges and distributions of species. In addition, no species in B.C. has its southern boundary as the 49th parallel. So we have international obligations as well as practical concerns in protecting this province's biological diversity. Success must be judged by how well a species is maintained in its natural range, not just in its B.C. range.

An inward provincial focus will have destructive consequences for our native flora and fauna. An example will illustrate this point; it is one of the most personally discouraging incidents in my former position as Director of Wildlife. We arranged with the State of California to capture and transport 15 juvenile Bald Eagles, to re-introduce them into a pesticide-free area there. As B.C. holds about 25% of the world population of Bald Eagles, such an action would have had no negative impact on the species in the province. A local organization formed a group called "Save the Eagles" and successfully lobbied to stop the arrangement with the simplistic and vacuous argument that we should not be shipping B.C. eagles to California! We therefore had to cancel the project to the detriment of the species, and to the impoverishment of the natural values of California.

The preservation of biological diversity is a provincial, national, international and global concern, and any strategy should emphatically recognize this. We have Burrowing Owls and Sea Otters re-established in B.C. because of Washington and Alaska; Nevada and Idaho have some Mountain Goats and Bighorn Sheep because of B.C. That is how it should be. There is no room for picayune ecological politics in addressing issues of biological diversity.

I am concerned that the general public feels we should protect only what exists now. Yet distributions, movements and behaviour of species are dynamic, and change over time. Global warming, for example, will profoundly affect what species occur in B.C. and where. Other changes occur naturally for unknown reasons. Caribou are shifting north and will apparently continue to do so in spite of habitat alteration by humans. Clearly we must plan for change, as well as plan for current patterns of biological diversity. Will the Barred Owl naturally out-compete the Spotted Owl in B.C.? I don't know and I don't want to debate it here. I do want to stress that we have to sell the concept of preserving the dynamic, natural processes of ecology and evolution to succeed in conserving biological diversity. It is not enough to preserve a snapshot of plant and animal life in 1991, as snapshots can not capture the complex, dynamic and variable natural processes that are the essence of biodiversity. This is difficult to explain to the general public and even more difficult to explain to politicians.

Whatever we decide is needed as reserved or special areas to protect biological diversity, we should try to fit within the existing system of protected areas and designations. Already there are too many designations; any attempt to create special areas based only on values unique to biodiversity will add to the present load of parks/wilderness areas, ecological reserves, wildlife management areas, migratory bird sanctuaries, forest recreation areas, federal sanctuaries, etc. It will be met with resistance by the decision makers and will slow down the critical work of preserving biological diversity. The present confusing myriad of designations in this province must be reduced to an understandable few, so we all know of what we speak.

There is much talk about protecting natural diversity from logging, mining and other forms of development and from flagrant misuse such as poaching and illegal hunting. But there is also a great need to protect the province's natural biological diversity from immigrant species. I refer to the demand to introduce non-native wildlife into this province. Such introductions of non-native species have high potential for disease, displacement and hybridization. We do not want to turn this biological treasure house into another Texas or Florida, with their numerous exotic species.

The Ministry of Environment has been extremely vigilant about introductions at considerable cost to how the Ministry is perceived by professional colleagues and some components of society. Great pressure is still being exerted to import many forms of wildlife for game farming, and the ministry has restricted this activity to three species, all under tight control: the Fallow Deer, Bison and reindeer. Escapes of species on game farms are inevitable, and to maintain the unique species we have, we must provide strong legislation and regulations to minimize the impact of these escapes.

Our internal policies should also reflect this concern for native biological diversity. We have transplanted native animals around the province to support the policy of trying to restore animals to their historic ranges. This means that we will attempt to transplant animals to areas where historical evidence shows they occurred before, but not to other vacant areas. This policy differs from past ones; many species on the Queen Charlotte Islands, for example, were purposefully introduced by the old B.C. Game Department (Red Squirrel, Black-tailed Deer,

Muskrat, Beaver, Racoon, etc.). The ministry no longer does this; requests to introduce Bighorn Sheep and Cougars to the Queen Charlottes have been rejected. We have recognized the need to be still more rigorous, however, and have tried to only transplant animals between ecologically similar parts of the province.

It is becoming more and more difficult to restore natural patterns of biological diversity by transplanting animals. Legitimate concerns from forestry and agriculture have made it difficult to implement certain transplants. Such opposition is likely to continue as available habitat comes increasingly into contact with settlement and with agricultural enterprise.

In conclusion, I am quite positive about the prospects for successfully preserving biological diversity of native forms in this province:

- Because B.C. has great biological diversity and abundance, the eyes of the national and international communities are always on us and they will demand that we be vigilant. (If you don't believe this, try removing a few wolves.)
- There are active, aware, concerned communities in this province. This conference is testimony to that.
- We currently have several initiatives underway that, if co-ordinated correctly, will contribute greatly to the protection of this province's diverse resources; e.g., Forest Resources Commission, Round Table, Parks Plan 90, Old-growth Strategy, Forest Wilderness System Plan (B.C. Ministry of Forests 1989, 1990, 1992; B.C. Ministry of Parks 1990).

The challenge in implementing a strategy is the same as understanding the concept. The parts are all there. We just have to fit them together.

Literature Cited

B.C. Environment. 1991. Managing wildlife to the year 2001: a discussion paper. B.C. Ministry of Environment, Wildlife Branch, Victoria, BC.

B.C. Ministry of Forests. 1989. Managing wilderness in Provincial forests. A policy framework. B.C. Ministry of Forests, Victoria, BC.

B.C. Ministry of Forests. 1990. Wilderness for the '90s. Identifying one component of B.C.'s mosaic of protected areas (map). B.C. Ministry of Forests, Victoria, BC.

B.C. Ministry of Forests. 1992. Towards an old-growth strategy (public review draft). B.C. Ministry of Forests, Victoria, BC.

B.C. Ministry of Parks. 1990. Parks Plan 90. Areas of interest to B.C. parks (draft working map). B.C. Ministry of Parks, Victoria, BC.

The Role of Protected Areas in British Columbia

Bill Wareham

Earthlife Canada Foundation, P.O. Box 47105, #19 - 555 West
12th Avenue, Vancouver, British Columbia, Canada V5Z 4L8
and Ducks Unlimited, Box 1170, Station A, Surrey, British
Columbia, Canada V3S 4P6.

Abstract
The goal of Earthlife Canada Foundation is to promote sustainable development
through information, co-operation and suitable policies. The foundation's
activities include the B.C. Endangered Spaces Project, which advocates
protected-area planning based on ecological criteria and viability. The needs for
research and community-level involvement are recognized as essential for the
success of endangered spaces protection.

Introduction

Earthlife Canada Foundation

The main objective of Earthlife Canada Foundation is to promote
the concept of sustainable development and to provide information that
will help to form suitable sustainable development policies. The
organization strives to disperse new information and build liaisons
among industry, government and the public.

The B.C. Endangered Spaces Project

Earthlife Canada Foundation's primary efforts are to operate the
British Columbia Endangered Spaces Project, designed to bring together
individuals from different sectors who are interested in issues involving
protecting natural areas. Many people are familiar with the Endangered
Spaces Project, launched in 1989 by World Wildlife Fund Canada.
World Wildlife Fund Canada promotes ways of ensuring that a

In *Our Living Legacy: Proceedings of a Symposium on Biological Diversity*, edited by
M.A. Fenger, E.H. Miller, J.A. Johnson and E.J.R. Williams, pp. 271-274. Victoria,
B.C.: Royal British Columbia Museum.

representative portion of each of Canada's major ecosystems has some form of protection.

The B.C. Endangered Spaces Project will go beyond World Wildlife Fund's objectives by developing a comprehensive program to assess options for protected areas in B.C. The objective of the B.C. project is to bring different sectors together to analyse what biological diversity we have in B.C. and to assess options for protecting these values. This must, of course, be done within the context of a provincial planning system that includes an assessment of resources, economics, scarcity factors and land-use demands.

How Much is Enough?

Whether or not it is wise to protect only a representative portion of an ecosystem is questionable. As has been proven around the world, representative areas soon lose the diversity they were set up to protect. The analysis proposed by the B.C. Endangered Spaces project will help to determine how to ensure that ecosystems remain functional. Protected areas are only part of the equation in maintaining biological diversity. Progressive land use that considers ecosystem needs must be practiced throughout the land base.

Protected Areas and the Land-use Strategy

Protected areas play a valuable role in ensuring that critical biological values are not lost. Protected areas are just one piece of the land-use pie. For the B.C. Endangered Spaces Project protected areas are the existing designations we work with today. They include ecological reserves, provincial parks, national parks, Ministry of Forests wilderness designations, national wildlife areas and provincial wildlife management areas.

Protected Areas Designation Criteria

Existing designations currently protect tourism, recreation and conservation values. Provincial parks also protect special features and landscapes. If the concern, however, is to protect biological diversity, protected areas must be used to their fullest extent to contain natural biological values. It may be a mistake to allocate areas as protected if they do not have significant biological value. To use protected areas to protect critical biological values that would be lost if we left them unprotected may be wiser. Whether our existing parks and wilderness proposals capture biological values is questionable. The B.C. Endangered Spaces Project is designed to help answer that question.

How Much Protected Area do We Need?

Do we really need more protected areas in B.C.? If we hope to maintain the level of biological diversity we currently have then we probably do. As the global population approaches 14 billion in the next century we will undoubtedly have more people in B.C. This means increasing demands on our resources and more of our natural areas will be developed. As British Columbians we must decide whether we want to hold on to what we have and try to maintain as much biological diversity as possible. Opinion polls indicate that we do. The B.C. Endangered Spaces Project will help provide information that will allow us to assess the trade-offs and options we have today. The information will help everyone make better land-use decisions in the future.

Many of the presentations at this symposium have outlined values for protecting natural areas. The reasons for protection include research, species conservation and spiritual values; but most of all, protected areas serve as models for functioning ecosystem processes. This is something we must understand if we are to adequately manage the rest of the land base in a sustainable, productive manner. The productive capability of second- or third-growth forests and of agricultural land depends on our ability to better understand natural processes. Protected areas are vital in providing us with this information. We must, therefore, ensure that we protect ecosystems that are part of the working land base on which our economy depends. There is less value in protecting rock and ice and high-elevation lands because those are not areas in which we farm and harvest forests. We need reference points to understand how the land we depend upon works. Right now, although we have some understanding, we have much to learn.

Based on this argument, protected-area planning should be based on biological criteria. Important criteria to consider are the functions of ecosystems, the diversity of communities and species, and the degree of fragmentation.

How do We Complete a Protected Areas System?

How can we do all this? First we have to want to do it. It's like getting married. You know it will be a lot of work, but it will be worthwhile if you can get through the tough parts. The reward is a healthy relationship.

Some key items must be addressed if we hope to create a viable system of protected areas that help protect biological diversity. These include public support, the understanding of industry, a process that allows objective debate and resolution of contentious issues and the availability of objective information. The B.C. Endangered Spaces Project is designed to provide objective information by establishing broad sectorial advisory groups that set terms of reference for a variety of issues and approve research projects on them.

What Research is Required?

The research projects were identified in a two-day colloquium hosted by the Endangered Spaces Project in December 1990. More than 80 people from diverse backgrounds suggested over 30 research topics they felt should be addressed if we are to move ahead with a protected areas plan. Descriptions of these research topics are currently being drafted and will be distributed for comment. Earthlife Canada Foundation will work to raise funds to complete these projects.

Community Involvement

A key factor in this entire process is that communities must feel they are part of the process and that they have a say in what happens. This means a well organized and well funded effort such as regional round tables. Round tables would offer stakeholders the opportunity to meet and learn about the issues. They would create forums for reaching consensus on land use. Currently, many communities either feel left out of the process or don't want anything to do with it. This is likely because little information is being dispersed at the local level. We all have a job to do in informing the people of the province about the values at risk and the options for the future.

Finally, if maintaining biological diversity is to be a reality in this province, each of us must look at how we can contribute to better decision making and better education about biological diversity. The B.C. Endangered Spaces Project will try to provide the information required for everyone to better understand issues related to protecting natural areas. The project invites your participation and support.

Crown-land Planning in British Columbia: Managing for Multiple Use

Thomas Gunton

School of Natural Resource Management, Simon Fraser University, Burnaby, British Columbia, Canada V5A 1S6.

Abstract

One of the principal issues in British Columbia is the management of land for competing uses. While much of the discussion on land-use management deals with technical issues and disputes in specific watersheds, increasing attention is being paid to the nature of the decision-making process. This paper assesses the current process for crown-land management in B.C. Criteria are developed for evaluating decision-making structures and these criteria are used to evaluate existing processes. It concludes that the current process should be reformed to better take into account the diverse objectives of land-use planning. In particular, the opportunities for diverse stakeholder involvement in the decision-making process should be increased.

Introduction

One of the principal issues in British Columbia forestry is the management of land for competing uses. The issue is so important that the B.C. government appointed a special Wilderness Advisory Committee (WAC) in 1986 to investigate the conflict between wilderness and extractive uses (Wilderness Advisory Committee 1986). More recently, the B.C. Task Force on Environment and Economy (1989; hereafter, "Task Force") identified land-use management as the focus for much of the environmental concern in B.C. Concerns about land use were also prevalent in the recent provincial government hearings on Tree Farm Licences and in the hearings conducted by the Forest Resources Commission.

In *Our Living Legacy: Proceedings of a Symposium on Biological Diversity*, edited by M.A. Fenger, E.H. Miller, J.A. Johnson and E.J.R. Williams, pp. 275-293. Victoria, B.C.: Royal British Columbia Museum.

Much discussion about land management has dealt with conflict in specific watersheds such as the Stein and Carmanah, but has also focused on the nature of the decision-making process. The WAC and the Task Force, for example, both recommended that the land-use decision-making process be revised. Indeed, sound planning depends on a good decision-making process which ensures that diverse concerns for land use are adequately taken into account.

The purpose of this paper is to assess the current process for managing crown land in B.C. The paper will begin with a brief summary of the current crown-land planning process. This will be

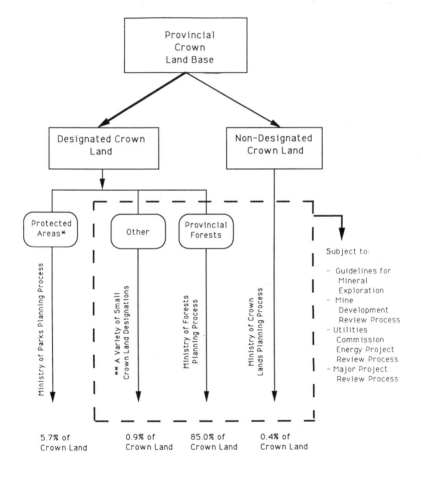

* Protected Areas include Parks and Ecological Reserves
** Includes Ministry of Highways, etc.

Figure 1. Schematic summary of crown-land planning processes in B.C.

followed by an evaluation and recommendations for improving the process to ensure that competing objectives are adequately incorporated into the decision-making process. The analysis is based on background papers prepared for the Forest Resources Commission and the B.C. Round Table on the Environment and the Economy. Those wishing more detail should consult Brenneis (1990), Duffy (1990), Gunton and Vertinsky (1990 a,b) and M'Gonigle et al. (1990).

The Crown-Land Planning Process

The crown-land planning process in B.C. is summarized in Fig. 1. The B.C. Ministry of Forests (MOF) is the dominant agency involved in crown-land planning. The legislative basis of MOF's jurisdiction is contained in the Ministry of Forests Act (British Columbia 1979a), which sets out the functions of the Ministry of Forests, and the Forest Act (British Columbia 1979b), which provides the legislative basis for managing crown land that is in provincial forests. Approximately 85% of the land base in B.C. is subject to the Forest Act.

The MOF planning process is summarized in Fig. 2 (B.C. Ministry of Forests 1983; Duffy 1990; Vance 1990). MOF is legally bound to prepare a ten-year resource analysis plus an annual report outlining management priorities for the next five years. These plans include silviculture programs to increase forest productivity. The MOF process calls for establishing regional priorities in consultation with government officials. To date, no comprehensive regional priorities have been established.

The next level of planning involves preparing resource-management plans for Timber Supply Areas (TSAs) and Tree Farm Licences (TFLs). Resource-management plans provide the broad, twenty-year management strategies for each TSA and TFL. These plans contain a summary of the major issues and objectives for TSAs and TFLs, quantitative targets for timber, range and recreation uses, and proposed management strategies for achieving these targets. MOF is obligated under the Ministry of Forests Act to incorporate timber as well as non-timber values into the planning process.

As that act states, MOF is required to "...plan the use of the forest and range resources of the Crown so that the production of timber and forage, harvesting of fisheries, wildlife, water, outdoor recreation and other natural resource values, are co-ordinated and integrated, in consultation and co-operation with other ministries and agencies of the Crown and with the private sector" (British Columbia 1979a: Section 4(c)).

277

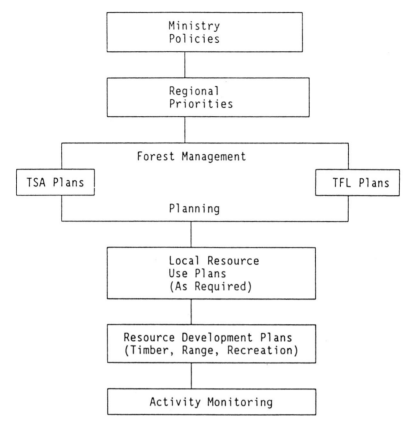

Figure 2. Schematic summary of the B.C. Ministry of Forests' forest planning framework.

The process for preparing the TSA plan is summarized in Fig. 3. Currently, preparation of TSA plans is managed by a steering committee normally comprising MOF staff and industry representatives. Sometimes representatives from other agencies are included. In response to a recommendation by the Forest Resources Commission, MOF decided to create district planning committees to prepare the plan (B. Sieffert pers. comm.). The district planning committees will be chaired by MOF staff with representatives from relevant government agencies. In future, the specific form of public involvement will be determined by the district planning committee. It is currently at the discretion of MOF staff. The policy is to invite public input when issues are first identified, and again on the draft TSA plan. Depending on the decisions of local staff, additional public involvement such as public meetings, workshops and task forces may be used. After public input is

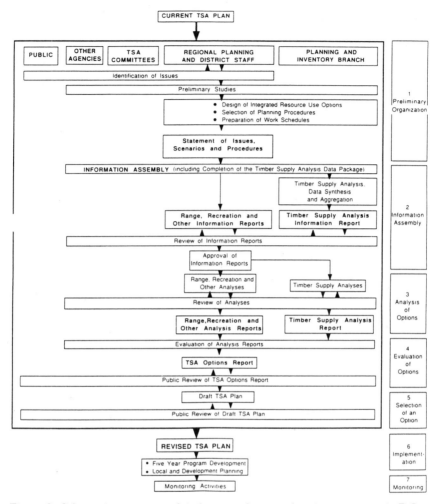

Figure 3. Schematic summary of timber-supply-area planning processes in B.C. (from Andrews and Higman 1986).

received, the draft TSA plan is submitted to B.C.'s Chief Forester, who has authority for its contents and approval.

The planning process for TFLs is similar to that for TSAs except that the TFL licence holder, not a steering committee, is responsible for preparing the TFL plan. The licence holder is obligated to notify the public when the initial issues and management objectives are identified and when the final draft TFL plan is ready to submit to the Chief Forester for approval. Input is also sought from relevant resource agencies. At each stage, the public has the right to review and comment on the documents.

The next level of planning, local resource-use planning, is undertaken for special resource problems within TSAs. At the discretion of MOF, a process may be initiated to prepare detailed integrated resource plans for these special issues, often under the guidance of special advisory committees representing the various stakeholders. Implementing recommendations for local resource use may be the responsibility of the district MOF manager, the Chief Forester or Cabinet, depending on the recommendation.

Resource-development plans represent the most detailed level of the planning process. They are specific plans that outline the planned resource development for the sections of each TSA and TFL expected to be harvested over the next five years. Locations of logging, logging guidelines and other specific proposals are included. These plans, prepared by MOF staff or forest companies at the district level, are updated each year. The plans are available for public review and comment. Implementing plans at the site level requires approval of pre-harvest silviculture prescriptions.

In recognition of the demand for non-timber uses, the MOF recently adopted a new wilderness policy that provides for setting aside special areas for wilderness preservation (B.C. Ministry of Forests 1988). The process consists of identifying potential wilderness areas in provincial forests. After input is sought from relevant stakeholders, recommendations are made to Cabinet, which has the authority to designate the proposed area as wilderness under section 5.1 of the Forest Act. The designated area is then managed for non-timber uses. Mineral development is permitted, subject to the Mine Development Review Process (Mine Development Steering Committee 1989). Currently, MOF is seeking public input on proposed wilderness areas.

The other major land-use agency is the Ministry of Parks (MOP), which currently manages 5.7% of the provincial land base. The goal of MOP is to identify representative diversity of natural environments in B.C. and to conserve them within an ecologically sustainable framework (British Columbia 1979c). To achieve this objective, MOP is responsible for proposing new additions to provincial parks and ecological reserves and for preparing management plans for parks and reserves under its jurisdiction. Stakeholder input is sought on new proposals that are then sent to Cabinet, which decides whether the proposed areas should be designated as parks. The Ecological Reserves program provides for setting aside unique ecological areas in special reserves designed to maintain their ecological integrity (British Columbia 1979d). MOP is currently seeking stakeholder input on a major new parks proposal (B.C. Ministry of Parks 1990a).

Evaluation of B.C. Forest Planning

Forest management is fundamentally a question of values. The Ministry of Forests Act explicitly recognizes this by acknowledging that forests must be used to support competing objectives. The key problem, therefore, is to ensure that these diverse objectives, ranging from timber harvesting to maintaining biological diversity, are adequately incorporated into the management process. The characteristics of an effective decision-making process for multiple-objective planning are summarized below. These characteristics will be used to evaluate the current crown-land planning process (see Fox 1976, Office of the Ombudsman 1988, Wondolleck 1988 and Gunton and Vertinsky 1990a,b).

Involvement of Stakeholders

Principle: Stakeholders, defined as those significantly affected by a decision, need to have the opportunity to participate effectively in decision making. In part, this is achieved by electing representatives through the formal political process. Given the long intervals between elections, the large number of issues and the sometimes tenuous links among elected representatives, their constituents and the bureaucracy, the formal electoral process can be improved by direct participation of stakeholders in individual decisions affecting their interest. Direct participation has several advantages. First, stakeholders can give decision makers important information on issues such as goals and objectives and impacts of alternative management strategies. Second, by allowing stakeholders to participate directly in the process, potential conflict and opposition may be reduced. This increases the likelihood of formulating and implementing a sustainable management strategy. Specifically, stakeholders should be told about the structure of the decision-making process, formally notified of the opportunities to participate, given the opportunity and resources to participate throughout the planning process, and informed of the final decision in a written report outlining the reasons for the decision and the response to various stakeholder interests.

Evaluation: Many stakeholders are poorly informed about the nature of the MOF land-planning process. Documents such as "Forest Planning Framework" (B.C. Ministry of Forests 1983) describe the existing planning process, but no single comprehensive document is available that describes the roles of various participants, the decision criteria employed and the processes followed. Even when stakeholders have been invited to participate in a special advisory committee

281

overseeing the preparation of a local resource plan, their roles have not been clear.

Notification is also inadequate. In the case of TSAs, forest industry representatives and relevant government agency representatives guide development of five-year TSA plans by being on the steering committee and the planning committee. MOF normally notifies the public at the beginning of the TSA planning process. The guiding principle, however, is that if stakeholders want to be involved in the process, they will let agencies know. Although some MOF staff actively solicit broad stakeholder involvement by direct contacts with potential participants, there is no explicit requirement for this. In the case of TFLs, the resource-management plan is prepared by the licensee and reviewed by relevant government agencies. Notifying other stakeholders, however, is not effective. TFL holders are obligated to notify the public when issues are identified at the beginning of the TFL planning process and when the TFL draft plan is completed. Stakeholders may be unaware of key decision points between initiating the planning process and completing the draft plan.

Opportunity for involvement could be improved. Forest industry stakeholders are fully involved in preparing TSA and TFL management plans. They are responsible for developing TFL plans and are involved in preparing them through their representation on TSA steering committees. Various government agencies are also represented in TSA steering and planning committees. They are not as actively involved in the TFL planning process. Both industry and government agencies are also represented in the more detailed local resource-use planning process, often guided by a special advisory committee.

Other stakeholders, however, are not given the same opportunity for extended involvement. In TFL planning, they are allowed to comment on issues identified at the beginning of the TFL planning process and on the draft TFL plan. They are not active participants as the plan develops. In TSA planning, they are provided with the opportunity to comment but are not normally directly involved through ongoing representation on TSA steering committees or on planning committees. At the discretion of MOF they may be invited to participate in special advisory committees that have been created to prepare a local resource plan. The public is encouraged to review and comment on more detailed resource-development plans outlining proposed harvesting guidelines for the areas of the TFLs and TSAs to be logged. There is no opportunity for extended involvement in preparing these detailed resource development plans.

Not all stakeholders have adequate resources to participate. While major stakeholders such as the forest industry and government agencies have resources to participate in the planning process others, such as representatives of environmental groups, do not have sufficient budgets to hire large staffs of experts or to sit for extended periods on planning committees. Overall, inequality among stakeholders' abilities to participate is significant.

Access to information is inadequate. In December 1989, MOF released a new policy explicitly providing for disclosing non-confidential information to stakeholders. Implementing this policy is constrained by insufficient resources to generate and provide the information requested. Some concern has also been expressed about the definition of confidential information. Unlike more formal hearings, stakeholders do not have a formal right to request that other stakeholders or government agencies undertake special studies to provide relevant information. The problem is most acute for TFLs where licensees are allowed to keep basic management information such as growth and yield data confidential on the grounds that it is proprietary corporate information. Also, Cabinet confidentiality prohibits releasing information that forms the basis of recommendations to Cabinet on key decisions such as the management options for the Carmanah. Critical information is therefore not subjected to a thorough scrutiny by diverse interests.

MOF decisions are also not available in a written form that clearly outlines the criteria used to make the decision. MOF policies stipulate that MOF staff must give written reasons for decisions to reject public recommendations, but these policies are not effectively implemented. Also, decisions by the Chief Forester on proposed TSA and TFL management plans or local resource-use plans are not justified in a comprehensive document outlining the criteria used, the reasons for the decision and responses to stakeholders' concerns. Consequently, it is difficult for stakeholders to understand why decisions were made and how their concerns were dealt with.

Explicit Decision-Making Framework

Principle: The decision-making process should be formally structured throughout legislation or regulations. This will ensure that the processes, information requirements, objectives and responsibilities of various participants are fully understood. Performance and accountability are easier to judge when the decision structure is explicit.

Evaluation: Although many aspects of the planning process, such as the requirement to prepare working plans, are contained in legislation, some planning procedures employed by MOF are in policy documents that lack legal force or certainty (e.g., the access-to-information policy, the public participation program, decision-making criteria and analytical requirements for plan preparation). Consequently, important aspects of the decision-making process are not formally structured in legislation or regulations, leading to a high degree of administrative discretion.

Methods of Analysis

Principle: The decision-making process should be based on sound analysis that provides decision makers with a comprehensible and comprehensive assessment of the implications of alternative management plans.

Evaluation: The analytical techniques used by MOF are reviewed by Gunton and Vertinsky (1990 a,b). The major deficiencies of the current analytical process are that data collection by relevant agencies is inadequately co-ordinated, issues are often identified on an ad hoc basis, and comprehensive data collection and analysis guidelines that reconcile assumptions and values in a systematic fashion are lacking.

Representative Decision Maker

Principle: Those responsible for making the final decision must be impartial, adequately represent the broad public interest, possess relevant expertise and be directly or indirectly accountable through the democratic process to those affected by the decision.

Evaluation: Ultimately, Cabinet is responsible for making land-management decisions. Cabinet and the legislature are designated representatives of the public interest. Although Cabinet retains its role in broad decisions involving classifying land into parks or provincial forests, much of the decision-making power has been delegated directly to civil servants such as the Chief Forester, who approves TSA and TFL management plans. The Chief Forester is required to consider timber and non-timber values. It is unlikely, however, that the Chief Forester can represent adequately all these various stakeholder interests. Although the interests of other agencies and stakeholders representing environment and wildlife concerns should be taken into account, MOF is a line agency with a dominant orientation. It is clearly unreasonable to expect its members to adequately represent all stakeholder interests.

Efficiency

Principle: The decision-making process should be efficient. Decisions should be reached in a timely manner at reasonable cost. Indefinite delays caused by excessive appeal processes or unclear delineation of decision-making authority should be avoided.

Evaluation: Overall, the preparation of resource-management plans is efficient. Plans are prepared in a timely and cost-effective manner following a strategic planning process. Broad strategic overviews are prepared for TSAs and TFLs and more detailed plans are prepared for specific issues and conflicts as required.

Equity

Principle: The decision-making process should ensure an equitable outcome by providing compensation to stakeholders left worse off by a decision. If the decision is in the public interest, the benefits should exceed costs, leaving adequate resources to compensate those made worse off.

Evaluation: The only formal provision for compensation is for holders of timber licences who experience a reduction of the annual allowable cut of more than 5%. Although special arrangements for compensation have been made in cases such as the creation of a national park on the Queen Charlotte Islands, no formal mechanism exists to compensate forest workers laid off as a result of management decisions or to compensate other user groups negatively affected by logging (e.g., trappers or fishermen). Compensation, however, is a complex issue beyond the parameters of this paper. Suffice it to say that the issue of compensation needs a comprehensive review.

Monitoring

Principle: Implementation of land-use decisions should be monitored to measure the success of the management strategy. The monitoring process should be clearly prescribed and results should be available to all interested parties.

Evaluation: In a recent Ombudsman report, concerns were raised about the adequacy of MOF monitoring (B.C. Auditor General 1989). Indeed, a recent study of logging on Lyell Island showed widespread infractions (Thompson 1988). Clearly, resources available for monitoring are inadequate.

Flexibility

Principle: The planning process should be flexible enough to accommodate diverse planning problems.

Evaluation: The MOF planning process is flexible. Unique planning processes can be created to accommodate diverse planning problems within a comprehensive planning framework. As the creation of the WAC and the Forest Resources Commission indicates, unique initiatives outside the MOF planning process have been undertaken.

Appeal

Principle: An appeal process should be available to allow stakeholders to challenge land-management decisions that contravene standard procedures or are inconsistent with prescribed guidelines or

285

objectives. The appeal process should be narrowly defined to ensure that the appeal is not used as a mechanism for simply transferring decision-making authority to an appeal tribunal or causing lengthy and costly delays. The appeal tribunal should have expertise in land-use management and administrative processes.

Evaluation: Provision has not been made for appealing planning decisions to any independent third party when process or policy has not been followed, other than the normal appeal to courts on questions of law and jurisdiction.

Mediation

Principle: Mediation and negotiation techniques should be used when possible to reduce conflict or achieve consensus. The process should encourage constructive conflict resolution rather than confrontation.

Evaluation: No adequate mechanism exists for training or providing expert mediators to assist in forest-land planning. Unfortunately, the decision-making process is characterized by confrontation rather than co-operation.

Reforming the Decision-Making Process

The crown-land planning process in B.C. has many attractive features. Nevertheless, as the previous section on evaluation indicates, the process can be improved. The following proposals are made to help achieve more effective decision making for land-use planning.

- The elements of the decision-making process, including the roles of participants, procedures, planning objectives, decision criteria and analytical requirements, should be put in a legislative framework providing a clear and consistent decision-making structure. For example, legislating policies on public participation and access to information would improve clarity and accountability. Care should be taken to ensure that the legislative framework provides sufficient administrative discretion and flexibility to adjust to different circumstances. Regulations approved by Cabinet, for example, are more flexible than legislation that must be passed by the legislature. The scope of discretion should be carefully specified in the legislative framework. Non-timber objectives should be explicitly incorporated into legislation to ensure that they have equal weight in management decisions.
- The principal mechanism for land management should be an officially approved TSA/TFL regional resource plan outlining

the allocation of land among alternative uses, zoning guidelines enforcing land-use allocation, management guidelines regulating activities in each zone and resource-development targets such as the annual allowable cut. The plan would be an extension of existing TSA/TFL resource-management plans. Regional boundaries, however, may have to be adjusted to facilitate effective regional planning. The approved plan would have both legal force and a clearly defined process for amendment. More detailed resource-development and operating plans would be prepared within the parameters of the TSA/TFL plans by MOF staff.

- The decision-making process should be outlined in a comprehensive and comprehensible document and an education program should be implemented to inform stakeholders of the nature of the process and their opportunity to participate.
- Notification requirements should be incorporated into legislation. Such requirements should ensure that people are notified about all important forest-land planning decisions including the various stages of TSA and TFL planning and of the various stages in any local resource-use plans and in resource-development plans. Notification requirements should be extended to cover non-forestry planning activities such as mining, energy development, parks and large development projects. Notification should outline the opportunities to participate and invite participants to become involved. Notification should also require that all major stakeholders expressing interest be separately contacted and informed of the process.
- Stakeholders should have a reasonable opportunity to participate in developing plans throughout all stages of the planning process. These rights should be incorporated into legislation so that stakeholders have sufficient power to be effective in negotiation and plan preparation.
- A B.C. Participation Fund should be created to help stakeholders that have a significant interest but insufficient funds to become involved in planning. The fund should be managed by a separate provincial agency independent of the planning process. Funds should be allocated according to explicit criteria developed in collaboration with provincial stakeholder groups. Criteria used by agencies that provide funding, such as the Ontario Energy Board, could help in developing funding guidelines to ensure reasonable and fair allocation of funding.
- Stakeholders need improved access to information. This could

be accomplished through comprehensive provincial freedom-of-information legislation and by providing sufficient resources to the MOF to respond to information requests.

- An independent, endowed forest-management research centre should be created to do independent research on behalf of stakeholders and the public. Such research would allow stakeholders to be more effective participants in the planning process.

- A provincial mediation centre should be created to provide independent expert mediators to assist planning committees in resolving disputes.

- Although efforts should be made to achieve a consensus, stakeholders still need a forum in which they can question assumptions and state positions prior to final plan approval. If sufficient interest is shown, proposed TSA/TFL plans should be reviewed prior to approval through a formal public hearing. The public hearing should allow stakeholders to present input and review and comment on the input of others including those submitting the plans. Because the more detailed resource development and operating plans would be prepared within the parameters of the TSA/TFL plans, they need not be subject to the same hearing process.

- TSA/TFL resource-plan approval should be provided in written decisions outlining the reason for the decision, the criteria employed, and the response to stakeholder input. All participants should receive the written decision and be informed of their opportunities to appeal it.

- A new Resource Appeal Board should be created to hear appeals from the public. The basis for appeals should be narrowly defined to include only questions that follow appropriate procedures in making plans and complying with officially approved planning guidelines. The Board would speed planning by keeping appeals out of the normal judicial process. Appeal procedures should not cause costly and lengthy delays.

- A common Geographical Information System and comprehensive guidelines for analysing land-use options should be prepared similar to those used by the U.S. Water Resources Council. The guidelines should be based on a multiple-accounts framework that considers all relevant factors. The guidelines should also summarize when different analytical methods should be employed. The process should first use inexpensive techniques such as biophysical overlays to provide initial assessments of potential land-use conflicts; more advanced techniques (e.g., multiple accounts) should only be necessary for complex conflicts (see Gunton and Vertinsky 1990a,b).

• Sufficient resources should be made available for monitoring TSA/TFL resource plans. Annual performance reviews should be published and stakeholders should have the right to challenge non-compliance through the Resources Appeal Board.

Reforming Decision-making Authority

The previous section contains proposals for improving the decision-making process. Regardless of the process employed, someone has to have the responsibility for making final decisions. Alternative structures for allocating decision-making authority are summarized below (see Gunton and Vertinksy 1990a,b).

Revised Status Quo

One option is to continue to allow final approval by the Chief Forester of TSA/TFL plans. Preparing TSA and TFL resource plans is the responsibility of the TSA district planner and the TFL licence holder as outlined in the previous section. The major differences from the current process are: (a) the Chief Forester or designate must conduct a formal public hearing prior to TSA/TFL plan approval if sufficient interest is shown and (b) the stakeholders will have more opportunity to be involved. The principal advantage of this option is that it improves the process while minimizing the potential disruption caused by major changes. The disadvantage is that it leaves the decision-making authority with MOF, an organization that does not adequately represent all stakeholder interests.

Regional Resource Management Committees

This option involves broader stakeholder representation on the steering committee, renamed the Regional Resource Management Committee (RRMC), and using it and an interagency planning committee for preparing TFL and TSA plans. Representation on the RRMC could be by election or appointment by the provincial government.

The principal advantage of this proposal over the revised status quo proposal is that it broadens the perspective in preparing TSA/TFL plans by including the public in the RRMCs. The disadvantage is that it may increase the uncertainty and difficulty of preparing TSA/TFL plans.

Independent Commission

Under this option, the TSA/TFL plans would still be prepared at the regional level under the direction of the RRMCs. The difference between this and the two previous proposals is that TSA/TFL resource plan approval would be transferred from the MOF to a new independent commission appointed by the Lieutenant Governor and representing various agency and stakeholder interests. Those appointed to the commission would have expertise in resource planning and administration. The commission would also have a professional staff, likely comprising transfers from existing government departments. Plans would be approved after hearings on proposed TSA/TFL plans in the respective regions. In cases where insufficient interest is shown, the commission could approve the plan without a hearing. Approving more detailed resource-development plans would still be the responsibility of the RRMCs or MOF staff at the regional level.

The hearing, following the normal hearing process used by agencies such as the B.C. Utilities Commission and the National Energy Board, would include full disclosure of all information used in making the decision, submitting information requests, submission of evidence and the final argument, the right to cross examine, and written decisions.

The principal advantage of this option is that the decision would be readied in public, would be an open forum in which all relevant stakeholders have an opportunity to participate and all stakeholders (including MOF) would be forced to justify their positions in public. The potential disadvantages are that the process could be more lengthy and costly.

Decentralized Authority

The difference between this proposal and the three summarized above is that the final decision for plan approval resides with the RRMC. The potential advantage of this is that final decision-making authority resides with those most affected by management decisions (i.e., those living in the region). The potential disadvantage is that legitimate provincial concerns may not be adequately taken into account.

Conclusions

B.C. has a comprehensive process for planning and managing the province's land base. The provincial government has made a concerted effort in recent years to improve this process by developing public participation programs and new integrated planning techniques. MOF, in particular, deserves credit for its efforts, especially considering its limited resources.

Despite these efforts, additional changes in the process should be considered. The process should be outlined more explicitly in a legislative framework that provides certainty and sufficient flexibility to adapt to changing circumstances. Objectives such as biological diversity, decision criteria and information requirements for decision-making should all be specified. An education program should be implemented to inform the public of the process. The public should be notified about proposed plans and invited to participate throughout plan development. Participants should be given sufficient resources and access to information.

Mediators should be provided to expedite planning and final TSA/TFL plan approval should be subject to a public hearing if warranted by sufficient interest. The decision makers should give detailed written decisions. Adequate resources should be available to monitor and enforce the recommendations. Comprehensive guidelines for analysing land-use options need to be prepared.

Regardless of the process used, the key question is: who has decision-making power? The above four options are presented to help resolve this. Each option suggests that a new Appeal Board be created, to hear challenges on questions of process and compliance. Each option also includes broader representation in the decision-making process. All four options are viable. The choice among them depends on the tradeoff between centralized and decentralized authority, and between due process and administrative efficiency.

The case for decentralized decision making is that regional residents are the people most immediately affected by regional planning decisions. Consequently, regional residents would have a stronger incentive than central authorities to manage regional resources properly. The case for more centralized decision making is that regional decisions have impacts on the province as a whole. Important provincial impacts could be ignored by regional residents.

All four options have elements of decentralization including regional steering committees, district planning committees and public participation. All four also have an element of centralization. Provincial government staff sit on district planning committees and regional steering committees and the process is conducted within the provincial legislative framework. Nevertheless, the options give different weights to provincial versus regional interests. The proposals also differ in the degree to which they involve the public in the decision-making process. The tradeoff is between the objectives of due process and the costs of more complex and uncertain decision making.

Clearly the current decision-making process needs to be improved. Failure to reform the system may lead to increased conflict that itself creates uncertainty and discourages economic activity. While the options in this paper involve significant changes, they are all viable means for improving forest-land planning to incorporate broader interests and objectives.

Acknowledgements

This paper is based on Gunton and Vertinsky (1990a). The support of the B.C. Forest Resource Commission and the B.C. Round Table on the Environment and the Economy is gratefully acknowledged.

Literature Cited

Auditor General. 1989. Annual Report, B.C. Prov. Gov., Victoria, BC.

Brenneis, K. 1990. Evaluation of public participation in the B.C. Ministry of Forests. B.C. Forest Resources Commission, Victoria, BC.

British Columbia. 1979a. Ministry of Forests Act. Revised Statutes of British Columbia, B.C. Prov. Gov., Victoria, BC.

British Columbia. 1979b. Forest Act. Revised Statutes of British Columbia, B.C. Prov. Gov., Victoria, BC.

British Columbia. 1979c. Parks Act. Revised Statutes of British Columbia, B.C. Prov. Gov., Victoria, BC.

British Columbia. 1979d. Ecological Reserves Act. Revised Statutes of British Columbia, B.C. Prov. Gov., Victoria, BC.

B.C. Ministry of Forests. 1983. Forest planning framework. B.C. Ministry of Forests, Victoria, BC.

B.C. Ministry of Forests. 1988. Managing wilderness in provincial forests: a proposal policy framework. B.C. Ministry of Forests, Victoria, BC.

B.C. Ministry of Parks. 1990a. Striking the balance: B.C. Parks policy. B.C. Ministry of Parks, Victoria, BC.

B.C. Office of the Ombudsman. 1988. Annual report. B.C. Office of the Ombudsman, Victoria, BC.

Duffy, D. 1990. A review of the British Columbia land allocation and management planning process. B.C. Forest Resources Commission, Victoria, BC.

Fox, I. 1976. Institutions for water management in a changing world. Natural Resources Journal 16: 743-758.

Gunton, T.I., and I. Vertinsky. 1990a. Reforming the decision making process for forest land planning in British Columbia. B.C. Round Table on the Environment and the Economy, Victoria, BC.

Gunton, T.I., and I Vertinsky. 1990b. Methods of analysis for forest land use allocation in British Columbia. B.C. Round Table on the Environment and the Economy, Victoria, BC.

M'Gonigle, M., T. Gunton, K. Brenneis, J. Campfens, J. Fox, W. Clissold-Hoyle, T. Makinen, J. Nyboer, C. Rankin, E. Van Osch and M. Valenius. 1990. Crown land use planning: a model for reform. *In* Law reform for sustainable development in British Columbia, *edited by* C. Sandborn. Canadian Bar Association, Vancouver, BC. pp. 35-45.

Mine Development Steering Committee. 1989. Mine development review process: an overview. B.C. Ministry of Energy, Mines and Petroleum Resources, Victoria, BC.

Task Force on the Environment and the Economy. 1989. Sustaining the land. B.C. Prov. Gov., Victoria, BC.

Thompson, T.M. 1988. Audit of Block 6 Number 39, Queen Charlotte Islands. B.C. Ministry of Forests, Victoria, BC.

Vance, J.E. 1990. Tree planning: a guide to public involvement in forest stewardship. B.C. Public Interest Advocacy Centre, Vancouver, BC.

Wilderness Advisory Committee. 1986. The wilderness mosaic. B.C. Prov. Gov., Victoria, BC.

Wondolleck, J.M. 1988. Public lands conflict and resolution: managing national forest disputes. Plenum Press, New York, NY.

Legislation for Biological Diversity: Directions for British Columbia

Colin Rankin

School of Resource and Environmental Management, Simon Fraser University, Burnaby, British Columbia, Canada V5A 1S6.

Abstract

Legal efforts to maintain biodiversity in four areas are briefly reviewed: general principles, conservation planning, protected areas and species protection. An effective strategy to maintain biodiversity will involve incentives and informal approaches as well as legislation and regulations. Examples are used to illustrate the possibilities for legislative reform.

Protected area legislation should include system planning and implementation provisions. Species protection measures include endangered species legislation and the legislated use of management indicator species. Land-use planning legislation should require clear conservation objectives in plans, an environmental assessment that addresses the impacts on biodiversity and independent review bodies such as an appeal board and a legislative commissioner.

In British Columbia, legislation is lacking in clear guiding principles for maintaining biodiversity. Criteria for weighing decisions between environmental integrity and economic development are absent from legislation, the public trust responsibilities of the government are not explicit and citizen standing is limited. While the province has a wide variety of protective designations, no formal mechanisms co-ordinate their application. Species protection measures are not explicit and protection of endangered species is left to ministerial discretion. Conservation planning takes place subsequent to development planning with little opportunity for public review or appeal. Recommendations to address these issues include the creation of a natural areas advisory council, appointing a legislative commissioner for the environment and legislation on endangered species, forest practice and land-use planning.

In *Our Living Legacy: Proceedings of a Symposium on Biological Diversity*, edited by M.A. Fenger, E.H. Miller, J.A. Johnson and E.J.R. Williams, pp. 295-306. Victoria, B.C.: Royal British Columbia Museum.

Introduction

I would like to offer a word of comfort to all the foresters at the Biodiversity Symposium. I have been working with a lot of lawyers lately and have found that if any profession is more maligned than foresters it must be lawyers. But I fear that lawyers, like foresters, are involved in the issues concerning biodiversity whether or not we want them to be.

Lee Doney talked about the challenge facing the Round Table, of translating the concept of biological diversity into public policy (see Doney, this volume). This is the focus of my presentation.

For a number of reasons, we have a tremendous opportunity in British Columbia:

1. Much biodiversity remains, as Jim Pojar has eloquently pointed out (see Pojar, this volume).
2. Over 90% of B.C. is Crown land. This is not the Florida that Larry Harris described (Harris, this volume), with isolated patches of public land surrounded by private land. We have clear legal avenues available to direct the use of public land should we think that maintaining biological diversity is important.
3. We can learn from the experiences of others.
4. With our legal environment we can change quickly — all it takes is a Minister's (or Chief Forester's) decision.

I will make a brief aside here about differences between Canadian and American approaches to legislation and roles of courts and the executive. If that difference can be summed up in two words, they are "may" and "shall". Americans are concerned with "legally enforceable" agreements, Canadians with ministerial and executive discretion. We have a fear of following the U.S. legal route into a snake pit of lawyers and delays — and this is a legitimate fear. We can and must provide direction and bounds to decision makers. We can do this through clear guiding principles and avenues for review of government actions.

In his talk, Jack Ward Thomas referred to the seesaw in the U.S. between the courts and land managers over National Forests and to the hazards of prescriptive legislation (see Ward Thomas, this volume). Land managers in B.C. have a lot of freedom to change right now, though if you are in the trenches you may feel otherwise. We must use our opportunity and explicitly address biological diversity in our actions or face far more restrictive and possibly misguided direction from legislators responding to public pressures such as those we ourselves felt yesterday at the demonstration in support of preserving the Walbran and Carmanah valleys.[1]

I will quickly review some of the issues discussed at this symposium, give some examples of how other jurisdictions have tried to deal with the issues and offer some suggestions for change in B.C. These suggestions are offered to stimulate discussion, not as definitive answers.

I will look at legal efforts in four areas: (1) general principles, (2) conservation planning, (3) protected areas and (4) species protection. As pointed out by other speakers, we cannot maintain biodiversity through protected areas or species-based approaches alone. We must also develop a co-ordinated approach involving all elements of diversity. The same holds true for legal efforts. An effective strategy to maintain biodiversity will involve incentives and informal approaches as well as legislation and regulations. For each of these areas, I will review (1) what we need, (2) some examples from other jurisdictions, (3) how well we are doing in B.C. and (4) how we can change. Realizing that this is an ambitious scope, I'll use examples to illustrate the possibilities for change rather than setting out a complete agenda for legislative reform.

General Principles

What Do We Need?

First, consider some of the assumptions of law. Law is concerned with rules: "Thou shalt not kill." "Thou shalt not steal." We deal with the environment by making individual elements of it property: "Thou shalt not kill my tree." If you "own" that tree, you have the "right" to chop it down. There are no conservation obligations associated with that "right". Or else an element of the environment belongs to "nobody". Anyone has the right to use air or the ocean.

The rules of law are concerned with relationships among people, not between nature and people. Destroying a component of biodiversity, whether a genotype, a species or an ecosystem, seldom directly affects a human interest protected by law. Thus an injury to nature will almost never be considered a reparable injury unless it can be shown that a person has suffered directly from the damage or will suffer from it. Furthermore, the damage suffered must be assessed in monetary terms, a requirement that can seldom be fulfilled. In the absence of a legally protected interest, nobody has the standing to bring the matter to court, as a general rule. So our present laws do not clearly set out our responsibilities toward maintaining the environment in which we live. These responsibilities can be made explicit in legislation.

One approach holds that we, as thinking beings who can consciously change the environment, have a responsibility for its stewardship. This "stewardship ethic" is not new. Aldo Leopold, for one, outlined a land ethic in the 1930s. Leopold (1949: 224-5) felt that we must examine each land-use question "...in terms of what is ethically and aesthetically right, as well as what is economically expedient. A thing is right when it tends to preserve the integrity, stability and beauty of the biotic community. It is wrong when it tends to do otherwise."

An extension of the stewardship ethic is the "public trust doctrine", which holds that the present generation of humans has an obligation to future generations to preserve the diversity and quality of the world's life-sustaining environment. This requires each generation to give the planet to future generations in no worse condition than it received it and to provide future generations with equitable access to its resources and benefits. Each generation is thus both a trustee for the planet with obligations to care for it and a beneficiary with rights to use it.

Where is This Happening?

I will cite examples from New Zealand, the Yukon and Washington state. In New Zealand, a major reform of resource-management legislation is being proposed, concurrent with regional government reform. The overriding purpose of the comprehensive Resource Management Bill is "sustainable management" underpinned by a number of principles decision makers must consider.[2] These principles include biological diversity, the needs of future generations, the balance between public and private interests, maintaining natural character and features, and the ways in which protection, use and development contribute to community well-being. This combination of purpose and principles gives a clear basis for interpreting specific legislative measures or management actions.

The proposed Yukon Environment Act includes explicit public trust responsibilities of the government and rights of standing for those who wish to protect the environment.[3] A Sustainable Forestry Bill, presently before the State of Washington Senate, makes explicit the principles by which both private and public lands should be managed. The bill declares that it is in the public interest for the state to conserve the productivity of forest land and its associated resources by maintaining the integrity of biological processes while producing commodities and other services.[4]

How are We Doing in B.C.?

Natural resources are generally referred to in B.C. legislation in terms of the economic and social benefits they convey. There are few references to the need to maintain environmental integrity or natural processes for their own sake or for the sake of future generations. Criteria for weighing decisions between environmental integrity and human actions are generally absent from legislation. Such decisions are left to the discretion of Provincial Cabinet, individual Ministers or government agencies with little legal guidance.

Standing in the courts to take action on environmental protection is generally limited to those who are able to show (a) a greater or different right, harm or interest in the matter than any other person, or (b) a pecuniary or proprietary right or interest in the subject of the proceeding. The B.C. Environment and Land Use Act, for instance, directs a Cabinet committee to "maximize beneficial land use, and minimize... despoilation of the environment occasioned by that use."[5] The question of how much minimizing can be accommodated by the environment is left open to Cabinet interpretation. In another example, the Ministry of Forests Act directs the agency to "manage, protect and conserve the forest and range resources of the Crown, having regard to the immediate and long term economic and social benefits they may confer on the Province."[6] Using "forest and range resources" for various "natural resource values" is to be "co-ordinated and integrated".[7]

How Can We Change?

The maintenance of biological diversity should be made an explicit responsibility in B.C. resource-management legislation. Citizens of B.C. should have the right to take action in court to protect the environment.

Amending provisions to resource acts could read:

1. It is in the public interest of the province to maintain the integrity of biological and ecological processes for the benefit of present and future generations.
2. Citizens of B.C. have a right to a healthy and sustainable environment, a right to the protection of ecosystem and biological diversity and a right to expect reasonable action by the government of B.C. within the limits of its authority to protect or recover, or to require protection or recovery of, environmental quality.
3. Notwithstanding any other Act, any resident of B.C. may institute a private prosecution against any offence under this Act or any regulation under this Act.

Conservation Planning

What Do We Need?

Ensuring that functioning ecosystems remain for the future is the most challenging issue facing conservation biologists, land managers and legislators. The role of conservation planning in development and management actions is of critical importance and we in B.C. have much to learn in this area.

What do we need? First, we need to know what we have. We need to identify areas rich in biodiversity before they're gone. Then we need to know what kind of actions we must take, how the land base of B.C. can be managed to maintain biodiversity. We have been told that this can be done: the challenge is to apply these lessons to the B.C. situation.

In terms of legislation then, two things are necessary. The first is an effective conservation-planning toolbox that includes the means of determining and designating protected areas and regulations governing actions in areas outside of reserves. The second is an environmental assessment process that accounts for the impacts on biodiversity of development proposals and management actions.

Where is This Happening?

New Zealand has attempted to place conservation planning on an equal footing with development proposals by separating the conservation and development roles of government agencies such as the New Zealand Forest Service. Responsibility for balancing differing agency plans is given to an independent agency established for this purpose. As well, a Parliamentary Commissioner for the Environment has been appointed to audit the government's environmental record and to report to Parliament on the adequacy and application of environmental laws.

In New South Wales, Australia, an independent Land and Environment Court with broad access by the public has been created to ensure adherence to environmental laws.

The major public land-management agencies in the United States, the U.S. Forest Service and the Bureau of Land Management, are given explicit direction for conservation planning under several acts.[8] The National Forest Management Act, for example, contains specific directives to the U.S. Forest Service to maintain biodiversity at genetic, species and ecosystem levels. First, the act requires the Forest Service to maintain viable populations of existing native and desirable non-native species. Second, the act directs the Forest Service to maintain the diversity of tree species and to protect the resources and habitats upon which vertebrate populations and endangered species depend. Third, the act directs the Forest Service to preserve and enhance the diversity of plant and animal communities within each management area.

The National Environmental Policy Act (NEPA)[9] is probably the single most powerful U.S. federal law governing environmental quality. What is important to note about NEPA is that it is procedural and directs a process rather than a particular decision. The act gives government agencies and citizens a well defined opportunity to respond to development proposals. The assessment procedures of NEPA can also be amended to explicitly address specific environmental concerns. A National Biodiversity Act, presently being considered by the U.S. House of Representatives, proposes an amendment to NEPA that would require environmental impact statements to include impacts on biological diversity.[10]

How are We Doing in B.C.?

How we are doing depends a lot on who you talk to. Public concern regarding resource planning in B.C. is considerable. The Ministry of Forests, with legal responsibility for both timber planning and integrated resource management, is often perceived as having conflicting interests in balanced resource planning.

The government and various agencies have been attempting to respond to public concern with several initiatives and review bodies, such as the Forest Resources Commission, the Round Table on Environment and Economy, the Old-growth Strategy and others. Changes to assessment procedures, such as the Major Project Review Process, are also being considered by the province.

How Can We Change?

Two suggestions indicate some possibilities for change. First, we can make goals and responsibilities for conservation planning explicit in legislation and policy. The Ministry of Forests, for example, could be held responsible for maintaining biological diversity at regional, watershed and stand scales in Timber Supply Area plans.

Second, we need an independent review authority to ensure that government agencies are acting to meet conservation goals. The first task of such an authority could be to audit government legislation and policies for impacts on biological diversity. The New Zealand Parliamentary Commissioner for the Environment provides one model that could be followed.

Protected Areas

What Do We Need?

First we must recognize the limitations of protected areas in maintaining biodiversity. Biodiversity cannot be maintained through reserves alone. A reserve system designed with biodiversity in mind is, however, critical to maintaining that diversity.

Some reserves are needed to serve as representative protected areas, some as landscape links and some to protect special species or areas. We must evaluate the extent of biodiversity in our present parks and reserves, identifying gaps in the system and the means by which critical areas should be protected. Subsequent to designation, protected areas must be managed for their role in contributing to overall diversity and ecosystem values, not as if they were preserves functioning as isolated islands with distinct biological boundaries.

Where is This Happening?

Probably the most comprehensive review of areas protected for biological diversity is the strategy developed for the (in)famous Spotted Owl (*Strix occidentalis*) in the United States. The strategy provides an example of how the needs of a particular species, a system of protected areas and directed management actions can be combined in a regional approach to maintain critical elements of biological diversity.

Another example of planning and designating a broad range of protected areas in a co-ordinated manner is contained in the Alaska National Lands Interest Conservation Act.[11] The 1980 legislation gave protection to almost 42 million hectares of land in Alaska, establishing or adding to 60 protective units including 5 national parks, 10 national preserves, 2 national monuments, 20 national wildlife refuges and 25 wild and scenic rivers. The extensive planning process leading to the designations recognized the cultural and land ownership needs of Native Peoples, sought to provide for migratory wildlife and ecosystem integrity and considered international factors such as migratory species.

How Are We Doing in B.C.?

We have an abundance of protected area designations in B.C.: national and provincial parks, wilderness areas, ecological reserves, wildlife management areas and more (see Walker, this volume). There are, however, few formal mechanisms to co-ordinate the use and management of the various types of reserves. B.C. was the first province in Canada to pass an ecological reserve act in 1971, but the ecoreserve program is limited in resources and designation powers.

A major initiative to identify and designate parks and wilderness areas in the province has been undertaken through the "Parks and Wilderness for the '90s" program. Parks and wilderness areas will likely form the bulk of large protected areas in the province, hence this is a very important step forward in determining the conservation role for protected areas in the province.

How Can We Change?

First, we must act on our good intentions. We have lots of protected area designations but varying degrees of success in using them. We need mechanisms that force the bureaucracy to act on legislative intentions.

1. Create a Natural Areas Advisory Council to review and recommend to the government areas for protection under various categories. This would help to ensure that critical areas are not overlooked by government agencies and help co-ordinate the use of various protective designations. Such a committee was recommended by the Wilderness Advisory Committee in its report of 1986 and could be modelled on the Committee on the Status of Endangered Wildlife in Canada.

2. Identified areas for protection have frequently been lost to other uses before protection is implemented. Clear implementation plans and reports on progress toward achieving conservation goals are needed. U.S. legislation is usually very explicit about this. Agencies are required to report progress to legislative bodies at specific times. An annual report to the legislature on biological diversity and the state of protected areas would provide a public review and impetus to develop conservation systems. Such a report would best be prepared by an independent body such as a Natural Areas Advisory Council or a Legislative Commissioner for the Environment.

Second, we must develop mechanisms that consider the role of protected areas in the larger landscape.

1. Develop framework agreements outlining conservation goals for managing protected areas and adjacent lands and waters. Clear goals and working relationships with other land users are critical to effectively protected areas maintaining biological diversity. The Greater Yellowstone Co-ordinating Committee in the United States and the Fraser River Estuary Management Program in B.C. provide examples.

2. Support and use the Biosphere Reserve designation for appropriate protected areas and adjacent lands. This international designation provides a means of integrating a variety of reserve types and land uses in meeting collective goals.

Protecting Species

What Do We Need?

As with protected areas, we must acknowledge the limitations of species-based approaches. Relying on maintaining species populations alone is not sufficient to maintain biological diversity. Individual species or communities of species, however, can act as indicators and provide direction in a regional strategy aimed at maintaining functioning ecosystems.

We need to address the protection of the diverse range of species in functioning ecosystems. Our management actions in the past have often focused on "glamour species" such as endangered animals and large "shootable" mammals. Legislation aimed at protecting species often does not provide for protecting habitat even though fragmentation of habitat is the prime cause of species extinctions.

Where Is This Happening?

The most explicit species protection provisions are contained in U.S. legislation. Regulations require the U.S. Forest Service to maintain viable populations of existing native vertebrate species in planning areas. Furthermore, the agency must monitor population trends of "management indicator species" to determine the relationships between population changes and modifications to habitat.[12] The agency is presently moving beyond using individual species to using communities and guilds of species as management indicators.

The U.S. Endangered Species Act[13] contains strong habitat protection provisions and regulatory requirements for listing species. Legal responsibility towards endangered species for government agencies extends to recovery plans, with the goal of bringing populations to the point where protection under the act is no longer necessary. Proposals for amending the act presently call for protecting endangered ecosystems in addition to endangered species.

How Are We Doing in B.C.?

The Wildlife Act[14] gives the Ministry of Environment the mandate to protect and conserve wildlife species in the province. Plants, invertebrates and fish are not considered as wildlife under the act. The habitat on which wildlife lives is not under primary control of the ministry and provisions for designating areas as Wildlife Management Areas are not used extensively.

The Ministry of Forests is presently grappling with the management of old-growth species and other aspects of managing forests for biological diversity. The Ministry of Environment is attempting to shift from its traditional emphasis toward game species to managing for biodiversity. A proposal for Endangered Plant legislation is being drafted for tabling in the legislature.

How Can We Change?

    1. Resource-management legislation should include the following provisions to provide guidance in directing actions affecting

species: (a) the genetic viability of species shall not be compromised by actions taken under this act and (b) the population levels of all species, wild and domesticated, must be at least sufficient for their survival, and to this end necessary natural habitats shall be safeguarded.
2. Requirements for identifying, managing and monitoring management indicator species should be set out in legislation affecting natural habitats. Such indicator species should include (a) threatened and endangered species, (b) species sensitive to intended management practices, (c) game and commercial species, (d) non-game species of special interest and (e) ecological indicator species.
3. Endangered species legislation in the province needs strengthening, particularly for plants and invertebrates. A draft act for B.C. has been drawn up by Calvin Sandborn of the West Coast Environmental Law Association.
4. Establish a Real Estate Transfer tax on the subdivision or conversion of land for a fund to purchase critical habitat similar to the existing Habitat Conservation Fund. This mechanism is fair and well suited to support natural areas and natural diversity protection. Transfer taxes require that real estate development, one of the major sources of natural area destruction, pay its true social costs. It has gained wide acceptance in other jurisdictions, with the majority of states in the U.S. having some form of transfer tax.

Conclusion

On his deathbed, W.C. Fields, a lifelong agnostic, was discovered reading a Bible. "Just looking for a loophole," he explained. In a similar way, we can acknowledge the limitations of the law. We need explicit measures to maintain biological diversity in B.C. legislation. But without widespread understanding and acceptance of the need to change our ways, we will remain to the end, searching for loopholes.

Acknowledgements

I wish to acknowledge the support of the Law Foundation of B.C. for this research.

Notes:

1. Editors' note: on 1 March 1991, a public demonstration was held in Victoria to protest logging in the Carmanah and Walbran valleys on Vancouver Island.
2. The Resource Management Bill was before the New Zealand Parliament when there was a change in government in 1990. The legislation has been tabled pending review by the new National government. (Editors' note: this legislation was subsequently passed).
3. Government of Yukon, Discussion Draft of the Yukon Environment Act. (December 1990), Part 1 (4)(a): "Every person resident in the Yukon has the right to protect the environment and the public trust from adverse effects and/or from the release of contaminants by commencing an action in the Supreme Court against any person causing an adverse effect on releasing any contaminant into the environment."
4. State of Washington, Senate Bill 5616, 52nd Legislature, 1991, s. 101(1)(a): "Sustainable forestry requires: A long-term commitment to stewardship informed by advances in scientific knowledge and responsive to changing human values."
5. Revised Statutes of B.C. 1979, c.272, s.4(b)
6. Revised Statutes of B.C. 1979, c.101, s.3(b)
7. Ibid., s.4(c).
8. The National Forest Management Act, 16 U.S.C. 1976, ss. 1600-1687 authorizing amendments to the Forest and Rangeland Renewable Resources Planning Act of 1974 directs the U.S. Forest Service in Managing forest lands. Multiple-use planning provisions for the second largest U.S. land management agency, are established under the Federal Land Policy and Management Act, 43 U.S.C. 1976, ss. 1701-1784.
9. 42 U.S.C. 1969, ss. 4321-4370.
10. United States House of Representatives, 101 U.S.C., Bill 1268, National Biological Diversity Conservation and Environmental Research Act 1989.
11. 16 U.S. Congress 1980, ss. 3101 et seq.
12. National Forest Management Act Regulations, 36 CFR 219.9.
13. 16 U.S. Congress. 1973, ss. 1531-1543.
14. Statutes of B.C. 1982, c.57, index chap. 433.1.

Literature Cited

Leopold, A. 1949. A Sand County almanac and sketches here and there. Oxford Univ. Press, London, England.

South Okanagan Conservation Strategy

Debbi A. Hlady

B.C. Ministry of Environment, Lands and Parks, Integrated
Management Branch, 780 Blanshard St., Victoria, British
Columbia, Canada V8V 1X4.

Abstract

The South Okanagan Conservation Strategy is a five-year management plan that
was created to co-ordinate the efforts of two major projects: the Nature Trust of
British Columbia's South Okanagan Critical Areas Program and the B.C.
Environment's Okanagan Endangered Spaces Habitat Conservation Fund
Project.

Seven key objectives and associated conservation actions are recommended.
These focus on the identification and protection of threatened habitats and
species, encourage sustainable use of the land, and promote interagency and
public involvement. Integrated management opportunities and operational
activities are also described under a co-operative framework in which key
agencies are identified.

The Integrated Management Branch is a new branch recently
developed in the British Columbia Ministry of Environment to help co-
ordinate resource management and planning, both internal and external
to the ministry.

I will describe the development of the South Okanagan
Conservation Strategy, published in January 1991. The South Okanagan
Conservation Strategy is a local action plan created to co-ordinate the
efforts of two major projects: the Nature Trust of B.C.'s South Okanagan
Critical Areas Program and the Ministry of Environment's Okanagan

In *Our Living Legacy: Proceedings of a Symposium on Biological Diversity*, edited by
M.A. Fenger, E.H. Miller, J.A. Johnson and E.J.R. Williams, pp. 307-310. Victoria,
B.C.: Royal British Columbia Museum.

Endangered Spaces Habitat Conservation Fund Project. The Ministry's Habitat Conservation Fund is generated by a surcharge on consumptive use of wildlife: angling, hunting, guiding and trapping. We need to expand the fund or create a new one to finance biodiversity initiatives such as the Okanagan Endangered Spaces Project.

The South Okanagan Conservation Strategy has been developed with other agencies whose mandates emphasize maintaining biodiversity, primarily the Royal B.C. Museum, the University of British Columbia and the Canadian Wildlife Service. Together we have identified and set priorities on management activities to conserve natural ecosystems and associated flora and fauna in the South Okanagan and Similkameen valleys over a five-year period.

This project was initiated after the The Nature Trust of B.C. organized a set of workshops to assemble provincial experts to identify sites of ecological importance and to indicate those species of mammals, birds, reptiles, amphibians, invertebrates and vascular plants that are at risk from development. This information was collected to assist in setting priorities to acquire lands for conservation.

Seven key objectives and associated management activities have been identified in the strategy. These activities are described further under an implementation strategy, which sets out a co-operative framework in which 20 agencies are listed.

Objective 1 is to identify and protect the remaining threatened habitats (primarily sagebrush and antelope-brush grasslands) and wetlands (mainly small marshes and riparian thickets) by preparing and implementing habitat securement, management and restoration plans, and promoting research on conserving natural habitats. This objective set the stage for biophysical mapping (see Harper, this volume). A habitat management/protection plan has been completed for the White Lake area and another is underway for the Kilpoola – Mt Kobau area. These plans were developed to avoid or resolve land-use conflicts among agencies and landowners so that management for biodiversity can occur.

Objective 2 is to identify species of flora and fauna to be considered unique and endangered, threatened, rare or of management concern in the South Okanagan and Similkameen valleys, so that representative areas of habitat are secured and appropriately managed for the long-term welfare of the identified species. This was accomplished through The Nature Trust of B.C. Workshops. The list of species that are of conservation concern should be updated each year as status reports on them are completed.

As an example of some of the work done, it was discovered that 14 bat species inhabit the South Okanagan. Unfortunately, owing to people's fears and misconceptions, bats are often brutally exterminated and, globally, many species are endangered.

Under Objective 2, determining land status for protection purposes will also play a major role in maintaining biological diversity, since the land is a mixture of federal land, Crown land, provincial forest and private lands.

Table 1 summarizes the sites currently protected. With the exception of Okanagan Mountain Provincial Park, which is about 10,000 hectares, fewer than 3,000 hectares are protected in ecological reserves, parks, federal reserves, The Nature Trust properties and Ministry of Environment lands. These areas contain only one half of the 54 vertebrate species that are at risk.

Objective 3 is to protect and enhance the populations of South Okanagan species designated nationally or provincially as endangered, threatened, rare or of management concern. Specific management activities have been developed for each species and each species has been assigned to one of five categories of endangerment. Activities include species status reports, management plans, recovery plans, immediate habitat protection and monitoring of known populations. An example of this is the recent completion of the Burrowing Owl Recovery Plan, based on a successful re-introduction program initiated in 1983.

Objective 4 is to ensure that no additional species become endangered, threatened, rare, of management concern or extirpated. This will involve enhancing and rehabilitating degraded habitats, primarily by activities such as fencing to protect areas from being trampled and overgrazed by cattle. It will also involve the immediate protection of known hibernacula, nesting cavities and other critical sites through agreements with landowners and the acquisition of properties.

The last three objectives are: (5) to encourage interagency co-ordination and planning, to improve awareness and incorporate conservation of natural habitats and priority species into their programs and to encourage public participation in this process; (6) to encourage balanced use of public and private lands to allow sustainable use of the land while maintaining and enhancing the natural biological diversity of the South Okanagan; and (7) to promote public awareness of the values and importance of South Okanagan natural habitats and their unique flora and fauna.

Some of the recommended activities under these objectives are: working closely with other agencies and conservation clubs; supporting development of a grazing plan for the entire study area; annually reviewing protected sites to monitor the condition of habitats; releasing a communications plan to better inform the public about the natural biological values in the South Okanagan; promoting current Ministry of

Environment activities like Project Wild, Wildlife Viewing and the Wildlife Tree Committee; and reducing exotic plant populations that have infested natural grasslands (e.g., knapweed).

The Conservation Strategy outlines a list of nine major habitat types and associated priority species with a numerical summary, associated percentages, a list of habitats and sites currently protected and land tenure. These figures will be refined when 1:20,000 biophysical mapping is completed. The strategy also provides a list of research and thesis topics for people doing research in the South Okanagan.

Acknowledgements

I would like to thank those who provided support during the development of the South Okanagan Conservation Strategy, especially the steering committee members: Ron Erickson, Ken Redpath, Orville Dyer, Mike Sarell, Ted Lea, Rob Cannings and Dick Cannings. Thanks are also due to Bruce Pendergast for supporting the three-year biophysical mapping project. I would also like to thank Rick Page, Mike Fenger and Bill Harper for inviting me to present this paper.

Literature Cited

Hlady, D.A. 1990. South Okanagan conservation strategy 1990 - 1995. B.C. Ministry of Environment, Lands and Parks, Victoria, BC.

What Can We Do?

Conservation of Biological Diversity in the United States

Hal Salwasser

USDA Forest Service, P.O. Box 96090,
Washington, D.C., U.S.A. 20090 - 6090.

Abstract

The United States does not have a comprehensive policy or strategy for conserving biological diversity. Yet a strong foundation for conservation of biological diversity is provided by the aggregate of environmental laws and policies at federal and state levels, an abundance of zoos, botanical gardens and gene banks, and a large and diverse system of protected lands and waters, both public and private. The framework of scientific knowledge, inventories, conservation actions and education needs to be strengthened. But the foundation for a comprehensive conservation strategy already exists.

Biological diversity is the variety of life and its processes. It encompasses the spectrum of biological organization from genes to biomes and the spectrum of geographic locations from microsites to the biosphere. Significant losses of biological diversity could affect the future well being of human life. They will certainly affect the structure and function of the planet's ecological systems. Extinction of species and simplification of ecosystems diminish future resource options and reduce the availability of natural areas to provide life-supporting ecological services.

Major factors that affect biological diversity include: conversion of wild areas to agriculture, industry and other human uses, toxic chemicals, pollution and global climate change; overuse of species by humans; fragmentation of habitats and populations; restoration of species and ecosystems; and management of wild areas for sustainable uses of natural resources. The last two factors are useful in countering the potentially negative effects of the others.

Reasons for conserving as much of the variety of life as possible include its intrinsic values, its roles in providing current and future resources, and securing environmental quality. Biological diversity, however, is so complex and intangible that its conservation cannot be approached without focusing on

In *Our Living Legacy: Proceedings of a Symposium on Biological Diversity*, edited by
M.A. Fenger, E.H. Miller, J.A. Johnson and E.J.R. Williams, pp. 313-337. Victoria,
B.C.: Royal British Columbia Museum.

specific elements and processes. Some key elements include genetic resources, populations of species, biological communities and regional ecosystems.

In setting goals for biological diversity we must address specific, achievable objectives for the principal elements of concern in an area. These goals must be integrated in three major ways. First, they must be integrated into overall plans for resource management. Second, goals and actions must be integrated up and down geographic scales so that actions taken at individual sites or stands contribute to goals for watershed conditions, which in turn contribute to goals for regional ecosystems that can sustain both the desired environmental quality and resource products to meet human needs. And third, goals and actions must be integrated across the biological spectrum of genetic resources, species, communities, and ecosystems. This increased need for integration and coordination is a daunting challenge that is fraught with scientific, technological, and political barriers and uncertainties.

Introduction

The United States does not have a comprehensive policy and strategy for conservation of biological diversity (U.S. Congress 1987; Keystone Center 1991). It does, however, have a significant system of protected lands and major laws and policies at federal and state levels for protecting basic environmental quality and native species. These have been reviewed extensively in U.S. Congress (1987).

Conservation of biological diversity in the U.S. began with the traditions of native Americans and their reverence for all forms of life. Modern industrial societies would do well to reconnect with those early North American traditions. Game-protection laws in the colonies of the 1600s to 1800s extended some protection to selected animals, but the explicit and implicit policies of the nation up to the late 1800s largely worked against native biological diversity (Matthiessen 1959; Trefethen 1961; Borland 1975). Rampant destruction of forests, overgrazing of the western range and depletion of vast flocks of wild birds and herds of wild ungulates led to the conservation movement of the late 1800s and the nation's first efforts to conserve its biological diversity. People of the time did not talk about the issue in terms of biological diversity, but the actions they took built the foundation upon which the native flora and fauna of the nation depend to this day: national systems of federal parks, forests, wildlife refuges and, eventually, their analogs in states and counties.

Today, nearly one-third of the land area of the U.S. is protected in some form of federal ownership (Keystone Center 1991). These lands include large and diverse areas in national parks, national forests, national grasslands, wildlife refuges, defense installations and public lands (Fig. 1). Another one-third of the nation is held by states or private owners in wildland conditions that range from highly protected nature reserves to intensively managed tree farms and pastures. Approximately 20% of the nation's land is under agricultural use. The remaining land is in residential and industrial use, some of which includes local parks, greenways and urban forests. This is the land base upon which conservation of biological diversity in the wild must occur. Changes in how these lands are managed may be in order, but the nation can build a conservation strategy with the infrastructure of land-use designations in place.

As an overlay to the foundation of lands available for conservation of biological diversity, the nation has a complex array of federal and state laws and regulations for protecting various aspects of the environment. The most significant federal law affecting biological diversity is the Endangered Species Act of 1973, as amended (Kohm 1991). It provides strong protection to species and subspecies of plants and animals that are listed under federal regulations as being either threatened or endangered. But, as noted by a review of federal laws and policies regarding biological diversity (U.S. Congress 1987) and by Noss (1991) and others in Kohm (1991), the Act falls short of providing a comprehensive mandate for conserving biological diversity.

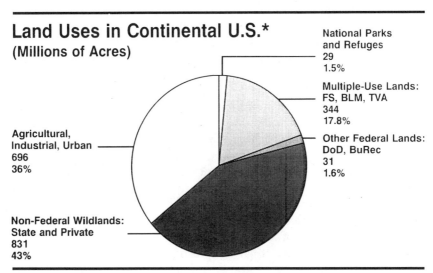

Land Uses in Continental U.S.*
(Millions of Acres)

National Parks and Refuges
29
1.5%

Multiple-Use Lands: FS, BLM, TVA
344
17.8%

Other Federal Lands: DoD, BuRec
31
1.6%

Agricultural, Industrial, Urban
696
36%

Non-Federal Wildlands: State and Private
831
43%

* Data from CEQ; USDI 1986

Figure 1. Relative area in different ownerships and land-use classes in the United States [after World Resources Institute et al. (1988) and USDI land statistics].

Several laws and policies augment the provisions of the Endangered Species Act regarding biological diversity. These include: the Wilderness Act of 1964, which has resulted in more than 36 million hectares of federal lands protected in wilderness condition as of 1991; the Land and Water Conservation Fund Act of 1965, which provides for federal acquisition of significant natural areas; the Wild and Scenic Rivers Act of 1968; the National Environmental Policy Act of 1969, which calls for comprehensive evaluation of the estimated environmental effects of all major federal actions; the Clean Water Amendments of 1972; the National Forest Management Act of 1976, which specifically mandates that management of national forests and national grasslands "provide for diversity of plant and animal communities...in order to meet overall multiple-use objectives..."; the Clean Air Amendments of 1977; the Soil and Water Resources Conservation Act of 1977; the Agriculture and Food Security Act of 1985, which worked to restore native vegetation to more than 16 million hectares of marginal farmlands and reverse the historical incentives for converting wetlands to agriculture; and the Agriculture and Food Security Act of 1990, which includes a forestry title and provisions for expanding the scope of federal assistance in stewardship of state and federal forestlands. There are numerous state laws and regulations and provisions in the mandates for international programs that aim at similar objectives. The above list is not comprehensive and is provided only to demonstrate the array of statutes that contribute in various ways to conserving aspects of the nation's biological diversity.

Added to the roles and missions of government agencies there is a strong private role in conserving biological diversity in the U.S. This is led by The Nature Conservancy with its programs in land stewardship, natural-heritage data systems and conservation education. Other private-sector entities contribute lands, technical expertise, financial resources and educational services to the cause of conserving biological diversity. These include private individuals, charitable foundations, business firms, industries involved in natural-resources management, local community action groups and a multitude of conservation and preservation organizations.

The preceding material provides an overview of the institutional framework for conservation of biological diversity in the U.S. Let us now consider how to strengthen the effectiveness of that framework and specifically how to develop a conservation strategy for federal agencies responsible for land and resource management. The background concepts and elements of this strategy follow largely from the report of the Keystone National Policy Dialogue on Biological Diversity on Federal lands (Keystone Center 1991). What I have chosen to emphasize in this paper reflects my own experience and, to no small degree, my view of the role of one particular federal agency, the USDA Forest Service.

The Nature and Context of Biological Diversity

Humans have a great influence on the variety and health of life around them. For many this means a responsibility to care for the richness of life on Earth. This has been stated in many ways across the centuries, though rarely better than by Aldo Leopold (1953: 147): "If the land mechanism as a whole is good, then every part is good, whether we understand it or not. If the biota, in the course of eons, has built something we like but do not understand, then who but a fool would discard seemingly useless parts? To keep every cog and wheel is the first precaution of intelligent tinkering." We now call these biotic cogs and wheels biological diversity, or biodiversity.

Biodiversity is the variety of life and its many processes. It includes all life forms from bacteria, fungi and protozoa to higher plants, insects, fishes, birds and mammals. That could be as many as 5 to 30 million different species worldwide (Wilson 1988; Table 1). Biodiversity also includes countless millions of races, subspecies and local variants of species and the ecological processes and cycles that link organisms into populations, communities, ecosystems and, ultimately, the entire biosphere.

Ideally we would restore or perpetuate every part and process of a diverse biota in every possible place. But it is not possible, or even desirable, to save every cog and wheel of this variety of life given the increasing pressure of human population growth and its artifacts. Nevertheless, conserving major portions of biological diversity, including a healthy global ecosystem, is vital to providing a future with productive, sustainable natural resources. Growing human populations, with their major demands for resources, their attitudes about long-term environmental health and their propensity for spewing toxic chemicals and pollutants into the environment, mean that we must continually make choices about which aspects of the variety of life are of highest priority for conservation and how we should blend their perpetuation with other social goals (Ehrlich 1990). In this fundamental sense, conservation of biological diversity is like conservation of any other resource or thing of value: we must often choose to do one thing at the expense of another or select options that (a) only reduce the rate of loss or (b) only approximate desired ecological conditions (Cairns 1989).

The big questions we face on the future diversity of life on Earth

317

Table 1. Estimated variety in species of plants and animals world-wide (after Wilson 1988).

Kind of Organisms	Number of Described Species
Viruses	~ 1,000
Monera	4,760
Bacteria	3,060
Blue-green algae	1,700
Fungi	~ 47,000
Algae	26,900
Higher plants	~ 248,400
Bryophytes	16,600
Ferns	10,000
Dicots	170,000
Monocots	50,000
Protozoa	30,800
Invertebrates	~ 900,000
Worms	36,200
Molluscs	50,000
Insects	751,000
Chordates	~ 44,000
Bony fishes	18,150
Amphibians	4,180
Reptiles	6,300
Birds	9,040
Mammals	4,000
Total of all described species	~ 1,400,000
(Estimated total of all species	5 - 30 million)

include: What exactly is biodiversity? What are the threats to its future? What parts and processes of biological diversity are of highest immediate concern? Why should we be concerned with perpetuating those parts and processes? How should we go about conserving biodiversity while satisfying the other needs of people for food, shelter, clothing, recreation and livelihood? We'll be many years getting comprehensive answers to these questions. But we need not wait for scientific consensus before embarking on a path to conserve what we already know to be some of the most important parts and processes.

Setting a course for action without all the knowledge we might desire means we must make some assumptions. Here are mine:

- Biodiversity, along with soils, water and air, is a basic building block for all nature and hence all life.
- No single institution or land owner — public or private — can individually perpetuate the full array of biological diversity or even certain major elements, such as migratory birds and mammals.
- Government land- and resource-management agencies, by nature of their trust responsibility over large areas of land in North America, together with private owners of the remaining lands, can and must play significant and complementary roles in an overall conservation strategy.

The Nature of Biodiversity

Aside from generally being the variety of life and its processes, what specifically is biological diversity? Because it is extraordinarily complex and much of it is hidden from easy view, we need some "handles" to describe its most significant measurable parts and processes (Fig. 2).

The foundation of biodiversity is *genetic variation*. Genetic

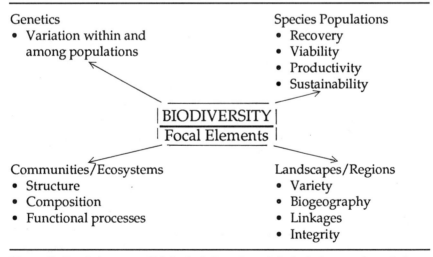

Genetics
- Variation within and among populations

Species Populations
- Recovery
- Viability
- Productivity
- Sustainability

| BIODIVERSITY |
| Focal Elements |

Communities/Ecosystems
- Structure
- Composition
- Functional processes

Landscapes/Regions
- Variety
- Biogeography
- Linkages
- Integrity

Figure 2. Focal elements of biological diversity might include: genetic variation in selected species; viability of endangered or sensitive species; the richness, structure and function of species and processes in biological communities; and characteristics of regional landscapes such as continuity, variety of ecosystems and patterns of different biological communities.

Potential Bioregions

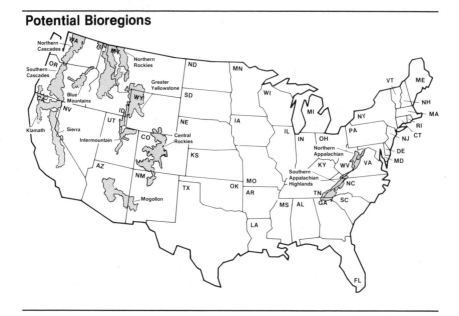

Figure 3. Some large areas of contiguous public lands that could form a network of regional ecosystems for conservation of biodiversity.

variation affects a species' physical characteristics, productivity, resilience to stress and long-term evolutionary potential. Genetic diversity means that species contain characteristic levels of genetic variation within and among their populations, including that of extreme populations (see Ledig, this volume).

A more easily recognized element of biological diversity is distinct *species*. Some species, such as White-tailed Deer, Rainbow Trout and Red Maple are abundant and widespread. Others, such as Kirtland's Warbler, Dune Tansy and Gray Wolf, face endangerment or even extinction. Species diversity means that an area contains its native species in numbers and distributions that contribute to both natural genetic variation and a high likelihood of long-term continued existence throughout their geographic ranges. Ecologists consider populations that possess such bright prospects to be viable (Shaffer 1981; Soulé 1987).

Associations of species in an area are another recognizable element of biological diversity. We refer to these associations as *biological communities* and usually recognize them as distinct stands, patches or sites, such as forests, meadows, fencerows, ponds, riparian areas and wetlands. Communities form the biotic parts of ecosystems. The variety of species in an ecosystem, usually referred to as its species composition, is intertwined with its structural and functional integrity and hence with the diversity of its ecological processes. Biologically diverse communities contain sufficient compositional, structural and functional variety that they are assured a high prospect of continued presence and ecological influence in an area (Franklin 1988).

Finally, at the geographic scale of *regional landscapes* (see Fig. 3) diversity includes variety in the kinds of biological communities and a biogeography (that is patterns, sizes, shapes, juxtapositions and interconnectedness) that provides for free, natural interchange of individuals throughout the area. Many species, especially those with specialized habitat affinities or those that are migratory or wide-ranging, can only be sustained in viable numbers and distributions in very large wildland areas. Examples of regional ecosystems that are capable of sustaining most if not all of their native biological diversity include the Southern Appalachian Highlands, Sierra Nevada and Adirondacks. Biologically diverse regional ecosystems contain a complement of species and biological communities with sufficient genetic variation, population viability and compositional, structural and functional integrity that the full range of natural processes, including evolutionary potential, is sustained. Such ecosystems might be thought of as being intact or possessing full biological integrity.

This classification of the parts and processes of biological diversity is admittedly artificial; there are other valid ways to categorize the variety of life. But classification helps us comprehend the infinitely varied and dynamic thing we call biodiversity.

Threats to Biological Diversity

Why is there such concern for biodiversity? Simply put, because we're losing parts and processes of the variety of life at alarming rates and we're not sure of the long-term consequences for future environmental health or quality of human life (Myers 1988; Wilson 1988). These losses of biological diversity result from many threats and they occur in every region of the planet (Ehrlich 1988; Table 2).

The best research and resource management will have only minor lasting effects on biological diversity if social and political systems do not address the global "megathreats" of human population growth, poverty, profligate consumption, pollution and political instability. These megathreats are reality; we cannot wish them away or ignore them. On the other hand, people can only effect change on a global scale through progressive local actions.

Therefore, we'll address a local and regional focus in this paper:

Table 2. Threats to biological diversity.

Global Megathreats	Local or Regional Threats
Population growth of humans	Overuse of species by humans
Poverty	Toxic chemicals
Profligate consumption	Habitat fragmentation
Pollution of air and water	Desertification
Political instability	Simplification of ecosystems and gene pools
	Spread of exotics
	Conversion of wildlands to agricultural, residential or industrial uses

how to conserve those elements of biological diversity that can be influenced by plans and actions we control. If we follow the maxim that even a journey of a thousand miles begins with one step, this is a good beginning.

The immediate factors that affect the variety of life in a forest, farm, rangeland or other relatively undeveloped area include toxic chemicals and pollutants, overuse of plant and animal resources by humans, habitat loss and fragmentation, global climate change, simplification of ecosystems, reductions in genetic variation and the introduction or spread of undesirable exotic species. These are the factors our conservation programs must address.

Values of Biodiversity

Before we embark on a program to deal with these factors, however, let's give some thought as to why we should expend the energy in the first place and what we should expend it on. What parts and processes of biodiversity should we be concerned about immediately?

It is axiomatic that we cannot perpetuate all biological diversity while extracting an increasing portion of its productivity for growing human uses. Nor can we perpetuate desired species, communities and ecological processes simply by trying to stop human progress or by preserving a few isolated natural areas and trying to prevent change in them, whether natural or human-induced. And, though initially appealing, we have now learned that trying to maximize the variety of life on every hectare or in every watershed is not the solution (Samson and Knopf 1982).

So, what should we care about first and why is that more important than other concerns? For example, should we focus on sustaining and enhancing all genetic variation, recovering all endangered species, restoring all riparian areas, perpetuating all old-growth forests or conserving all sundews, swamps and salamanders? Probably not — more of everything is not possible. Regardless of the ecological notion that all species are of equal value, the answers to these questions will be greatly influenced by economics, aesthetics and ethics, as well as biology and ecology.

Perhaps a look at how the variety of life serves people will offer some insight into immediate priorities, though this is not meant to imply that utilitarian values are the sole or even primary reasons to conserve biological diversity. Nevertheless, air, foods, medicines and recreational resources all contain elements that derive from biological diversity. We enjoy hiking in diverse forests, visiting biologically rich seashores and savouring the bounty of diverse and productive fish and wildlife populations. Diverse communities of plants, animals and micro-organisms also provide indispensable ecological services. They recycle wastes, maintain the chemical composition of the atmosphere and play a major role in determining the world's climate. Moreover, many people believe humans should revere all life on Earth and bear an obligation for its stewardship (Callicott 1986, 1991; Naess 1986; Norton 1986; Ehrenfeld 1988).

Of course, the potential values and uses of biological diversity far exceed our current knowledge (Oldfield 1984). We know only a small fraction of the species on this planet, especially in tropical ecosystems, despite decades of scientific effort (Lugo 1988). Nevertheless, we can begin by addressing those elements of biological diversity that we already know to be important and provide a "safety net" for the remainder that we do not know about or understand. To do that, we will need an overall conservation strategy for biological diversity that complements our approaches to managing other natural resources.

Building a Conservation Strategy for Biological Diversity

Building a successful strategy for conserving biological diversity will require the recognition that meeting human needs and responding to unpredictable changes are fundamental. These can only be accommodated by using large regional ecosystems as the context for long-term planning and for co-ordination of public and private actions (Salwasser 1991).

Management of genetic resources, species populations, ecosystems, bioregions and human activities to perpetuate the variety of life while meeting short- and long-term human needs must integrate many goals and considerations. It must also employ a full spectrum of conservation actions, from protection, restoration, enhancement and sustainable harvesting to research, inventorying, assessment, planning, monitoring, interpretation, marketing and education. Conservationists have been engaged in such efforts for many decades. But they have tended to focus on specific resources and specific places, seldom with biodiversity as an overall concern and seldom on a large enough geographic scale. The following are some essential aspects of a comprehensive conservation strategy.

Understand the Larger Context of Local Actions

We are not currently organized to conserve biological diversity because we lack the mechanisms and operating strategies to deal with many elements of biodiversity at large geographic scales across institutional boundaries. Addressing specific elements such as endangered or featured species, rare biological communities and specific places, such as the Gifford Pinchot National Forest or the Adirondack State Park, is an essential part of an overall conservation strategy. But if we operate as if these are ecologically isolated entities we are deluding ourselves. They are not isolated and any science or management strategy that treats them as such is a dead end for biological diversity. Regardless of whether you are working on one species or hundreds, one hectare or one million hectares, you are part of a larger context. The larger context influences your ability to achieve your goals and you influence the ability of others in the larger area to achieve theirs.

So, point one of a conservation strategy is to act locally but always know that you are part of something much larger (Crow 1991).

Embrace Complexity

Point two is to embrace the complexity of the issue. There are no simple solutions for biodiversity. This should excite and challenge us. To approach such a comprehensive goal as perpetuating biological diversity, we will need to better understand the structure, functions and processes of regional populations and ecosystems, and their responses to the long-term, cumulative effects of environmental change, human activities and resource management, then make prudent decisions on how to rehabilitate, sustain or enhance ecosystem productivity for diverse values and uses (Cairns 1988a). This means that we must begin a dialogue and experimental actions to diversify our theories, concepts, approaches and methods for carrying out conservation of lands and resources. But we must set the course for a biologically diverse world before we have all the knowledge we may desire (Soulé 1986; Cairns 1988b). This must start with a broadly agreed upon public policy for biological diversity to give us focus and the sense of urgency needed to mobilize attention (U.S. Congress 1987).

Work with the Political Aspects of Biodiversity

Point three is to recognize that biodiversity is a political issue, as well as a scientific concept. Taking action to conserve biological diversity will entail changes in social priorities and hence changes in current conservation policies and plans. Without recognition of the compelling need for action, making these changes will be difficult. Beyond the need for a national policy for biological diversity, we need clear, concise, well understood regulations and programs for its protection, restoration, enhancement and sustainability.

Many federal and state agencies, together with private entities such as The Nature Conservancy, are strengthening their contributions to a biologically rich future. But overall, the policies and actions for conserving biodiversity in the U.S. are fragmented and unco-ordinated (U.S. Congress 1987). We need to integrate them into a comprehensive conservation strategy. This does not necessarily require new legislation: much can be done by building from existing laws, regulations, policies and institutions.

Co-ordinate Plans and Actions

Point four is to increase co-ordination. National and state strategies must foster co-ordination among all agencies and organizations — public and private — that share interest or responsibility for natural resources (U.S. Congress 1987). We need to reduce barriers — real and perceived — that have been created between functional disciplines such as timber and wildlife, between science and management, between parks and multiple-use forests, between federal and state agencies and between government and the private sector. To maintain biodiversity, all must play complementary roles.

On private lands we must provide technologies and assistance for willing landowners and have sufficient incentives to favour necessary conservation actions.

Encourage Participatory Planning and Management

Point five is that agencies must open their planning and resource-management processes even more than has occurred under federal and state environmental regulations and planning processes. The alternatives to such openness, such as resolving controversy through litigation, increasingly centralized technology-driven planning and solutions by smoke-filled rooms of experts, waste time and talent and run counter to our democratic tradition and spirit of solving problems through community consensus (Wondolleck 1988; Cairns 1989). The latter two options, though holding great appeal to technologists, have never worked for long anywhere they have been tried — Soviet agriculture is perhaps the best example of technology-driven planning gone awry.

Many of the rarest elements of biodiversity occur on public lands — some are indigenous to these lands, others find their last refuge there. We develop and change policies for managing public lands and resources by democratic means. Our society used to delegate nearly complete authority for managing public resources to professionally trained rangers, biologists, game managers and foresters. Those days are gone. We have an increasingly knowledgeable and concerned citizenry. They want a voice in making the tough choices we face regarding their resources (Wondolleck 1988). The recently completed Keystone dialogue on biodiversity on federal lands is a national example of providing an open forum on shaping a conservation strategy (Keystone Center 1991). Many examples of similar regional dialogues exist.

Integrate Science with Management

Point six is to enhance the role of science in management. Because biodiversity conservation requires regional approaches and integrated ecosystem management as opposed to simple, functional, stand or population management, we need new knowledge and technologies. Research and technology development must embrace large-scale, long-

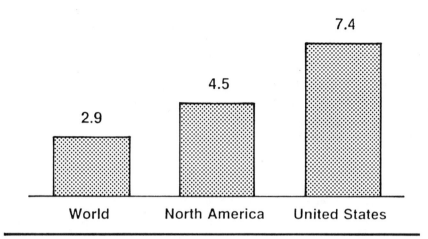

Figure 4. Relative area protected in national parks, nature reserves and wilderness areas in the United States, North America and the world (after World Resources Institute et al. 1988).

term biodiversity issues such as population viability, ecological resilience, landscape linkages, ecological restoration, multi-resource yield functions, habitat fragmentation, biodiversity indicators, cumulative effects analysis and monitoring. There is an implication here for education and science: both must become more — not less — interdisciplinary and both must focus increasingly on adaptive approaches to solving wicked problems that defy technical solutions (Holling 1978; Walters 1986). Research and development need to become integral parts of management strategies and their implementation. We can no longer afford the illusion that the only role of science is to yield pure knowledge and that development and application of that knowledge are someone else's business.

Market Biodiversity

Point seven is that marketing biological diversity is a worthwhile goal. Most people know what spruce trees, trout fisheries and prairies are. Some even know the roles of management in their conservation. But biodiversity is another story. Scientists have yet to reach a consensus on what it is and hardly anyone is willing to admit that he or she knows how to conserve it while meeting other human needs. So we have an enormous job of interpretation, education and marketing to gain awareness, understanding and support. It is conceivable that investments in interpretation and education programs will have a higher long-term return for biodiversity than investments in more science and technology.

Integrate All Actions and All Land Uses

Finally, the keys to success in conserving biodiversity are integration and on-the-ground action. We will need to use all lands from the most highly protected to the most intensively managed and muster all the management tools we can. Conservation of biodiversity can start with reserved areas such as parks, wilderness and natural areas, but it cannot stop there. Only about 7.4% of the U.S. is in parks, wilderness and nature preserves (Fig. 4). And the U.S. is at the top of the world in such land protection.

Multiple-use lands will also be critical to the solution (Fig. 1). It has been argued that they are more important to the solution than parks and preserves, and I agree (Norse et al. 1986; Wilcove 1988). However, even they will not be enough. We cannot meet our resource-production

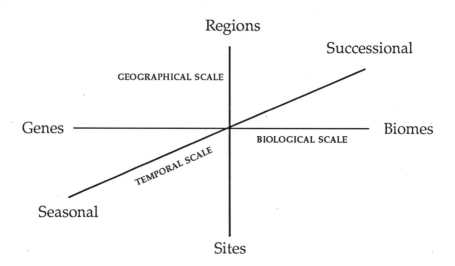

Figure 5. Planning for conservation of biological diversity must integrate actions with consideration of factors along three dimensions: temporal, geographic and biological.

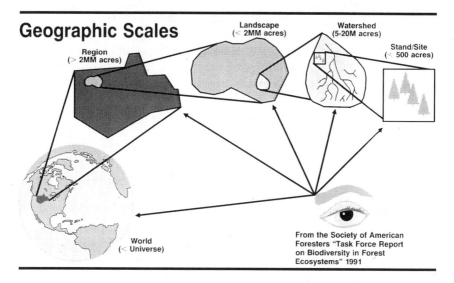

Figure 6. Geographic integration blends considerations from stand to landscape to watershed to regional scales.

needs and save major elements of biological diversity just on public multiple-use and reserved lands. Private lands, including those intensively managed for agricultural and wood products, must also play significant roles. High productivity from any piece of land or water will allow other parcels to be managed for other purposes, including restoration or enhancement of rare or sensitive elements of biodiversity.

A comprehensive, long-term, regional perspective on biodiversity implies that plans and actions will be integrated along three dimensions: biological, temporal and geographic (Fig. 5). We can get a feel for the need to integrate these different dimensions by thinking in terms of blending plans and actions for specific elements of diversity (the focal elements from Fig. 2), elements ranging from stands to watersheds to regions (the geographic scales in Fig. 6) with attention to consequences for periods of time that may range from the next few years to the next century.

The need for increased integration is a greater challenge than anything conservation science or management have yet tackled.

Some Priorities for Action

Let us now turn to some specific actions that managers and scientists can take to build or strengthen a biodiversity conservation strategy. Based on the state of knowledge about biological diversity in 1991, the Keystone dialogue recommended numerous actions for conservation plans, programs and on-the-ground conditions (Keystone Center 1991). I have selected what I consider to be the most pressing of the Keystone actions and put them in my own words. Many of these are currently underway in the U.S., though they do need to be more consistently applied:

- Aggressive and successful actions must be carried out to *recover and conserve threatened or endangered species,* including programs involving zoos, botanical gardens and gene banks. These are the elements of biodiversity we already know to be in the greatest peril. The goal should be a net decline in the number of listed species and subspecies in the nation and in major regions of the nation.

- Management of habitats, human activities and artifacts, and wild populations of plants and animals, must assure that *populations of native species are viable and well distributed throughout their geographic ranges.* The goal is to secure the places and functions of all native species in regional ecosystems before they reach the point where federal listing as threatened or endangered calls into play the extreme measures of crisis conditions (Salwasser 1988). Especially important in achieving population viability is the perpetuation of demographically resilient local populations, the characteristic genetic variation of the entire species and the full range of the species' roles in ecological processes. The principles of conservation biology are particularly useful in this task (Soulé and Wilcox 1980; Soulé 1986b, 1987).

- Management of lands, human activities and artifacts, and habitats must assure that a *network of representative native biological communities* is maintained across the landscape. Especially important are communities or assemblages of species that are rare or imperiled in the region or nation. The landscape must provide for the full range of ecological processes characteristic to the area, as well as the essential resources for all species, including conditions needed for normal movement of plants and animals throughout the landscape.

- *Natural elements of structural diversity* such as snags, caves, fallen trees and seeps must be maintained in sufficient qualities, amounts and distributions across the landscape that their roles in sustaining the diversity, productivity and resilience of regional ecosystems are assured.

- The genetic resources of intensively managed wild populations such as timber, fish and game species should be under special management programs to sustain the *natural levels of genetic variation within and among populations and the genetic integrity of representative and extreme populations* (Ledig 1986, this volume; Millar 1987).

- *High-productivity sites* such as flat ground with deep, loamy soils and wetlands, and featured species such as pines, firs, oaks, Elk and trout should be managed in ways that *sustain the production of resources desired by people*, thus meeting human needs with minimal impacts on more fragile sites and sensitive species.
- *Management activities* that affect soils, water, air and biota should be designed to *perpetuate the long-term health, diversity, productivity and renewability of the ecosystem*. Resource-management systems should employ the natural restorative powers of ecosystems to the maximum extent feasible. Actions that are known to degrade long-term ecosystem diversity, productivity or resilience should be avoided if possible or mitigated promptly when not. The kinds, amounts and distribution of living and dead organic matter to be left in the ecosystem for long-term renewal of the land following resource harvest should be routinely reviewed along with how much of the productivity of the system is to be removed for human uses.
- Biological communities, water and soils that have been degraded by natural events or past human actions should be under special programs of *restoration and renewal*, embracing the concepts and methods of restoration ecology and management (Cairns 1986, 1988a, 1988b, 1988c, 1989; Jordan et al 1987; Bonnicksen 1988; Jordan 1988).

These are some of the major actions and conditions that would strengthen conservation of biological diversity in the wild. They are all possible to accomplish given sufficient commitment and priority on resources. They characterize the direction federal land-management agencies are taking in the United States. But making them happen on the ground requires some changes in our resource-management systems, planning processes and working relationships. For example:

- *Landscapes* must be planned and managed to meet specific objectives that will yield both desired future ecological conditions and desired economic and social goals, while balancing conflicts between uses and values (Foreman and Godron 1986).
- Entire *regional ecosystems* at the geographic scale of millions of hectares must become the areas in which *multiple uses and multiple values* are planned and scheduled to conserve native

biological diversity as well as to meet people's needs for resources. The complementary contributions of different ownerships must be blended by using incentives to encourage entrepreneurial activity with minimal infringement of private property rights (Salwasser et al. 1987). Overall conservation plans must employ the full range of land-use classes available, such as wilderness, wildlife refuges, national parks, natural research areas, stream-side management zones, greenways, resource-production areas and special-interest areas. Managers must meet the needs of people for essential resources and the needs of our economies for natural-resource products in ways that also assure that representative examples of all native habitats, biological communities and ecosystems and their ecological processes are perpetuated in the region. Concepts such as Biosphere Reserves and wildlife-movement corridors need to be considered in planning land uses (Gregg and McGean 1985; Harris and Gallagher 1989).

- Unprecedented degrees of *citizen involvement and co-ordination between agencies and affected citizens* will be required. Consensus building and negotiated problem solving must replace litigation and technology-driven planning as primary approaches to conflict resolution (Wondolleck 1988). People who are affected by resource-management and conservation strategies must feel a strong commitment to being part of the solution rather than perceiving themselves as injured parties or as aggrieved critics.

- All agencies and affected interest groups and enterprises must contribute to *common inventories* of the basic conditions of soils, water and biota, sharing data and other information as appropriate to their missions and property rights. For elements of biological diversity, the inventories should build from the foundation of state heritage programs and multi-resource inventories conducted by various state and federal agencies to allow prudent choices to be made on priorities for investments and protection (Scott et al. 1987, 1991; Jenkins 1988).

- Agencies, universities and affected interest groups and enterprises should improve their co-operation in *long-term, interdisciplinary ecosystem-scale research and development*. Managers need dramatic improvements in scientific understanding of biological diversity; they also need a wide array of tools to plan and evaluate the expected effects of management options, and expanded choices for sustainable harvest and management of resources.

- Agencies, universities and affected interest groups and

enterprises should cooperate in a *comprehensive program of interpretation, education and demonstration of co-ordinated, adaptive ecosystem management for multiple values and uses.* The result should be broader understanding and support for the complementary roles played by different agencies and ownerships in the overall conservation strategy.

- Agencies, universities and affected interest groups and enterprises should co-operate in *monitoring carefully selected indicators of ecosystem diversity and valued resources.* The monitoring program should be guided by the use of decision-analysis tools to ensure that the most vital information is being collected in useable quality and in a timely fashion (Maguire 1988, 1991). A major result should be more timely response by management using current information to adjust plans and actions.
- The levels of *trust* that once existed between the scientific community and land managers, their customers, clients and publics need to be restored.

Conclusion

Conserving biological diversity is an ideal — as is this perspective. It may be achievable only partially in the U.S. and perhaps even less so in other nations (Cairns 1989). But we should aim beyond what we think we can achieve. As Goethe said, dream bold dreams, they're full of power and magic.

All nations and states must build their strategies for conserving biological diversity on the foundation of land uses and legal mandates in place. Every nation will find the need and opportunities to strengthen their foundation and craft a framework to address the many dimensions of biological diversity. But in any nation or state, conserving biological diversity while trying to meet people's needs for resources will not be easy. Goals for recovery and conservation of native species and biological communities and for desired ecological conditions will have to compete effectively with goals for specific resource uses. Tradeoffs will be inevitable because there is only one land base and the competition for different values and uses will only get fiercer with time.

Nevertheless, conservation of biological diversity is not a lost cause. And it does not mean the end of resource production from our wildlands. It will require some adjustments that will probably cause short-term changes in land uses and in production of some resources. If we make those adjustments with prudence they should enhance our long-term prospects for environmental and economic sustainability.

In the U.S., we have made some definite progress toward this goal in recent years. But we also have a long way to go. I hope that Canadians can learn from our American successes and failures, and I trust we can learn from theirs as well. After all, we are partners on the same journey. It is one Earth. And its future richness and productivity depend on all of us taking this journey that we call conservation of biological diversity.

Literature Cited

Bonnicksen, T.M. 1988. Restoration ecology: philosophy, goals, and ethics. The Environmental Professional 10: 25-35.

Borland, H. 1975. The history of wildlife in America. National Wildl. Fed., Washington, DC.

Cairns, J., Jr. 1986. Restoration, reclamation, and regeneration of degraded or destroyed habitats. *In* Conservation biology: the science of scarcity and diversity, *edited by* M.E. Soulé. Sinauer Associates, Sunderland, MA. pp. 465-484.

Cairns, J., Jr. (Editor). 1988a. Rehabilitating damaged ecosystems (2 vols.). CRC Press, Boca Raton, FL.

Cairns, J., Jr. 1988b. Restoration ecology: the new frontier. *In* Rehabilitating damaged ecosystems, Vol. 1, *edited by* J. Cairns, Jr. CRC Press, Boca Raton, FL. pp. 1-11.

Cairns, J., Jr. 1988c. Restoration of damaged ecosystems and opportunities for increasing diversity. *In* Biodiversity, *edited by* E.O. Wilson. National Academy Press, Washington, DC. pp. 333-343.

Cairns, J., Jr. 1989. Restoring damaged ecosystems: is predisturbance condition a viable option? The Environmental Professional 11: 152-159.

Callicott, J.B. 1986. On the intrinsic value of nonhuman species. *In* The preservation of species, *edited by* B.G. Norton. Princeton Univ. Press, Princeton, NJ. pp. 138-172.

Callicott, J.B. 1991. Conservation of biological resources: responsibility to nature and future generations. *In* The challenges in the conservation of biological resources: a practitioner's guide, *edited by* D.J. Decker, M.E. Krasny, G.R. Goff, C.R. Smith, and D.W. Gross. 1991. Westview Press, San Francisco, CA. pp. 33-42.

Crow, T. 1991. Landscape ecology: the big picture approach to resource management. *In* The challenges in the conservation of biological resources: a practitioner's guide, *edited by* D.J. Decker, M.E. Krasny, G.R. Goff, C.R. Smith, and D.W. Gross. 1991. Westview Press, San Francisco, CA. pp. 55-66.

Ehrenfeld, D. 1988. Why put a value on biodiversity? *In* Biodiversity, *edited by* E.O. Wilson. National Academy Press, Washington, DC. pp. 212-216.

Ehrlich, P.R. 1988. The loss of diversity: causes and consequences. *In* Biodiversity, *edited by* E.O. Wilson. National Academy Press, Washington, DC. pp. 21-27.

Ehrlich, P.R. 1990. The population explosion. Simon and Shuster, New York, NY.

Foreman, R.T.T., and M. Godron. 1986. Landscape ecology. John Wiley and Sons, New York, NY.

Franklin, J.F. 1988. Structural and functional diversity in temperate forests. *In* Biodiversity, *edited by* E.O. Wilson. National Academy Press, Washington, DC. pp. 166-175.

Gregg, W.P., Jr., and B.A. McGean. 1985. Biosphere reserves: their history and their promise. Orion 4(3): 41-51.

Harris, L.D., and P.B. Gallagher. 1989. New initiatives for wildlife conservation: the need for movement corridors. *In* Preserving communities and corridors, *edited by* G. Mackintosh. Defenders of Wildlife, Washington, DC. pp. 11-34.

Holling, C.S. (Editor). 1978. Adaptive environmental assessment and management. John Wiley and Sons, New York, NY.

Jenkins, R.E. Jr. 1988. Information management for the conservation of biodiversity. *In* Biodiversity, *edited by* E.O. Wilson. National Academy Press, Washington, DC. pp. 231-239.

Jordan, W.R., III. 1988. Ecological restoration: reflections on a half-century of experience at the University of Wisconsin-Madison Arboretum. *In* Biodiversity, *edited by* E.O. Wilson. National Academy Press, Washington, DC. pp. 311-316.

Jordan, W.R., III, M.E. Gilpin, and J.D. Aber. (Editors). 1987. Restoration ecology. Cambridge Univ. Press, Cambridge, England.

Keystone Center. 1991. Final report: of the Keystone Policy Dialogue on biological diversity on federal lands. Keystone, CO.

Kohm, K. A. (Editor). 1991. Balancing on the brink of extinction: the Endangered Species Act and lessons for the future. Island Press, Washington, DC.

Ledig, F.T. 1986. Heterozygosity, heterosis, and fitness in outbreeding plants. *In* Conservation biology: the science of scarcity and diversity, *edited by* M.E. Soulé. Sinauer Associates, Sunderland, MA. pp. 77-104.

Leopold, L.B. (Editor). 1953. Round River: from the journals of Aldo Leopold. Oxford Univ. Press, New York, NY.

Lugo, A.E. 1988. Estimating reductions in the diversity of tropical forest species. *In* Biodiversity, *edited by* E.O. Wilson. National Academy Press, Washington, DC. pp. 58-70.

Maguire, L. 1988. Decision analysis: an integrated approach to ecosystem exploitation and rehabilitation decisions. *In* Rehabilitating damaged ecosystems, Vol. 2, *edited by* J. Cairns, Jr. CRC Press, Boca Raton, FL. pp. 105-122.

Maguire, L. 1991. Using decision analysis to manage endangered species. *In* The challenges in the conservation of biological resources: a practioner's guide, *edited by* D.J. Decker, M.E. Krasny, G.R. Goff, C.R. Smith, and D.W. Gross. 1991. Westview Press, San Francisco, CA. pp. 139-152.

Matthiessen, P. 1959. Wildlife in America. The Viking Press, New York, NY.

Millar, C. 1987. The California forest germplasm conservation project: a case for genetic conservation of temperate tree species. Conservation Biology 1: 191-193.

Myers, N. 1988. Tropical forests and their species: going, going ...? *In* Biodiversity, *edited by* E.O. Wilson. National Academy Press, Washington, DC. pp. 28-35.

335

Naess, A. 1986. Intrinsic value: will the defenders of nature please rise? *In* Conservation biology: the science of scarcity and diversity, *edited by* M.E. Soulé. Sinauer Associates, Sunderland, MA. pp. 504-515.

Norse, E., K. Rosenbaum, D. Wilcove, B. Wilcox, W. Romme, D. Johnston, and M. Stout. 1986. Conserving biological diversity in our national forests. The Wilderness Society, Washington, DC.

Norton, B.G. (Editor). 1986. The preservation of species. Princeton Univ. Press, Princeton, NJ.

Noss, R. F. 1991. From endangered species to biodiversity. *In* Balancing on the brink of extinction: the Endangered Species Act and lessons for the future, *edited by* K. A. Kohm. Island Press, Washington, DC. pp. 227-246.

Oldfield, M.L. 1984. The value of conserving genetic resources. USDA National Park Service, Washington, DC.

Salwasser, H. 1988. Managing ecosystems for viable populations of vertebrates: a focus for biodiversity. *In* Ecosystem management for parks and wilderness, *edited by* J.K. Agee and D.R. Johnson. University of Washington Press, Seattle, WA. pp. 87-104.

Salwasser, H. 1991. In search of an ecosystem approach to endangered species conservation. *In* Balancing on the brink of extinction: the Endangered Species Act and lessons for the future, *edited by* K. A. Kohm. Island Press, Washington, DC. pp. 247-265.

Salwasser, H., C.M. Schonewald-Cox, and R. Baker. 1987. The role of interagency cooperation in managing for viable populations. *In* Viable populations for conservation, *edited by* M.E. Soulé. Cambridge Univ. Press, New York, NY. pp. 159-173.

Samson, F.B., and F.L. Knopf. 1982. In search of a diversity ethic for wildlife management. Trans. N. Am. Wildl. Nat. Res. Conf. 47: 421-431.

Scott, J.M., B. Csuti, J.D. Jacobi, and J.E. Estes. 1987. Species richness: a geographic approach to protecting future biological diversity. Bioscience 37: 782-788.

Scott, J.M., B. Csuti, and F. Davis. 1991. Gap analysis: an application of Geographic Information Systems for wildlife species. *In* The challenges in the conservation of biological resources: a practioner's guide, *edited by* D.J. Decker, M.E. Krasny, G.R. Goff, C.R. Smith, and D.W. Gross. 1991. Westview Press, San Francisco, CA. pp. 167-180.

Shaffer, M. L. 1981. Minimum population sizes for species conservation. Bioscience 31: 131-134.

Soulé, M.E. 1986a. Conservation biology and the real world. *In* Conservation biology: the science of scarcity and diversity, *edited by* M.E. Soulé. Sinauer Associates, Sunderland, MA. pp. 1-12.

Soulé, M.E. (Editor). 1986b. Conservation biology: the science of scarcity and diversity. Sinauer Associates, Sunderland, MA.

Soulé, M.E. (Editor). 1987. Viable populations for conservation. Cambridge Univ. Press, New York, NY.

Soulé, M.E., and B.A. Wilcox (Editors). 1980. Conservation biology: an evolutionary-ecological perspective. Sinauer Associates, Sunderland, MA.

Trefethen, J. B. 1961. Crusade for wildlife. The Stackpole Co., Harrisburg, PA.

U.S. Congress. 1987. Technologies to maintain biological diversity. U.S.

Government Printing Office, OTA-F-331, GPO Stock No. 052-003-01058-5. Washington, DC.

Walters, C. 1986. Adaptive management of renewable resources. MacMillan Publishing Co., New York, NY.

Wilcove, D.S. 1988. Protecting biological diversity. National forests: policies for the future, Vol. 2. The Wilderness Society, Washington, DC.

Wilson, E.O. 1988. The current state of biodiversity. *In* Biodiversity, *edited by* E.O. Wilson. National Academy Press, Washington, DC. pp. 3-18.

Wondolleck, J.M. 1988. Public lands conflict and resolution: managing national forest disputes. Plenum Press, New York, NY.

World Resources Institute, International Institute for Environment and Development and United Nations Environment Programme. 1988. World resources 1988-89: an assessment of the resource base that supports the global economy. Basic Books Inc., New York, NY.

Biodiversity Research in British Columbia: What Should be Done?

Ken Lertzman

School of Resource and Environmental Management, Simon Fraser University, Burnaby, British Columbia, Canada V5A 1S6.

Abstract

Faced with clear threats to biological diversity, we are challenged to take action. One necessary kind of action is research in applied conservation science. As in other fields, the role of research is threefold: understanding what is known, probing the boundary between the known and the unknown and determining what is *knowable*. This framework should be applied to five basic kinds of questions about biodiversity: What is there? Where is it? How does it work? How is it changing? and How can we keep it? Each question is fundamental to traditional biological disciplines, but they must be combined to build a research agenda that will support effective, long-term biodiversity conservation. In particular, research must address the problems of understanding and managing ecological processes that occur at large spatial scales and over long time scales.

A conservation research agenda also needs to recognize that we cannot learn enough fast enough to act from certainty. This means we must recognize that:
- what we do now in the best of faith may be wrong;
- we need to learn from our mistakes and we will need to change our policies and approaches in the future as we do so;
- to ensure that we do learn as much as we can as fast as we can, we must adopt an *adaptive management* strategy in our approaches to reserved areas and areas where conservation objectives must coexist with other needs; and
- we must act conservatively to maintain options.

In *Our Living Legacy: Proceedings of a Symposium on Biological Diversity*, edited by M.A. Fenger, E.H. Miller, J.A. Johnson and E.J.R. Williams, pp. 339-352. Victoria, B.C.: Royal British Columbia Museum.

Introduction

Some data presented in this symposium are alarming, if not depressing. Various speakers have demonstrated convincingly the enormous magnitude of the crisis in biological diversity that faces us. Worse, they have indicated that the rate of loss of diversity is accelerating and in some cases may be difficult to detect (see Scudder and Ledig, this volume). We have also been shown that British Columbia is extraordinarily rich in biological diversity and that this poses both special problems and opportunities for conservation (see Pojar, Tunnicliffe and McPhail, this volume).

The title of this session is "What can we do?" One answer to this question is "More research!" We need to do more research, not as procrastination to avoid taking real action or to satisfy intellectual curiosity, but because there are basic questions that need to be answered if we are to effectively protect biodiversity. There is a desperate need in B.C. for focused, applied research of a high quality in conservation science. Outlining the agenda for this research is the object of this paper.

What *should* be the role of research when faced with the type and magnitude of the problems facing us in conserving biological diversity? The role of research here is no different than in any other line of inquiry. The business of research is, first, to understand what is known and to organize that understanding into a coherent framework of theory, hypotheses and data. Secondly, this framework can be used to probe what is unknown and then further modify or replace the body of understanding. (Carefully defining the boundary between the known and the unknown is often the most difficult and worthwhile contribution of a researcher.)

A third basic task of research that is often forgotten by scientists but can pose tremendous problems for managers is determining *what is knowable*. This becomes especially critical when the problem is applied and short-term policy decisions await the answer to specific questions. For instance, I have been asked if, in a period of a month, I might be able to make quantitative estimates of the degree of fragmentation in old-growth forests on the B.C. coast. This is a worthwhile research objective, one that is important for decision making, and one that is achievable in principle but impossible in practice. For short-term purposes, the answer is not *knowable*. For questions such as this, which can be answered in time but not with the immediacy we might want, "What do we do in the meantime?" is as important a research issue as answering the original question. I will return to this point in the final section of this paper.

What do I mean by the term "biological diversity"? It is generally defined at three levels of organization in the biological hierarchy (Table 1): (1) genetic variation within and among populations of a species; (2) variation in numbers and relative abundances of species within and among communities; and (3) variation in the type and functioning of ecosystems across a landscape. In many cases, however, the development and persistence of diversity at these three levels are intimately linked to three aspects of ecosystem structure and function (Franklin 1988; Perry et al. 1989; Franklin et al. 1990). These three aspects are structural diversity, spatial diversity and functional diversity (Table 1). In defining a research agenda and in setting conservation policy these linkages need to be remembered both for the problems they present and the opportunities they provide. They reveal problems because a conservation strategy focused only on species or genes without considering ecological processes is doomed to failure. They reveal opportunities, because in many cases, ecological processes or structures are more amenable to direct management than are species or genes.

Table 1. Defining biodiversity at several levels in a hierarchy: diversity at the traditional levels of organization in biology is linked to ecosystem structure and function.

genetic	structural diversity
species	spatial diversity
ecosystem	functional diversity

An Outline for a Biodiversity Research Agenda

A biodiversity research agenda can be organized around five general questions, each associated with a traditional biological discipline (Table 2). For each I will highlight some of the main research issues.

What is There?

"What is there?" questions are the basic questions of systematics and ecological genetics: What species are there? How do they vary? To what can we attribute that variation? These questions are fundamental to all other components of biodiversity research. We need to be able to name and inventory the basic stock of biological systems before we can proceed effectively with either conservation research or management. As we have seen in other papers in this symposium (e.g., McPhail,

Table 2. Basic questions in biodiversity research and the biological discipline with which they are associated.

Question	Discipline
1. What is there?	systematics/genetics
2. Where is it?	biogeography
3. How does it work?	population, community and ecosystem ecology
4. How is it changing?	long-term studies in population, community and ecosystem ecology
5. How can we keep it?	conservation biology adaptive management

Chanway and Marshall), our knowledge at this most basic level is very poor for most taxa in B.C. Any survey of insects in remote parts of the province or in poorly known systems (like forest canopies) reveals species that are new to science. At the same time, small, poorly known organisms are extremely important in ecosystem function (Wilson 1987; Perry et al. 1989).

It is frightening that the most serious problem for "What is there?" research is one of human resources. Systematics is no longer considered "sexy science" by granting agencies and many scientists (Miller, this volume). Obtaining research funds for systematic research is difficult, few systematists are being trained and fewer are being hired. One regularly hears stories of excellent systematists leaving the country or finding other kinds of employment because no work can be found in their field. Concurrently, obtaining identifications for many of the lesser known taxa is becoming difficult.

There is also a tremendous paucity of information about biodiversity below the species level, such as patterns of genetic variation within and among populations over a species' range — especially for non-commercial species! As Ledig (this volume) has described, loss of local genetic variants constitutes a hidden catastrophe in biodiversity that is largely unrecognized by conservationists, managers or policy makers. For most species in B.C., we don't know enough about genetic variation to assess the magnitude of the problem. Loss of local genetic variants and homogenization of genetic structure are, however, issues of increasing importance (Allendorf and Leary 1988; Pielou 1990).

Where is it?

Biogeography is the science of the geographic distribution of organisms. It examines broad patterns of distribution and the processes that account for them. Because biogeographic processes usually occur over long periods of time and at large spatial scales, they are often not amenable to rigorous experimentation so are out of vogue in the world of big science. Yet many aspects of the modern ecology of organisms are conditioned by their biogeographic histories. For instance, if a small isolated population is a relict of a distribution that was much larger in the past, it may contain much more genetic variation than if it represents the edge of an expanding distribution (Ledig, this volume).

Island biogeography in general and the "equilibrium theory of island biogeography" in particular have received more attention than other aspects of biogeography in the conservation literature. Principles of island biogeography have often been the basis of often conflicting discussion about reserve design (Simberloff 1988). Issues of conservation importance in B.C. involve both true oceanic islands and continental habitat islands varying over several orders of magnitude in size. An understanding of the principles that apply to islands over this range of spatial scales is badly needed. When issues such as the impacts of climate change become recognized as conservation problems, such understanding will form an important basis for decision making.

Biogeographic data are a key component of both biodiversity research and conservation planning in general, and are an essential complement to "What is there?" data.

How Does it Work?

Why does the number of Snowshoe Hares fluctuate? Why are there 23 species of birds in this forest and only 15 in that one? Why can one community be invaded successfully by an exotic species, when another community cannot? In practice, answering such "why" questions comes down to asking "How does it work?" This is the realm of population, community and ecosystem ecology. In fact, many of the classic questions in ecology have clear applications for conservation biology (see Table 3). However, research on these general questions often does not produce data of direct applicability to conservation. The road to general understanding in ecology is a slow one with many hairpin turns and stop signs.

While basic ecological research sets the stage for conservation science and must continue unabated, a real need for research with a

343

Table 3. Some classic questions in ecology that can be basic to conservation problems.

---

1. What regulates populations?
2. What determines the number of species present in a community?
3. What characteristics of a community or ecosystem make it stable or resilient with respect to a given disturbance?
4. What controls the cycling of elements in an ecosystem?

---

focus on applied conservation problems exists. For instance, a central issue in conservation biology is the long-term viability of populations in reserves subject to different constraints (Soulé 1987). Our ability to determine the viability of real-world populations under real-world conditions, however, is often not good. The processes determining population viability are fundamental ecological processes: research on these processes need not be compromised in quality or in basic importance to apply directly to practical conservation problems.

A further issue with regard to these basic ecological problems is important. In many cases, if maintaining genetic variation, species or communities is the management objective, the components of the system most amenable to management will be ecological processes or structures rather than the genes, species or communities themselves. For instance, frequent, small-scale, low-intensity disturbances play an important role in maintaining community composition in many forest types. Maintaining these communities can be accomplished only by maintaining an appropriate disturbance regime. Similarly, conservation objectives may sometimes be combined with goals for commodity production by focusing on retaining key structural features such as snags (Franklin et al. 1986; Bunnell and Kremsater 1990).

How is it Changing?

Not only do we need to characterize biodiversity and the ecological relations that maintain it, we need to assess the rate at which it is changing. Recognizing the change is inherent in ecological systems irrespective of human intervention, and understanding the problems of assessing and interpreting change are key issues in designing conservation strategies that will work over the long term. Not only do ecological systems change over time, they often change at time scales that are difficult to *detect* but are still ecologically important (Magnuson 1990). Short-term data sets may be poor indicators of patterns or mechanisms of long-term change, especially for long-lived organisms (such as trees and whales) or other slowly changing systems. The importance of long-term studies has emerged over the last decade as a major "new" perspective in ecology (Davis 1981, 1986; Franklin and DeBell 1988; Magnuson 1990).

What does this mean for biodiversity research in B.C.? It means that although we need quick answers to pressing problems, we must be cautious of short-term studies. In conjunction with studies to provide quick answers, we need long-term research to address questions that cannot be answered any other way. Furthermore, we need to recognize that our understanding of ecological systems (and their conservation needs) will change as longer-term data become available and that this expectation must be built into our policy frameworks.

A key methodological issue related to recognizing the importance of change in ecological systems is our ability to *detect* change. For instance, if we were concerned with the potential impacts of resource development on a given animal population, we might ask "What is the likelihood that we could actually detect a decline in abundance if it occurred?" In other words, what is the power of the statistical tests we are using? A discussion of statistical power, beyond the scope of this paper, is provided in Toft and Shea (1983) and Peterman (1990). The main point here is that in many cases where researchers have come to a statistical conclusion of "no difference", it was not justified. While the implications of this finding for resource policy are obvious and disturbing, such methodological and statistical aspects of conservation problems rarely receive the attention they deserve.

Long-term change in populations and communities often reflects change in the spatial mosaic in which they are embedded. For instance, local changes in abundance may be caused by restriction or elimination of migration or dispersal routes. These landscape-level problems are central to conservation biology and will be discussed in the next section.

How can we keep it?

"Conservation biology" is the application of the above disciplines (and often others, such as soil science) to applied conservation problems. It attempts to answer the question, "How can we maintain, protect or restore biodiversity at all its levels?" Research in the conservation biology of threatened and endangered species has a long tradition, but more recently there has been a substantial focus on communities, ecosystems and ecological processes.

Many of the central problems in conservation biology deal with integrating the effects of processes occurring at large spatial scales or over long time scales. Spatial problems (Table 4a) are among the most difficult ones facing biodiversity research in B.C. They involve issues both of reserve design and of managing areas outside reserves. It is not feasible to create reserved areas big enough to maintain many large, mobile species or the communities in which they participate (Newmark 1987). Under scenarios of changing climate, this will be true for less mobile species as well. Thus, what we do outside reserves will make the difference between the success and failure of our reserve system (Bunnell and Kremsater 1990; Salwasser, this volume).

Three key lines of research stem from this conclusion. The first is research on reserve design criteria. This is made more difficult because evaluating reserve design must be done in the context of the processes underway outside the reserves. Such context dependence in reserve design is itself a major research issue. The second line of research deals with how to blend conservation objectives and commodity production objectives outside of reserves, so they can support reserve functions. This requires a substantial amount of social, economic and ecological research. All else being equal, the conservation strategy that minimizes socio-economic costs is the best. Our ability to implement an effective conservation strategy will depend on our commitment to the human environment in which it is embedded. The third line of research focuses on the problem of allocating scarce resources. If we have only so much money, time or people to invest in conservation, what is the best mix of investment in reserves, corridors between reserves and "special management areas"? This is a difficult problem that has received little attention in B.C. [see contrasting views on corridors in Noss (1987) and Simberloff and Cox (1987)].

Some significant research problems deal primarily with temporal issues (Table 4b). Perhaps the most prominent is the problem of population viability (Brussard 1991). How do we evaluate the likelihood that a population of a given size will persist over time when confronted with threats of random nature? How can we best use calculations of that likelihood to influence decisions on the size and number of reserved areas or the need for corridors connecting reserves? These are difficult questions to answer even in a theoretical framework, let alone under the vagaries of real-world populations and landscapes. Yet they are critical. We pay at least double for investments in conservation strategies that won't work (for example reserves that are too small, too few or too isolated) because of the opportunity costs.

The impacts of anticipated changes in climate are another major temporal problem. How do we plan for an environment 100 years from now that will be very different from the one we have today, and about which there is great uncertainty? Bet-hedging strategies that do an adequate job under a number of conditions but may not be perfect

Table 4. Research problems in conservation biology.

A. Spatial Problems

- reserve design problems: how big? how many? what shape? how connected?
- connectedness issues in general: dispersal, gene flow, recolonization, corridors
- fragmentation
- landscape level issues in general: technical + substantive

B. Temporal Problems

- population viability analysis
- anticipating the impacts of modern climate change
- diversity as a result of dynamic processes; e.g., disturbance regimes, past changes in climate

under any one are the obvious answer. But what does this mean for on-the-ground conservation planning? Ledig (this volume) suggests the need for *ex situ* gene preservation, but what about other kinds of biodiversity?

These examples of temporal problems have clear spatial components and their effective solution must be an integrated one. This is exemplified by the last example of temporal problems in Table 4b: recognizing that diversity is a result of dynamic processes. Natural disturbances are a critical component of such dynamic processes. They operate at time scales varying from thousands of years or more (such as the processes resulting in stickleback speciation: McPhail, this volume), to the yearly creation of open space in rocky intertidal systems by waves and logs (Paine and Levin 1981). In many ecological systems, the diversity we see is largely a consequence of the particular regime of disturbances experienced by the systems. Planning reserves to account for the certainty and necessity of natural disturbances is necessary for such systems (Pickett and Thompson 1978).

The successes and failures in managing natural fires in Yellowstone National Park have become almost an archetype of large-spatial-scale, long-temporal-scale conservation problems (Christianson et al. 1989; Romme and Despain 1989). Despite an unusually good understanding of the dynamics of natural disturbances at a variety of spatial and temporal scales (Romme 1982; Romme and Knight 1981), managers were unable to avoid a situation where, in one summer, fires burned a substantial proportion of the park. In fact, ecologically, there were good reasons to be grateful for this event, and it was not at all out of line with the scale of such disturbances in the past. Sociologically and politically though, it was disastrous. What has evolved is a recognition of the need to manage "the greater Yellowstone ecosystem" — the park and a very large area surrounding it — as the integrated system that it is. Many parks in the B.C. interior face similar decisions as have been made for the greater Yellowstone system, including the problems of multiple-management jurisdictions and objectives.

In coastal B.C., the issue that ties together difficult spatial and temporal problems best is that of the remaining large, unlogged watersheds. These are potentially significant as core areas in a reserve network, they are few relative to their original number and many are under severe pressure for harvesting. Consequently, setting aside large watersheds will place greater pressure on other areas. Is the trade-off justified? Will harvesting a small proportion of a watershed have broader effects over the whole watershed or will impacts be mainly local? Is it better or worse from a conservation point of view to reserve an equal area spread contiguously over several adjacent watersheds? Is a small, reserved watershed with a large "special management" buffer around it better or worse than a large, reserved watershed without a buffer? I suspect that the success of biodiversity conservation in the coastal forests of B.C. will be strongly influenced by how we answer questions such as these, but much of the research necessary for good decision making remains to be done.

Adaptive Management

"Tell us what you want and we can manage for it!" has been a central theme of public land management across North America (Bunnell 1976). Even today, with the current importance placed on public involvement in decision making, we assume that if we *did* know what people really wanted, we *could* manage for it. This is not true now and I doubt that it ever was.

If all we wanted from our forests was timber, we could probably do a good job of managing for it. If all we wanted was biodiversity, we might be able to do that too, though it would be harder. But we want both and, as I have stressed in this paper, we have many poorly understood problems that are central to achieving this dual objective. Even more dismaying, many of the problems probably fall into the "not knowable" category of Table 1. Over the time frame in which many key decisions must be made, we cannot know with certainty much of the information on which those decisions should be based.

Obviously we need to get on with the various components of the research agenda discussed above, but what do we do in the meantime about important decisions that may foreclose options for the future? We must acknowledge that decisions are being made and must be made on the basis of incomplete information. Just as we have made mistakes in managing for timber and are learning from them, we will make mistakes in conservation and must learn from them. Given the certainty that we will make mistakes, we must structure our conservation experiments to: (1) minimize the consequence of mistakes when we do make them; (2) learn as much as we can from our mistakes and successes; and (3) apply what we learn so that better decisions are made in the future.

These principles, embodied in the research and management strategy called Adaptive Management, view management actions as an opportunity for experimenting and learning (Orians 1986; Walters 1986; Hilborn 1987). Successfully carrying out Adaptive Management is not trivial. Many technical issues relating to experimental design need to be resolved before the approach can be used effectively.

Finally, recognizing our uncertainty about biodiversity and the processes that maintain it means that we must be more cautious than we would be otherwise so we do not foreclose options unknowingly. Until a species or process has been proven to be important, it is assumed to be unimportant, and until a land-management practice has been proven to be a threat, it is assumed not to be. We need to question carefully where the burden of proof should rest. When no proof is possible, we need to consider seriously which tactics maintain options and which foreclose them.

Conclusions

I have discussed key research issues related to five basic questions about biodiversity: What is there? Where is it? How does it work? How is it changing? and How can we keep it? A research agenda that will support effective, long-term biodiversity conservation must address all five questions. Problems and opportunities are inherent in each.

A research agenda also needs to recognize that we cannot learn enough fast enough to act from certainty. This means we must recognize that: (1) what we do now in the best of faith may be wrong; (2) we will need to change our policies and approaches in the future as we learn more; (3) to assure that we do learn as much as we can as fast as we can, we must adopt an adaptive management strategy in our approaches to reserved areas and areas where conservation objectives must coexist with other needs; and (4) we must act conservatively to maintain options.

Long-term conservation of biodiversity in B.C. will require unprecedented commitment, co-operation and honesty from all involved. Research on applied conservation biology depends on successfully integrating all the biological sub-disciplines mentioned in Table 2. Successfully applying that research will depend on the commitment of people with a broad range of skills, interests and professions. Helping to build the understanding and interest necessary for that commitment is the final job for the researcher.

Acknowledgements

I would like to thank C.S. Holling, C. Walters, D. Ludwig, R. Peterman and L. Gass for having inspired me over the years to think about what it means to recognize uncertainty and act adaptively. More recently, F. Bunnell and J. Pojar have contributed substantially to the ideas presented here, by both word and example.

Literature Cited

Allendorf, F.W., and R.F. Leary. 1988. Conservation and distribution of genetic variation in a polytypic species, the Cutthroat Trout. Conservation Biology 2: 170-184.

Brussard, P.F. 1991. The role of ecology in biological conservation. Ecological Applications 1: 6-12.

Bunnell, F.L. 1976. The myth of the omniscient forester. The Forestry Chronicle 52: 150-152.

Bunnell, F.L., and L.L. Kremsater. 1990. Sustaining wildlife in managed forests. Northwest Env. J. 6: 243-269.

Christiansen, N.L., J.K. Agee, P.F. Brussard, J. Hughes, D.H. Knight, G.W. Minshall, J.M. Peek, S.J. Pyne, F.J. Swanson, J. Ward Thomas, S. Wells, S.E. Williams, and H.A. Wright. 1989. Interpreting the Yellowstone fires of 1988. BioScience 39: 678-685.

Davis, M.B. 1981. Quaternary history and the stability of forest communities. In Forest succession: concepts and application, edited by D.C. West, H.H. Shugart, and D.B. Botkin. Springer-Verlag, New York, NY. pp. 134-153.

Davis, M.B. 1986. Climatic instability, time lags, and community disequilibrium. In Community Ecology, edited by J. Diamond and T.J. Case. Harper and Row, New York, NY. pp. 269-284.

Franklin, J.F. 1988. Structural and functional diversity in temperate forests. In Biodiversity, edited by E.O. Wilson. National Academy Press, Washington, DC. pp. 166-175.

Franklin, J.F., T. Spies, D. Perry, M. Harmon, and A. McKee. 1986. Modifying Douglas-fir management regimes for non-timber objectives. *In* Symposium Proceedings, Douglas-fir: Stand Management for the Future, *edited by* C.D. Oliver, D.P. Hanley, and J.A. Johnson. College of Forest Resources, University of Washington, Seattle, WA. pp. 373-379.

Franklin, J.F., D.A. Perry, T.D. Schowalter, M.E. Harmon, A. McKee, and T.A. Spies. 1990. Importance of ecological diversity in maintaining long-term site productivity. *In* Maintaining the long-term productivity of Pacific Northwest forest ecosystems, *edited by* D.A. Perry, R. Meurisse, B. Thomas, R. Miller, J. Boyle, J. Means, C.R. Perry, and R.F. Powers. Timber Press, Portland, OR. pp. 82-97.

Franklin, J.F., and R. DeBell. 1988. Thirty-six years of tree population change in an old-growth *Pseudotsuga-Tsuga* forest. Can. J. For. Res. 18: 633-639.

Hilborn, R. 1987. Living with uncertainty in resource management. N. Am. J. Fish. Mgt. 7: 1-5.

MacArthur, R.H., and E.O. Wilson. 1967. The theory of island biogeography. Princeton University Press, Princeton, NJ.

Magnuson, J.J. 1990. Long-term ecological research and the invisible present. BioScience 40: 495-501.

Newmark, W.D. 1987. A land-bridge perspective on mammalian extinctions in western North American parks. Nature 329: 430-432.

Noss, R.F. 1987. Corridors in real landscapes: a reply to Simberloff and Cox. Conservation Biology 1: 159-164.

Orians, G.H. 1986. The place of science in environmental problem solving. Environment 28: 12-41.

Paine, R.T., and S.A. Levin. 1981. Intertidal landscapes: disturbances and the dynamics of form. Ecological Monographs 51: 145-178.

Perry, D.A., M.P. Amaranthus, J.G. Borchers, S.L. Borchers, and R.E. Brainerd. 1989. Bootstrapping in ecosystems. BioScience 39: 230-237.

Peterman, R.M. 1990. Statistical power analysis can improve fisheries research and management. Can. J. Fish. Aq. Sci. 47: 2-15.

Pickett, S.T.A., and J.N. Thompson. 1978. Patch dynamics and the design of nature reserves. Biological Conservation 13: 27-37.

Pielou, E.C. 1990. Depletion of genetic richness is not "harmless" consequence of clearcutting. Forest Planning Canada 6(4): 29.

Romme, W.H. 1982. Fire and landscape diversity in subalpine forests of Yellowstone National Park. Ecological Monographs 52: 199-221.

Romme, W.H., and D.G. Despain. 1989. The Yellowstone fires. Scientific American 261(5): 37-46.

Romme, W.H., and D.H. Knight. 1981. Fire frequency and subalpine forest succession along a topographic gradient in Wyoming. Ecology 62: 319-326.

Simberloff, D. 1988. The contribution of population and community biology to conservation science. Annu. Rev. Ecol. Syst. 19: 473-511.

Simberloff, D., and J. Cox. 1987. Consequences and costs of conservation corridors. Conservation Biology 1: 63-71.

Soulé, M.E. (Editor). 1987. Viable populations for conservation. Cambridge Univ. Press, Cambridge, England.

Toft, C.A., and P.J. Shea. 1983. Detecting community-wide patterns: estimating power strengthens statistical inference. American Naturalist 122: 618-625.

Walters, C.J. 1986. Adaptive management of renewable resources. MacMillan, New York, NY.

Wilson, E.O. 1987. The little things that run the world (the importance and conservation of invertebrates). Conservation Biology 1: 344-346.

The British Columbia Conservation Data Centre

Andrew P. Harcombe

Wildlife Branch, B.C. Ministry of Environment, Lands and Parks, 780 Blanshard St., Victoria, British Columbia, Canada V8V 1X4.

Abstract

A Conservation Data Centre is being established in British Columbia to produce a centralized, computer-assisted inventory of rare and endangered species and of natural communities. It will be operated initially as a co-operative project between the provincial government and three non-government organizations, with the long-term goal of being fully supported as a government program. The Centre's aim is to compile and organize data related to conserving elements of biological diversity. Objective analysis of selected data will assist in setting clear conservation priorities. The Centre is networked with more than 60 other similar systems in Canada, the United States and Latin America, using standard methods and terminology that encourage interaction across jurisdictional boundaries. The system is dynamic, improving over time as new information is added and old information checked, and is designed to use a broad array of data sources and experts. Its success depends upon co-operation from all parties involved in conserving the province's biological resources.

Efforts to manage the biological resources of British Columbia have traditionally focused on consumptive uses such as game fish, big game, furbearers and timber. Over the past few years, this management focus has expanded to encompass a more holistic approach, attempting to follow a broader ecological philosophy. Examples of such efforts are the ecological classification and interpretation work of the B.C. Ministry of Forests and expansion of the non-game program in the B.C. Ministry of

In *Our Living Legacy: Proceedings of a Symposium on Biological Diversity*, edited by M.A. Fenger, E.H. Miller, J.A. Johnson and E.J.R. Williams, pp. 353-358. Victoria, B.C.: Royal British Columbia Museum.

Environment. Recently completed drafts of the provincial fish and wildlife strategies have strongly emphasized biological diversity. As with many initiatives, however, implemention of change lags behind the desire to make changes.

A major stumbling block continues to be the availability of basic data. In the ongoing old-growth strategy project, a basic question was asked: how much old-growth forest remains and where is it? The answer was not directly available. Relevant data often exist but tend to be scattered and are readily accessible only if you know the right people. Also, because data may be requested at different times, yet on similar topics, people often become encumbered with repetitious searches.

In the summer of 1989, at a workshop in Victoria, the Nature Conservancy of Canada (NCC) in co-operation with The Nature Conservancy of the United States (TNC) demonstrated methods for organizing key information about biological diversity in a single centralized system. This presentation, given to several provincial ministries, triggered strong support and interest that has led to a provincial government initiative to implement a similar inventory system dealing with rare biota.

The Nature Conservancy has a two-level approach to conserving natural diversity, described as *coarse filter* and *fine filter*. Using the coarse filter allows adequate examples of the major ecosystem types to be conserved in a network of managed areas so that most species will be conserved. The species that "fall through" the coarse filter will be the rare ones — species that do not occupy all their suitable habitats or are members of communities poorly represented in the landscape. To conserve them, one must consider them individually and carefully monitor them over time to ensure their continued existence. This is the fine-filter approach. Hence the methodology revolves around a data system containing records on common ecosystems and on rare species and habitats. Items that are tracked are referred to as *elements*.

The backbone of the data system is a permanent and dynamic atlas and data bank that include data on the existence, characteristics, numbers, condition, status, location and distribution of the elements of natural diversity. Its intent is to centralize and standardize data, concentrating on rare elements within a particular jurisdiction (B.C. in this case). Map, manual and computer files keep the information organized and easily accessible. Records will be indexed by parameters that will include standardized name, location, endangerment status, watershed, administrative districts and land ownership.

In the United States, these diversity data programs are referred to as Natural Heritage Programs. The preferred name now, however, is Conservation Data Centres. The first program started in South Carolina in 1974; it proved so successful that TNC has expanded the programs into all 50 states, 3 Canadian provinces and 12 Latin American nations (Grove 1988). All systems were created as joint ventures between TNC and local government during the start-up period (the first two or three years). In Canada, TNC is co-operating with the NCC in promoting and establishing the provincial systems. The intent is that these programs will be in high enough demand that government will incorporate them wholly into their own programs. To date, no programs have been cancelled, since demand from users has provided extremely strong support. TNC supplies the methodology, trains staff, offers technical support and co-ordinates data exchanges and other forms of multi-jurisdictional collaboration.

The system uses a standardized methodology and terminology to process information about natural elements. This enables sharing and exchanging data as well as summarizing range-wide information on species. For example, data on rare species in nearby jurisdictions such as Washington, Idaho, Montana and Alaska are critical for putting our own efforts in perspective. Elements may be ranked provincially, nationally and globally; their relative importance can then be assessed. The quality of a particular element occurrence can also be considered. Hence a rare species population may be healthy and abundant or depauperate and decreasing.

The B.C. Conservation Data Centre (CDC) will focus on elements that include biotic communities (which may be common or rare) and plant and animal species and other natural features that are rare at the national or provincial level. Individuals can identify and compare critical elements to ensure that conservation efforts focus on the most threatened. In B.C., staff can upgrade the resulting data system as new information becomes available. Monitoring extensive commonly occurring ecosystems (coarse filter) will follow a classification that combines ecoregions (Demarchi et al. 1990), site associations of the biogeoclimatic ecological classification (Pojar et al. 1987) and biophysical habitat units (Demarchi and Lea 1989). Rare elements that are tracked will include the terrestrial vertebrates named in the Red and Blue lists of the Wildlife Branch, rare fish identified by the Fisheries Branch and rare plants identified by the Endangered Plant Committee. Initial staff efforts will concentrate on compiling existing element data that are presently scattered in many locations. Consolidating existing sources will provide "one-stop" access to ecological information pertaining to these elements. The inventory is a cumulative process through which information is continuously updated and refined.

The system also has a component for managed areas. This component records each occurrence of tracked elements when they fall within the boundaries of a protected area. These areas are also included in the map files. Where available, data on communities are also maintained for managed areas. In B.C., managed areas will include ecological reserves, provincial parks, national parks, wildlife management areas and wildlife reserves.

The critical driving force of the entire program is co-operation. People and organizations with data must be willing to share information and expertise so that the best possible information is included in the data base. This completeness will ensure justifiable conservation efforts. In terms of data gathering and data use, the network of co-operators is unusually broad, involving universities, government agencies, corporations, federal agencies, private conservation groups and private individuals. The centralized repository approach is a necessity for clients of the information; hence the CDC must gain a reputation for shared support and for professional and careful use of this information. The CDC is not in the business of producing large reports or releasing data sets that may pre-empt other professionals' efforts to analyse and publish similar information.

Often people fear that data in the system, once centralized and accessible, could be misused. Although this could occur, mechanisms of data release can "hide" sensitive data, such as the specifics of geographic location for certain exploitable taxa to all but a specified group of resource managers (e.g., Peregrine Falcon nest sites, orchids). Refusal to provide any information, however, will ensure that current "ignorant" decisions leading to further damage of critical elements will continue. The people who make decisions must have access to good information; otherwise, who will know about or be concerned with the Nuttall's Quillwort in Beacon Hill Park, the Marbled Murrelet in our old-growth west coast watersheds or the sticklebacks in Enos Lake? We also assure our co-operators that a "need to know" must be established before we disseminate data. All reports from the system are channelled through the data manager. It will be at least one year, however, before enough information is in the system to provide useful reports. (The centre is now producing reports routinely - Eds) The order in which we acquire information will be governed by geographic priorities and the status of the element.

Uses for the information in the Conservation Data Centre include:
- Land protection: good information will help focus attention and dollars on the province's most critically threatened elements of natural diversity. Managers and land owners can be informed of the presence and importance of these critical elements so that they can voluntarily protect them.
- Environmental Impact Assessment (EIA): a professionally

staffed and centralized data base that is readily accessible to help with informed decision making will be invaluable for EIAs.

- Resource management: information about areas currently protected may be used to improve existing management practices and policies leading to wiser stewardship of what is already protected.
- Endangered species review: information collected and analysed by the CDC is valuable for establishing and reviewing priorities on provincial lists of endangered species and habitats.
- Research and education: information gaps identified will help direct further field work. Many educational uses of the information are evident.

The data will be of tremendous help to various client groups in B.C., including:

- regional management staff, to help ensure that the long-term protection of important elements is adequately considered during various stages of the referral process (reactive);
- other resource managers, to assure that rare elements are considered in development plans and activities (proactive);
- government and non-government conservation agencies, to help priorities for land acquisition and protection; and
- naturalists, to identify priorities and data gaps where better information is required by managers and researchers.

The CDC project started in early February 1991 when four Nature Conservancy staff were hired. The project, a joint venture of the B.C. Ministry of Environment, the Nature Trust of B.C., NCC and TNC will initially concentrate on vertebrates, vascular plants and ecosystems. (The Nature Trust is co-ordinating the provincial fund-raising for the private share of CDC funding. NCC is assisting, and is helping to co-ordinate this and other Canadian initiatives. TNC is providing all technical and staff support.) For the initial three-year period, direct financial costs are being provided by the Ministry of Environment and three non-governmental partners. Private funds still being raised by these non-governmental partners have already provided over half of the needed funds.

Andrew Harcombe will co-ordinate this project for the Wildlife Branch in close consultation with Bill Munro, Chris Dodd, Ted Lea, Don Eastman and others. The project has been strongly supported by the Ministry Executive and endorsed by the Royal B. C. Museum, Ministry of Forests, Ministry of Parks, Canadian Wildlife Service, University of British Columbia (Department of Zoology and Faculty of Forestry), Federation of B. C. Naturalists and B. C. Wildlife Federation. Staff training has just been completed and one of the staff's first tasks will be to arrange access to the many experts, both amateur and professional, available in B.C. Knowledgeable individuals who willingly contribute information and biological collections form the backbone of data sources.

The CDC employs four staff: a botanist, a zoologist, an ecologist and a data manager.

Acknowledgements

Much of the background information on the CDC has been obtained from manuals and handouts provided by the Nature Conservancy staff. Sue Crispin (Toronto) and Robert Jenkins (Arlington, Virginia), in particular, have been helpful in discussions on program methodology and philosophy as a background to designing this initiative in B.C.

Literature Cited

Demarchi, D.A., and E.C. Lea. 1989. Biophysical habitat classification in British Columbia: an interdisciplinary approach to ecosystem evaluation. *In* Symposium Proceedings, Land Classification Based on Vegetation: Applications for Resource Management, *compiled by* D.E. Ferguson, P. Morgan and F.D. Johnson. Forest Service- USDA, Gen. Tech. Rep. INT-257, Moscow, ID. pp. 275-276.

Demarchi, D.A., R.D. Marsh, A.P. Harcombe, and E.C. Lea. 1990. Environment. *In* Birds of British Columbia, Vol. 1, *by* R.W. Campbell, N.K. Dawe, I. McTaggart-Cowan, J.M. Cooper, G.W. Kaiser, and M.C.E. McNall. Royal B.C. Museum, Victoria, BC. pp. 55-144.

Grove, N. 1988. Quietly conserving nature. National Geographic Magazine, December 1988: 818-845.

Pojar, J., K. Klinka, and D.V. Meidinger. 1987. Biogeoclimatic classification in British Columbia. For. Ecol. Manage. 22: 119-154.

Public Expectations

Panel Discussion:
Opinions from Users of the Environment

Fred Bunnell

Faculty of Forestry, University of British Columbia, Vancouver, British Columbia, Canada V6T 1Z4.

Fred Bunnell (Moderator):

Jim Pojar's quote from America's first hippie, Walt Whitman, was, "I am large, I contain multitudes." I guess we all do. And in a pluralistic democracy, the democracy must contend with a multitude of opinions. That may be all we're exposed to in this session, but I hope we can do a little better. I hope we can find some of the common elements to help us capture not just the diversity of opinions, but the diversity of responses.

We have a panel that is diverse. Several of these individuals are scarred but sustaining veterans of the Old-growth Strategy. There's Vicki Husband, who's a representative of just about any environmental cause that's worthwhile; Andy MacKinnon from the Research Branch, Ministry of Forests; Jim Walker, Assistant Deputy Minister of the Ministry of Environment; Geraldine Shirley, who has the task of representing at least part of the opinions of the First Nations; Don McMullan, Chief Forester of Fletcher Challenge Canada Ltd.; Hal Salwasser, head of the New Perspectives program in the U.S. Forest Service, who is big enough that if this gets unruly, he can help; and Derek Thompson, Director of Planning and Conservation Services, Ministry of Parks. I've asked them each to briefly indicate some of the kinds of opinions they hear as representatives of their agencies or groups. Then the comments go back to the floor. This is your session. The microphones are out there and it'll either all unfold as it should or unravel as it might.

In *Our Living Legacy: Proceedings of a Symposium on Biological Diversity*, edited by M.A. Fenger, E.H. Miller, J.A. Johnson and E.J.R. Williams, pp. 361-385. Victoria, B.C.: Royal British Columbia Museum.

Derek Thompson (Ministry of Parks):

It's always intimidating to follow you. But at least I don't have to follow the rest of the erudite crew up here. I want to apologize to those of you who are fed up with hearing from me on old growth and Parks 90 and things like that. I'm neither a biologist nor a professional forester. I'm that most dangerous of things according to Jim Pojar, a professional planner, and worst of all, a professional parks planner.

On Wednesday, we finished a sort of epic — 105 public meetings around the province where we heard in no uncertain terms from a lot of the public. So I'm going to talk mostly about what the public seems to be saying to us.

These town hall meetings were amazing. We had a little more than 11,000 people come and more than 700 people actually made presentations. So my head is full of a lot of contrary opinion. People were from all walks of life and, as I said, they told us in no uncertain terms what they thought. We biased the sessions deliberately towards small communities — towards single-industry towns and native villages — and heavily away from the major urban centres. I find the message they've given us exhilarating, tiring and frustrating all at once. I want to make five key points.

First, love and pride for our land comes across from everybody, the feeling that British Columbia is something special, but also facing the difficulty of sustaining life in small communities. Everybody wants British Columbia to be better in some way.

Second, there's real confusion about how to do it. Public attitude shows a lot of misunderstanding and a great deal of misinformation. In one town, for example, I learned some unique features about the habitat needs of caribou and the benefits to them of logging, according to one young gentleman. Level of knowledge is low, but the desire to gain knowledge is great and it's very genuine. A lot of people feel frustrated that they don't have the information they think they need.

Third, there's a great deal of distrust and mistrust, particularly of professionals and particularly of professionals from Victoria and Vancouver who seem to think they know what should be done and want to impose it on smaller communities. It's quite humbling, really. In particular, when looked at from a ministry like Parks, people are concerned about what they see as competition among agencies, especially competition among different conservation agencies. They're very concerned about it and they want it to stop.

Fourth, very real concern about our environmental future is matched by concern about community futures. How can we work it out?

And fifth, people expressed almost universal preparedness and desire to change things. Unfortunately, however, as always, what we're going to change and how we might change it is very clouded. I heard a great deal of rhetoric about land-use policy and new ways of doing business but not a clear understanding about how that might be achieved. People expressed no absolutes, just a desire to change things and to be part of that change. That's a very simple summary of thousands of hours of meetings and submissions.

I guess for provincial parks, some of the messages that came across most clearly confirm thoughts we had already got from household surveys and talking with our own users. First, the two most important values the public believes it gains from provincial parks are protection of the environment in general and protection of special wildlife values and habitats, not the recreational opportunities. Second, even when people know their communities will suffer job losses and it may be them who will lose their jobs, they are willing to set aside areas for parks and risk the economic consequences. This is not a huge majority of people, but it is a significant number of people.

What does this mean for all of us here today? The professionals who spend time dealing with these issues have to do a better job of informing people and of working with groups like the Federation of B.C. Naturalists and inventory activities like the Conservation Data Centre Andrew Harcombe was talking about (see Harcombe, this volume — Eds). We professionals must understand that we have a long way to go to regain public trust and to gain public understanding about what we do. We have to work towards more truly local and public processes such as the one we have started with Parks and Wilderness in the '90s. And that means listening to people actively, not just listening to them, letting it bounce off and carrying on the way we did before. Government agencies must find ways of working together more effectively on common causes and give a little to each other's individual goals and objectives. Thank you.

Fred Bunnell:
Thank you, Derek. After Derek's summary of thousands of hours of hearings into a few minutes, I'll remind the other panelists that if you talk less, you give the floor more time. I want to make sure that the floor does have time. So how brief can you be, Hal?

Hal Salwasser (New Perspectives, U.S. Forest Service):
I'll let the floor have all my time.

Don McMullan (Fletcher Challenge Canada Ltd.):
Fred is an old class-mate of mine so I know a lot about him that I'd love to tell this audience, but I won't.

When I was preparing for this symposium yesterday morning I got a phone call from a young woman from Simon Fraser University who wanted to talk with me about biodiversity. As we started to talk, we had a tough time focusing on what the subject was. And she said, "Boy, it's a really squishy topic isn't it?" And I thought to myself that the topic really is squishy.

Last November, we brought consultant Nils Zimmerman into our corporate office in Vancouver and sat him down with about 20-25 of our senior vice-presidents and general managers of logging. Nils talked with them for two hours on biodiversity. These guys had never even heard the term before. Eyes glazed over, heads rolled back and people scratched themselves. At the end of two hours they walked out, and they walked out with tons and tons of questions, questions like: what about...? what if...? don't we already...? what evidence...? what cost...? who will pay? what to do? and so on. Like our executive management team, over the last couple of days you, the audience, have been welcomed, you've been told about global values of biodiversity, you've been exposed to principles of biodiversity, you've been told about the diversity of B.C.'s ecosystems, you've been shown diversity at work, you've been given strategies, and you've been asked, what can we do? And now, most controversial of all, you are to be told by this panel what the public expects. I don't know, Fred, did you pick this topic?

Fred Bunnell:

I just hoped for an overview. I liked Derek's summary because he's been exposed to so much public opinion recently. What I was after was, very briefly, a statement of the kinds of opinions that are coming to you, how you feel you need to respond and then give the audience time to ask questions.

Don McMullan:

The simple answer is that if the public doesn't know what biodiversity is or what it means, how can there be any real expectations? I think that's the easy answer, but I think it's the wrong answer. Yet how can we here today be so bold as to speculate what the public expects? It's especially risky because, as you've heard, we don't have species inventories, we don't understand species' roles within ecosystems, we don't know crucial thresholds, we don't know resiliencies and we don't have agreement on a whole range of related and central issues.

I believe that as a starting point the public wants healthy and productive ecosystems, and they want them as one of their legacies. But another public that you haven't heard about in the last two days wants a healthy and productive economy. I believe both are interdependent. And this interdependence of economy and environment is the foundation of the concept of sustainable development.

I see here today an opportunity to collectively work in the public interest on two important fronts. One is strengthening our economy over the long haul and the other is strengthening our environment over the long haul. We can harness the drive of our free-market economic system to achieve the goal of sustaining our environment. We can do this by developing a program that demonstrates to British Columbians that maintaining biodiversity is good for business and for the environment. How can we do this? By first developing an estimate of the economic value of maintaining biodiversity. It's a prerequisite to allocate sufficient resources of people and budgets to maintain biodiversity. It's also a prerequisite that the private sector allocate its talent and resources to the same program. Think what rapid advances we can make in this province if we were to achieve public understanding and support, government and private-sector commitment, and financial and program support. We need action on four fronts: (1) knowledge, (2) the economy and the value to it of maintaining biodiversity, (3) education, both for scientists and for lay people, and (4) the parliament.

If it's true that biodiversity is essential to healthy and productive ecosystems, then I cautiously suggest that the B.C. public expects something along the following lines. I've got nine points:

First, get cracking on essential scientific studies. Let's get on with it. Second, don't stop the world until all the facts are in. Third, build as much as possible on existing knowledge and experience. Fourth, ensure that the recommendations fit conditions and situations in B.C. Fifth, don't impose short-term drastic change. Sixth, concentrate our resources, talk about team work, involve the best people we've got on the task of maintaining biodiversity. Seventh, educate resource managers and the public under the principle of KISS — "Keep It Simple, Stupid" — because it's a very, very complex topic. Eighth, avoid destructive and confrontational rhetoric. And ninth, without a doubt the last thing I'm sure the public expects, don't ask for lots of money to do it.

Fred Bunnell:

Thank you, Don. I appreciate your summary of the breadth of perceptions within the forest industry.

Geraldine Shirley (First Nations of South Island Tribal Council):

We are pleased to be invited to participate in this symposium and lend our support as aboriginal people with our perspectives on all the issues that have been discussed in these past two and a half days. In our view, many plans taking place in B.C. conflict with nature. I can give many examples, and many of you probably know these examples, examples such as pesticide use near creeks, clearcutting, developments not properly planned and without proper sewage systems, and building going on where it shouldn't, like the proposed ferrochromium plant. In all this, we should have aboriginal input.

There seems to be a consensus at this symposium that we should find better ways to control what is being done to our environment. We must not build evidence against each other, but look at establishing a process and enforce that process, and do it together. What is the reason for this? Our children and future generations. We need planned land and water use because many things that have happened in B.C. were done without planning. One example is sewage dumping in the Saanich Peninsula.

We aboriginal people have guidelines that are passed down from generation to generation. I will share with you what we focus and centre on, and that is respect — respect for everything around us: the animals, the forests, habitat, water and environment. All things are for good use. We aboriginal people look at nature spiritually, for our well being. We would like to share our teachings with all people, not just among aboriginal people, because the respect is not just for nature but also for people. We respect, for example, logging companies and the people who work for those companies. But the companies lose our respect when they indiscriminately clearcut with no care for the area. We need reforestation, we need to protect fish streams and animals, we need to think about their place here. In closing, before future developments are made, we aboriginal people of B.C. would like to be consulted for we believe that we can contribute to the holistic quality of lands and waters in this province. Thank you so much.

Fred Bunnell:

Thank you, Geraldine. Despite all the dimensions of diversity we talked about and the compelling arguments for the practical values of it, I don't think anybody ever explicitly said it represented respect for nature, although it seems that's just what it must do. So in that case, there's a great deal of shared view on your comments. Your words also harkened back to one of the comments Hal Salwasser made in his talk. For this to work there will have to be unprecedented public participation.

Jim Walker (Ministry of Environment):

I'll try to address what I think the public wants with respect to fish and wildlife resources. I think it's fairly easy to determine what they want. The public wants everything, the same as if you ask them about health care or whatever. This is one of the conundrums we face, that we have to resolve. We've had people sit around and talk about integrated management or multiple use or co-ordinated planning or sustainable development, and someone will say, "You know, we don't really have to have all this confrontation." If some of us simply sit down and communicate better, we can have all these things. To use a wildlife analogy, as one sparrow said to the other, "Don't swallow that." One thing we have to do to bring debate to a reasonable level is to recognize that this province has to make some hard choices. Biodiversity is simply another factor in quality of life. How much quality of life do you want? How much biodiversity do you want? That may be kind of a paradox, but I'd like to use it to illustrate my point. With the wildlife resource in this province, we have to make decisions fairly soon. If you are satisfied with White-tailed Deer and Raccoons and Canada Geese, there's no problem. We can all go and have a beer because those creatures are doing very well. If we're going to preserve populations of Grizzly Bear or if we want populations of wolves or Caribou or some other animals, that's going to take a lot more. That will cost some people some wood and other people some space.

Those are the types of decisions we must make in B.C. I fear we're not in the right mode to make them because we face a unique situation in B.C., which is that too many of the people who have as much, say, as you and me live in urban areas, and they don't come to conferences like this. They're sitting at home with their feet propped up watching Marty Stouffer's "Wild America" and they think they can have what they want with simply a nod in the direction of wildlife management — and they can't. It's time for hard choices.

I think we're going to have a problem sorting out the conflict of demands in this province. I never cease to be amazed by just how many different groups there are, and I'm not talking about just public environmental groups. There are also industry groups and other groups. You name a species or a watershed and I can give the address of a group that's been formed to save it. We used to say in the Wildlife Branch that of the last two totally pristine unlogged watersheds on the west coast of North America, thirteen are on the west coast of Vancouver Island.

One thing we have to do is see that better information is put out to the public. People in government have fallen down. We haven't been able to contend with the pervasive influence of things like American television, which, as I say, tells people they can have whatever they want as long as they get a moderate level of management. That's not true.

The other thing we have to do is start engaging the public in some of the debates. Biodiversity isn't just a technical issue, it's not a geographical issue, it's not area-based, it's not a logging issue. It's a technical and moral issue, and people have to decide what they want, ethically and morally. I often use the example of the Wolf Control Program mainly because I was caught in the middle of it. The Wolf Control Program has two angles. It has a technical side, which was never an issue. It was used as an issue, and the word "extermination" was used, but extermination was never an issue. The troublesome angle about whether or not we should kill one animal to increase the population of another is a very valid question, a question that we in the Wildlife Branch continually debate among ourselves. It's one that society has to debate. An interesting off-shoot is what we're going to do with biodiversity. We have some areas right now in this province where some natural populations are going to be significantly affected by other natural populations — nothing to do with the hand of humanity. One is Caribou in the Quesnel area that may go extinct because of wolf predation. So an interesting question for the manager of biodiversity is, are we going to do anything about it? In the Queen Charlotte Islands, there are populations of seabirds that aren't hunted, so they are of no benefit to hunters — they're simply seabirds that nest on the ground. They're being decimated by Raccoons. Do we do anything about the Raccoons? These aren't technical questions. These are moral and ethical questions that all of us who are interested in wildlife or in natural resources have to engage ourselves in.

The last point I'd like to make is that, because the government hasn't come up with answers to these problems and is more or less in disarray as to how to resolve them, whether they be questions about wildlife or land use, the answer has been to form small groups in local areas in the hopes that they can come up with answers. I find that approach somewhat questionable. If a bunch of dispassionate bureaucrats in Victoria have been unable to resolve the situation, how can we expect the people who live in an area and have everything vested in it to be any more dispassionate when they try to make a decision? So one of the things a land-use strategy has to address is the fact that we simply cannot afford to put together a group of local people and ask them to make a decision every time we have a question about biodiversity or land use. The key thing for a land-use strategy is that the public has to be asked what they want in this province. In a specific region, the people must decide what balance they want among logging and wildlife and fisheries and cultural heritage or whatever it be, and then let the managers manage. The public can't try to manage on top of the managers. People have to be involved. The public should have a big say in what the annual allowable cut is in this province. They have to have a say. But they should not have a say in every cutting plan in this province, and that's what is happening. Perhaps lack of credibility by government has partially caused this, and government should try to address it. But we have to give some real thought about proper public input in land planning in the future and answer some of the questions I have raised and be certain we engage the moral and ethical debate as well as the technical. Thank you very much.

Fred Bunnell:
Thank you. Andy?

Andy MacKinnon (Research Branch, Ministry of Forests):
I'm here to briefly present the Ministry of Forests' opinion from the view of a Ministry of Forests' researcher. I've seen some members of our executive here. I'm sure they'll correct me if I misstep in representing the ministry's position.

While the Ministry of Forests is not legislatively mandated to manage for biological diversity, most of us in the agency consider that it is implicit in our ministry's commitment to sustainable development. However, as Colin Rankin noted so well in his talk, we tend to be judged by what we do rather than what we say. I'd like to take my three minutes to briefly note the number of biodiversity initiatives the Ministry of Forests is involved with now.

We have initiated more than 30 biological diversity and old-growth forest initiatives this year and I can provide details on these. But I'll just touch on a couple that I think might be of interest. It's perhaps fortunate that we have a researcher as a representative on this panel because a lot of what we do in trying to manage for biological diversity falls under the heading of either research or adaptive management. I can't imagine that anyone, anywhere, at any time, has ever successfully managed for biological diversity. It would be too difficult to tell whether you were retaining all the aspects of biological diversity that you were trying to, given the fact that a number of speakers have already told us that we don't know what 90% of what we're trying to manage for is. As a result, we try to apply our best knowledge to the management systems we're dealing with and follow up with what Ken Lertzman referred to as adaptive management. That is, we make our plans, we implement them, and then we try to see if they're achieving what we want. In that way I think researchers are in a rather unique position. We've been able to get access to more funds, and people are interested in what we're doing now, whereas they never seemed to be before (some of them were actually yelling at us). So that's a step in the right direction, I suppose.

We have to try to maintain biological diversity at a couple of levels, and I'd like to simplify it a bit. You can look at it from a landscape level, where you're considering ecosystems and some wide-ranging species, or you can look at more of an individual stand level, where you're perhaps considering less mobile species and concerns about genetic diversity. At a landscape level, the Ministry of Forests for the past 20 years or so has been classifying and describing the forested and range ecosystems of B.C., arranging them in a number of different climatic zones. We did this for one purpose which had to do with forest management, primarily at the stand level. Recently, we found that this has been useful in stratifying the province to develop a systems plan for ecological reserves and for coming up with definitions for old-growth forests. So a lot of information we have about old-growth forests in B.C — about their structure and their biological diversity, at least in terms of the plant species — comes from information on climatic and geographical zones gathered for forest management. In short, Ministry of Forests initiatives at both landscape and individual stand levels have been taking place.

Along with the excellent Parks Plan 90 that has been touring the province, the Ministry of Forests Wilderness Plan 90, suggesting 59 wilderness study areas that will add to the area of B.C. under protection status, has also been going around. The Ministry of Forests was mandated in 1978 to manage provincial forests for wilderness as well as timber. Several speakers have noted that we're not going to be able to achieve any of our objectives of maintaining biological diversity in protected areas alone, no matter how hard we try in B.C. We have to change the way we do some things on managed landscapes and this is something else we're looking at quite closely in this province — the way our forest practices impact on ecosystems and some of the alternatives that may be available. So a lot of money is going into things like silvicultural systems research right now, looking at different ways of managing ecosystems. We've been looking at the potential application of some of the new forestry principles and practices in B.C. within the constraints of the different ecosystems and climatic regimes we have here. Possibly we can use some of those principles to develop some new forestry practices of our own in this province. The Ministry has a land management report on principles and practices in new forestry for British Columbia if anybody's interested in that topic. (Hopwood, D. 1991. Principles and Practices of New Forestry. B.C. Ministry of Forests, Land Management Report #71.)

It's also important to look at how some research initiatives might be applied because, as everybody knows, research is research and sometimes that's all it is. In the protected areas we were talking about, we have the Ministry of Forests' Wilderness Plan. We also have the Old-growth Strategy project involving the Ministry of Forests and other agencies, mandated to ensure that we have representative, functional old-growth forest ecosystems of the various types that occur in B.C. So those are areas where we're actually seeing on-the-ground application of some of these ideas.

And finally, as Tom Gunton noted, we need to look at incorporating some of these biological diversity concerns directly into our management plans, ideally at different planning levels but especially in some of the larger-scale planning. In the Prince Rupert Forest Region, for example, a biodiversity committee is working with local Timber Supply Area (TSA) planning processes to make sure that concerns and strategies for biological diversity are incorporated explicitly in things like TSA plans. This also might be done at the local resource-use planning level like the biodiversity study at Tofino Creek that will deal with planning options and different levels of biological diversity in the Tofino Creek watershed. Those are some of the things that are going on, and I'll leave it at that, Fred.

Fred Bunnell:
Thanks, Andy. Vicki?

Vicki Husband (Friends of Ecological Reserves):
Thank you for letting me be last. It's a help because I always seem to write my speeches at the last minute.

I echo Geraldine's thoughts about respect for nature and I was pleased to hear Randy Chipps on the first day set a tone that contrasted with the political statements we'd heard previously. It put the right tone in the right place because this is what it's all about and sometimes we seem to have lost our way.

What does the public want? Well I have one immediate suggestion. I think we have got to entice Hal Salwasser up here to head the Forest Service. I was very impressed with what he had to say and with his understanding of what's going on. I thought that was really important.

But I wonder about some things. My real problem in land use has been dealing with the Ministry of Forests. The Research Branch has been wonderful, but I wonder whether the Timber Harvesting Branch is really listening. All things considered, it's surprising we're even here. Maps from the Ministry of Forests and Fish Forestry guidelines only look at fish and timber. They don't look at wildlife or recreation. There

371

are problems with that. It really does depend on your objectives and I think that's what the public is concerned about.

We do want a land-use strategy. In Mike Apsey's speech to the Council of Forest Industries, he envisions a defined commercial forest land base of about 30% of the land in the province. That's virtually every valley bottom and all the productive forests. That's also where the most productive wildlife habitat is and it's a lot of other things. He also says that, and it's easier to quote him, "What we've got is a veneer of public involvement as an overlay on top of a land-use decision-making system that is so flawed it can't possibly work." I do share that feeling. How are we going to make it work? That's our major issue.

My feeling in dealing with government is that I want the Ministry of Environment at the decision-making table. An interesting publication has come from the Canadian Bar Association, B.C. Branch, looking at forest-law reform and detailing quite clearly that the Ministry of Environment is given a responsibility and a mandate to protect the environment, wildlife habitat in particular, and yet they're not given opportunity to do that because they can only give recommendations to the Ministry of Forests. Obviously this hasn't been working. The Ministry of Environment doesn't have the staff, the resources, or the clout. I just visited Smithers and met with five people looking at fish and habitat issues there. The area they're supposed to plan for covers 33% of the province, all the way from the Yukon border down to Bella Coola, west to the Charlottes and east to Burns Lake. It's obviously an impossible task. What are we going to do about it? I'm told that we have 16 wildlife habitat managers in the province overseeing 75 million cubic metres of cut. In the U.S. it's 700 overseeing 55 million cubic metres of cut on national forests. So we've got to decide. Are we going to put more money into really protecting biological diversity? Where should we put it? What research needs to be done? If we want to protect biological diversity we have to change the way we're doing things.

I'm a director of the Friends of Ecological Reserves and we have a new study out done by Keith Moore that profiles the undeveloped watersheds of Vancouver Island, looking in depth at five pristine and four slightly developed watersheds on Vancouver Island (see Foster and Miller, both this volume - Eds). Keith gives a lot of credit to all the ministry staff up and down the Island who helped him — Environment, Parks, Forests — everyone was extremely helpful. But he says, "I was surprised, even shocked, how little systematic, good inventory information exists for most of the undeveloped watersheds on Vancouver Island. Most government staff agree that present inventory information is inadequate to be making major land-use decisions in the undeveloped watersheds. They recognize the value of better information and would strongly support field inventory programs to gather a variety of necessary information." So there's a real agreement. The present lack of information, however, is related to shortage of staff, money and the pressure of other priorities, and these do not appear likely to change. Better information is probably a long way off. Wildlife inventory, where it exists, is limited to deer and Elk. Other species such as bear, Cougar, wolf and Marten are mentioned only in passing if at all. Waterfowl information, even in the Tahsish, the best studied of the watersheds, is limited. I don't blame the Wildlife Branch. They haven't had the staff and they haven't had the resources. Information on freshwater fisheries is limited. Information on rare plants, big or old trees, and other vegetation features is also sketchy. Information on recreational and tourist use is lacking. The greatest amount of information gathered has been on forest-harvesting operations.

So we have to change some of our priorities. It's really important. We all need information if we're going to make adequate decisions. What can we do in the transition when we don't have this information? The Old-growth Strategy has started by identifying some critical old-growth forest areas and trying to get a deferral on logging plans, at least until a land-use strategy could be put in place. But obviously information is not available. Jack Ward Thomas said that the legislative process towards protecting the Spotted Owl would not have worked if the U.S. hadn't had an Endangered Species Act and if there hadn't been litigation. We still suffer with no endangered species act, we don't have a wildlife habitat act, we don't have a forest practices act. We do need some legislation, but with it has to go monitoring and enforcement, and that's important.

We're all concerned about education. I welcome the new Conservation Data Centre described by Andrew Harcombe and we will provide copies of our studies to the Centre so they go into the public database (see Harcombe, this volume - Eds). That's where they should be. We've got a lot to put together. At one time we were talking about how all things are not created equal and we people, women especially, are not equal in our society. But a level playing field and equity are really important. We need to be able to participate in some kind of equal way and that has not been made available for the public. And the public can't be everywhere because they just don't have the resources. But we do need to get information out, so a freedom of information act is vitally important too.

We also need a better process. It all comes down to a process, and for someone like Don [McMullan] who dealt with the old-growth deferral process that was imposed with impossible terms of reference and all of those issues, the process has to come from the different groups that are concerned about the use of an area. They have to mould the process that will work in that area. Whether it will work in a given situation I don't know, but at least we have to give it a fighting chance and every sector of the public has to be involved. How we get the message to the politicians that we want a different process, that we do not want just timber emphasized in this province, is a problem. Should the Ministry of the Environment and the Ministry of Forests share the land base so decision making is done by two agencies and not one, or should it be more agencies? But change is necessary and that's definitely what we're hearing from the public. The public wants a meaningful place at the decision-making table and, as far as we can, it should be an informed public. Thank you.

Fred Bunnell:
Thanks, Vicki. It's out to the floor now. Where's the stampede?

Rosemary Fox (Sierra Club):
I'd like to comment first on something the representative from Fletcher Challenge said, if I understood correctly, which is that a demonstration of the economic benefits of biological diversity should be provided. It's quite evident from what we've heard at this symposium that we're dealing with a great deal of uncertainty and a great lack of knowledge. Not everything can or should be measured in terms of economics. Other values are important to society that are just not measurable in dollars, and these should be given equal emphasis to economic values. That's not to deny that we need a healthy economy, but it's not measured only in terms of dollars.

Second, I'd like to emphasize again the need for a level playing field that a number of people have referred to. One land-use interest group not represented in the deliberations of the last two days and maybe not represented here at all is the mining industry. We're talking about biological issues, and it's really important that the mining industry be brought to the table so that we can establish a dialogue and not be in the position of being commanded that no land area can be protected until ten years at the very minimum has been set aside for exploration. The Forest Service is proud to say that it allows mining in its proposed wilderness areas and this is a major deficiency if we're going to aim for biological diversity.

Someone behind me keeps on saying, "Ask a question," but I believe this is an opportunity for public input and for you to know what public expectations are. My expectation as a member of the public is that the government should adopt as a basic premise of land management the need to protect biological diversity. It's important from my values to recognize that biological diversity does not mean replacing one set of species by another but rather maintaining the components of the natural ecosystem. This is my hope for the future. Thank you.

Fred Bunnell:

That was directed to you, Don. I don't think anybody can touch the mining industry.

Don McMullan:

I won't touch the mining industry. Your comments on evaluation are very correct. What I was trying to say is that I believe we can collectively attach a value to maintaining biodiversity and that's very important because if you can do that, even in a general sense, then you can start competing for dollars in the legislature. You can start saying we need more dollars for this, we need more dollars for that, because we have values which can be rated for protection. From my point of view in the forest industry, if we can demonstrate that maintaining biodiversity will maintain the productivity of the working forest within the total land base, that's one way that I can pursue better forest-management practices within my company, for instance. So, even in general terms, if we can make some estimate of how important it is and how much value is tied to biodiversity, then we can move ahead a lot faster, get more people on side a lot faster and get a teamwork approach.

Vicki Husband:

You can't calculate the value of biodiversity. It's priceless. You can't calculate it in how many species per cubic metre or whatever. I think we must put a very high price on it. I think it's the price of our survival and the survival of this planet.

Fred Bunnell:

Hal [Salwasser], this is the place for public comment but I'm just wondering, have you guys in the U.S. had to grapple with some of the hard moral issues that Jim Walker raised?

Hal Salwasser:

No, some people keep asking that, but to some extent Vicki's point is correct. You can't put an economic value on all these elements and processes of nature. On the other hand, it's not priceless because people constantly make decisions every day to do things, implicitly or explicitly. But you can come up with some assessment of the relative costs and benefits and tradeoffs among policy choices you have, and I think that's important to do. But you don't have to put it all in terms of dollars. You do need some way of communicating to people the possible consequences and relative merits and demerits of options though.

Fred Bunnell:

Thanks. There are quite a few people standing patiently. I hope you can phrase your opinion or question succinctly. (I'd like to call this to Vicki's attention. I did a quick count, and of ten people standing up, eight are women.)

Sharon Duguid (Earthlife Canada):

Rosemary echoed a lot of what I wanted to say, especially about the mining industry not being represented here, which is unfortunate. After what I witnessed at the Parks Plan meeting in Vancouver where they decided to dump on everything, it's really unfortunate they're not here. Speaking about economics, another group I would love to see here is big business, the big money that keeps B.C. going, the industries that are involved in trade. In terms of education about biodiversity, that's a group that is not educated at all. Banks, business, multinational corporations, they all need to understand what's happening to the province.

Unfortunately, Don, I'm going to have to address this to you, and again it's echoing Rosemary. Something you said that I don't really understand was that as we look at these issues, we need to keep the world going. How I interpreted that was that we can't just stop what we're doing in the logging industry and the forestry industry. I would like to pose two questions to you. One concerns areas that are highly controversial right now. I don't understand why a moratorium is not possible. If we're going to be working together for the areas, a moratorium does not necessarily mean that the forest companies will not have those areas for logging. It just means that for one or two years we defer a decision until we know what we're talking about. It's impossible to rebuild a tree that's been cut down and if we don't know enough I would far rather wait for two years to make that decision, and you'll have the cash flow in two years when it's an informed decision for logging. Why logging companies won't agree with moratoriums is beyond me, especially when the question marks are so big. And the second question I have for you is more an ethical question. I don't mean to sound patronizing, but I would like you to tell me what your definition of "compromise" is.

Don McMullan:

If I can go back to your first point about involving what you call big business, I think that's very important and I think that's what we had intended in our own corporation by bringing Nils Zimmerman in to talk about biodiversity. You're talking about the public and part of the public is senior management of companies. If you want to get business on side, you have to work very hard to involve them, to educate them, to tell them how to spell the word "biodiversity" if necessary and to invite people like Nils to speak to them. It's a long, slow process, but you have to do it — and you have to start, but it won't happen overnight. If you go up to them and say, "You're wrong. Change your practices," you won't get anywhere. You have to go the circuitous route of educating them to bring them on side.

With respect to your questions, first the moratorium. Within the old-growth process we did establish a process that deferred logging in some valleys. Certainly industry and some of the communities involved were very upset with the way the process went because it did impact directly on operations, on families and on individuals. So there's no easy answer to that one. It would be nice to say you have another valley to log, but in some cases there were no immediate alternatives. The roads weren't laid out, plans weren't in place and approvals from government agencies weren't there. So there was no alternative.

The other question was about trying to reach a compromise. Compromise is something that requires an incredible amount of work and understanding and it takes a long time. You have to spend most of your time trying to listen and understand what the other person is saying, and none of us are very good listeners. I think compromise is trying to come up with something that separates what's best here and what's best there and what's right. That sounds a bit philosophical but I don't think I can do any better than that.

Vicki Husband:

On the whole idea of deferral, which I was also partially involved in, the problem is that we've allocated all our productive forest land and the annual allowable cut has gone up and up and up. I suggest that we have to do things with public pressure to reduce the allowable cut at least to the long-run sustainable yield, which isn't the same as the sustainable development we say we're moving towards. In some areas we are cutting 100% above the long-run sustained yield. The TSA in the Charlottes is one example. We are in this bind continually. We have to reduce the cuts so that options are more open.

Fred Bunnell:

I made a tactical error. I should have asked individuals at the microphones to introduce themselves, in part to be considerate to everybody else and in part because we're trying to record this. So would you please do that.

Kat Enns (Larkspur Consulting):

I have a question for the biologists if they're still here and haven't gone squirrelling off to write their biodiversity papers. I would like to know if there's any link in their backgrounds between education and what they do. Do they have in their histories a really talented high school teacher who inspired them with a love of biology so that they studied and carried on with biology? A lot of people come to me wanting to study this topic, but they don't have a lot of tools. I have a question for Jim Walker. Is there any movement to change the high school curriculum so that we have students who get some background and gain an interest in this topic? Last year, when I was working on a Caribou habitat paper in the Chilcotin, I was verbally attacked by a contract logger who was really mad because his livelihood was at stake and his employees — his skidder operators and feller buncher operators — had a grade 10 education. They didn't have much fun in school and they dropped out early, but they made $40,000 a year. We hear all this talk about diversity studies, but a lot of us biologists have been ungulate biologists or bear biologists or carnivore biologist — not many of us have had the opportunity to work on non-passerine birds or insects or whatever with any kind of background or funding. So my question to Jim Walker is, is there a plan or any kind of link between your ministry and the Ministry of Education?

Jim Walker:

There are some links. Probably the best and the most positive is Project Wild, supported by our ministry. Over the last three years we've managed to get Project Wild into about a third of the school systems across the province. This was the elementary and secondary school system but not the university system. It's also something that we're evaluating now and will probably try to expand our involvement in. Project Wild is an excellent ecological program. It tells people how to think about the environment, it doesn't tell them what to think. It tells people about ecological processes. I think that's where we need more effort. I know the Ministry of the Environment has a program of environmental education that it's hoping to move on soon. One thing we did learn, and I am not saying thisfacetiously, is that educators tell you that once a person reaches 40 you're wasting your time to try to re-educate him or her. I suspect from some of the personal experiences I've had, I would go along with that. But there definitely is a real need to try to get through to the people. One of the common themes in this whole panel has been that people are not informed, that they don't know how to think about the environment.

We have to make certain that the next generation has better ideas about this province and its ecology, and programs like Project Wild will do it. We have tried, along with the Ministry of Forests, Department of Fisheries and Oceans and the Council of Forest Industries, through the Coastal Fisheries Guidelines, to go out to every camp on the coast and educate people about stream habitat with the use of videos and with on-the-spot instruction. It's difficult but they have had some success. So we have attempted to educate people on sites, but the big push has to be to educate the children of the next generation. And we are making moves in that direction.

The only other thing I would say is I did have some dealings with the Ministry of Education curriculum committee. I tried to get them to classify Farley Mowat's book "Never Cry Wolf" as fiction instead of non-fiction, but I didn't get anywhere.

Fred Bunnell:

Thanks, Jim. Bill, I think you're next.

Bill Wareham (Endangered Spaces Project):

I have two questions, one for Andy [MacKinnon] or Jim [Walker] , and the second for Jim and Derek [Thompson]. The first one to Jim is, looking at the Ministry of Environment's identified areas of critical habitat throughout the province, how well do you feel that they have been recovered or incorporated by the Ministry of Forests' wilderness areas that have been proposed and the Parks Plan study area proposals? And the second question: Derek mentioned that throughout the public process for Parks Plan 90 and the Ministry of Forests Wilderness Plan there was a phobia about people coming from Victoria or from high above to make recommendations on land use, and Jim, you commented that you didn't feel it was appropriate to set up local advisory groups. I was wondering if both of you, Derek and Jim, could comment a little bit more on that and what you see as a viable process, because I see Derek saying that the people want more local control and Jim, I see you having other ideas on that.

Jim Walker:

The last question first. I don't have a problem with local input. We need it. It's going to be a thing of the future. What I'm saying is that you have to determine at which level public input will be most effective. People in each region should have a say on what the cut is, how many Elk there will be, how much grass will be allowed, where settlements will be, etc. This should be done for each region of the province. That's the key to land-use planning — to set provincial objectives and underneath that extend them with regional objectives. After that it becomes a relatively simple matter, if they're ecologically consistent, for the manager to fit them into the larger picture. The public should comment on what the Ministry of Forests is going to do with annual allowable cuts. I don't think the public should be commenting on every cutting plan in this province and I don't think we can afford to do that. It's an ineffective use of public input. There is a need for the public to tell the politicians what they want this province to be but not to tell them how to cut trees down or how to manage Elk. And I think that's where we've gone.

Once again in B.C. it's the old pendulum. The other didn't work so we're way over here now. Everybody wants input into the 4,000 logging referrals we get every year. You have to determine the level where public input does the most good, and it's not needed in every single plan.

The second question is about how much input we've had into both the Parks Plan and the Ministry of Forests Wilderness Plan. We submitted our lists of areas. Not all of them are reflected on the map of the Ministry of Forests and I expected they probably wouldn't be because some of our concerns are things like wildlife or fish as opposed to some of the more recreational or other concerns. I'm not that perturbed. The process will go out into the public, the public will probably come up with a list as long as your arm in addition to the one the Ministry of Forests has, but when it comes back we'll see what the list is. Possibly it will more correctly meet our concerns. If it doesn't we can go back and recommend other areas to add. If those areas aren't put on the list we also have the opportunity to go back and recommend ecological reserves or wildlife management areas. So all of our areas aren't on the list, but the process isn't finished yet. Things are unfolding as they should.

Derek Thompson:

Regarding Jim's earlier comment about dispassionate bureaucrats, I don't think I've ever known a bureaucrat who's less dispassionate. As for regional, local involvement, I would err somewhere between the extremes that Jim has outlined. In the planning we do in Parks, we get the public involved and we intend to continue to get the public involved at a much lower level than simply ranking provincial objectives and things like that.

Yes, the ministries do have to manage, but with considerable suasion from the public that will involve increasingly, in Parks anyway, decisions down to the level of where the campgrounds will go and where the protected zones will be within these parks. So, as usual, we have an argument with one another. I would add two other things to what Jim has said. One on the issue of the representation. He makes a very accurate observation that what Parks and Wilderness for the '90s have put out is an initial inventory of areas. We're getting an enormous number of other areas identified too. The other ministries will clearly be involved in review and overview. But even then it won't be finished.

And one final thing to the person who was speaking about education. Education is very crucial. We recognize that. We're not doing a good enough job on our own interpretive naturalist programs in this province. We recognize the need to get into the school systems to propel kids and we're identifying the elementary grades as a key group to be reaching in the future.

Hal Salwasser:

I want to comment on the point Jim was making about professional management. This is one of the things on which professional managers in B.C. could perhaps take a lesson from us in the U.S. In both state and federal wildlife and forestry agencies, we have lost some credibility with the public because through this public involvement process it was discovered that we were not giving the public full, objective information on the potentials and the trade-offs, the proposed management practices and some of the costs and consequences. The result is that the public is insisting that it have higher involvement closer and closer to the field. This is extremely important. It's going to take us some years to rebuild the trust and credibility that we have lost in this process, and I can give you specific examples to verify my statement. It's going to take us some time to restore a level of public confidence in the professional managers that will allow them to say, "OK, we buy off on the policy and we'll let you carry it out according to your professional judgement."

Fred Bunnell:

Thank you. One of the things I was wondering about, Geraldine, was the feeling of the aboriginal people. You're dealing with a great many issues right now but you have very few people trying to deal with them. Do you have any observations on public participation you'd like to offer?

Geraldine Shirley:

Definitely. We're into educating people to understand the support

381

and roles that they must play in the negotiating process with the province. We've never been in these plans before, which causes hardships with trying to plan because no one is listening. So for us to gain trust we have to inform the people. We started an education process through our tribal council and we have been to many of the schools and the university and colleges in this area to inform them what we are looking at and what needs to be done and what they need to do too.

Fred Bunnell:

Thank you. I think it's you, Audrey.

Audrey Pearson (University of British Columbia):

I'd like to comment first about education. When I was eighteen years old, Ken Lertzman was my teaching assistant in forestry biology and he showed me all these slides of what a real live biologist does and I thought to myself, oh wow, I can be a real live biologist too. And Fred Bunnell gave me truth and enlightenment and hope.

Fred Bunnell:

I'm not going to accept all the responsibility.

Audrey Pearson:

So, in short, yes, teachers do matter, because here I am. I'd like to address my question to Don McMullan but also to the members from the government agencies. Since I began as an undergraduate wanting to be a real live biologist when I grew up (I'm almost there), I've heard some things consistently. We desperately need information, we need basic research, we need systematics, we need people, we need resources, but there's no money. Would you support a tax on stumpage or a research tax similar to stumpage that could be used towards generating a long-term stable research money source to use to answer some of the questions we've been dealing with today, since user pay is pretty sexy with government these days? The wealth the forests generate could go towards its stewardship as well.

Don McMullan:

The easy answer to that, of course, would be to say that the industry is bleeding to death nationally right now. But that's not what you're looking for, Audrey. The question we address here today, biodiversity, involves all users, and we shouldn't focus on the forest sector or the mining sector. Maybe if we're really going to commit ourselves as a society to maintaining biodiversity in this province we should go right across the board and say, "OK, all users of the land base should contribute fairly." How to do that? We could have user fees that we establish, or a royalty, or require that you buy a ticket before you go into a park, or something like that, but that's probably going to cost more than it's worth. So I think you've got to go back to what I said: you've got to convince the people in the legislature that this is really important and try to allocate funds through the legislative process. That's the only way to get everybody to contribute fairly.

Lynda Laushwey (Island Watch Society):

With all due respect to the scientists who made eloquent presentations in the past couple of days, I must say that I am concerned. It seems to me that one of the major reasons we're in the sorry state we're in today is because we are a generation who believes that we have the right to control and manipulate nature. I've heard a number of people suggesting that possibly the solution to the problem now is to further control and manipulate nature. This concerns me greatly. Nature survives very well all on its own without our intervention. The real, beautiful example of biodiversity is nature left on its own. It's done this for a millennium. We can't survive, however, without nature and that's the bottom line. To change so that we survive as a species, we have to look at our value systems, our attitudes and our own personal problems, because I don't believe our desperate drive to control and manipulate nature comes solely from economic models. It's a much bigger issue that needs to be addressed. It's the malaise of our society.

And further, in response to the panelist's statement that the world can't stop before all the facts are in, my concern is that the world *may* stop before all the facts are in. The people and public I'm connected with are very cynical at this point. They've lost faith that our provincial and federal governments are willing to care for our environment, and this is a very sad position to be in. This is nothing personal against any of the representatives here, because I know they're all very caring people. What we do at this point I'm not exactly sure, but I know that the reason part of this cynicism exists is because we have the attitude that we'll have round tables, we'll have discussions, we'll have research, but in the meantime it's business as usual. This has to stop. There has to be some concrete representation to the people of this province, the public, that there really is going to be a change. You just have to take an aerial look at Vancouver Island and you don't need to be a scientist or a genius to realize what the serious problems are. We simply can't continue. We can't wait for all the facts to come in. As for the confrontations everyone is concerned with, these are the acts of people who have exhausted all other channels. They are desperate people, and unless there is some concrete representation that business is not going to continue as usual, I'm afraid the confrontations are going to continue too. And I'm sorry about that. Thank you.

Mark Wareing (Western Canada Wilderness Committee):

Quite a statement to follow there. One aspect I was going to ask Hal Salwasser about is forest rotation length. One of the ways we can immediately implement changes to bring about protection of biodiversity as far as the Western Canada Wilderness Committee is concerned is by lengthening forest rotations. I haven't heard the rotation length questioned but it's a fundamental aspect of accountable stewardship. We feel that while we continue to be committed to 80- to 100-year rotations, there is no hope for maximizing biodiversity or maintaining a forest function. So I'd like to ask Hal what he feels about the impact of these 80- to 100-year rotations and what we should do about them.

Hal Salwasser:

Rotation lengths emphasize the production of wood fibre. In the southern part of the U.S. they're about 20 to 30 years, and on some acres that's going to be an essential management tool, part of the overall sustainability issue. To sustain biological diversity as part of the whole mix, for environmental values as well as sustaining a diversity of human lifestyles and a resilient environment, you will have some places where the rotation length is probably whatever nature decides it to be, augmented unfortunately by things that come in from the air that we can't stop, and you will have some areas that are managed, yet managed with a primary purpose of sustaining something like a native forest condition that is commensurate with taking some products out. Those places may have rotation lengths two or three times longer than what we currently envision for managed forest lands, or they would be managed in such a way that rotation length is not exactly an appropriate concept. They're managed with some selective harvests and maybe under something that the technical people call "unity area control" where you define the characteristic of the landscape you're trying to achieve, and that dictates what you can take out. The idea that you're going to leave some things on the land as biological legacies for the health of the land, for future productivity and resilience — biological diversity — obviously means those things aren't going to the mill. So yields on those kinds of acres will have to come down. People can figure out many ways to do that and their solutions will vary by forest type and soils, etc.

Fred Bunnell:

When I agreed to do this, I didn't realize the full implications of it. One of them is I have to ask you to be quiet to listen because I have to summarize this. But I'll take one more question.

Renée Jackson (Victoria, B.C.):

I don't have a question. Will you take a statement?

Fred Bunnell:

I will listen to your statement and I expect the rest will too.

Renée Jackson:

I've often heard in these last two days the phrase "users of the environment". I think of myself as part of a whole living system and that's important to me. I, unfortunately or fortunately, am part of that whole living system and one of the species who, like the modern gathering swine, drive their polluting cars down to the waters that are rising to drown us. I haven't given up, but Derek, you spoke of really listening to the public. Great, but can I suggest that after you really listen to someone and they feel listened to, you can then start telling them some of your concerns. Because I am also convinced that the general public really needs to be better informed and they need to be because they're in this just as much as we're in this, and all the plants and animals are in this.

Fred Bunnell:

Thank you. I think we should all thank the panelists. It's a job I don't like, as a panelist, because it's alternatively boring then terrifying when somebody directs a question at you. Thank you.

Summary and Wrap-Up

Fred Bunnell

Faculty of Forestry, University of British Columbia,
Vancouver, British Columbia, Canada V6T 1Z4.

In some ways it's an honour to summarize. The myth about Pojar and me being twins isn't true. My mother explained it to me. She said, "Fred, you're not twins. You are much older than he is." Nothing about wiser, just older. It's too bad we're not twins because I really admire and respect the wonderful places to which Jim's mind roams. And I do like his quotes, particularly the multitudes one we've heard a multitude of times.

I think I've summarized more than my share of symposia and conferences. Usually I find it somewhat trying because you have to go to every session and you have to stay alert and you should think. Then instead of going out carousing at night you show some discipline and review your notes of each session and extract the major points. I find all of those activities compete directly with my preferred behaviour.

I found it less trying this time for three reasons, two of which I'll tell you now, the other I'll discuss later. First, because I find the subject matter fascinating, it's easy to listen. Second, I tried something different this time. I still listened and I still tried to think, but instead of reviewing the points session by session, I want to try to place them into the context of the history of this gathering. That gives me more freedom, which I dearly love.

This symposium had its beginnings less than a year ago here in Victoria and not very far from this auditorium. We were gathered a short distance away in the museum tower [Fannin Building] to discuss and develop a research project on Marbled Murrelet. It became clear that the project was going to be extremely costly. At lunch we wondered aloud just how, if it was going to cost us this much for a single species, we were ever going to grapple with biodiversity. And I

In *Our Living Legacy: Proceedings of a Symposium on Biological Diversity*, edited by M.A. Fenger, E.H. Miller, J.A. Johnson and E.J.R. Williams, pp. 387-392. Victoria, B.C.: Royal British Columbia Museum.

believe it was Bill Pollard of MacMillan Bloedel who said we really needed to pull some minds together on this subject, whatever biodiversity was. Bill's concern was simple. It's easy to be down on what you aren't up on. He wanted to learn a little more. So the initial demands of the symposium came from the forest industry and a desire to define the scope and dimensions of this biodiversity thing.

At one of those gatherings that I affectionately refer to as executive meetings, we scribbled on the pub napkins and started to mock up a program. Another meeting, another pub, more napkins and we had an outline. Very shortly after, the Forest Service took the lead role or at least the major role in making this gathering real.

Naturally, it's changed from those first notes. It's evolved. But if a symposium can have roots and grow, the roots of this one were simple questions, not trivial, but simple. Just what is biodiversity, what is the situation in British Columbia and what are we going to do about it? Does anybody have any ideas we could borrow or steal? We were optimistic that we would move towards answers to those questions.

This may seem bizarre, but as I thought of the history of this symposium, it recalled to my mind a story about Chauncey Depew introducing President Taft. Taft is distinguished among American presidents as having been the largest they ever had. God only knows what the long-winded Chauncey was up to, but his introduction went on at length about a man pregnant with vision, pregnant with hope, pregnant with intellect, pregnant with courage, pregnant with charity. At 300-plus pounds, Taft could have been pregnant with almost anything. He was truly a man large and containing multitudes. When the seemingly endless introduction was over, Taft eased his bulk up to the podium and said, "If it's a boy, we'll call it Courage. If it's a girl, we'll call it Charity. But if it's only gas, we'll call it Chauncey." Now when we conceived this gathering, we were pregnant with enthusiasm, with hope, maybe vision, but we also knew that if we brought enough professionals and enough academics together we too could have a meeting pregnant with gas.

So let me return to those three simple but non-trivial questions that began this gathering less than a year ago. I want to very quickly review some of the things I heard and then try to summarize in slightly different words.

The first question, what is biological diversity? We went through an awful lot of that. Chanway noted processes — diversity as pieces and processes. Chris Pielou told us how we might measure pieces. As a sometimes quantitative ecologist, I enjoyed her humour. As one who tries to do applied research, I enjoyed the simplicity of her approach. As potentially depressing as Larry Harris's talk could have been, I confess I enjoyed the message for two reasons. One is my bias. Larry was emphasizing a notion I've been raving about for years, that the space around the space you preserve is important as well. Second, there was some optimism that gentler management in surrounding areas can increase our options. Wade Davis reminded us of points that Suzuki and Pojar also made, that cultural diversity is important too. And Tom Ledig — well damn, Tom, it's frightening enough to think of the rates of extinction of things that I can see, the rates that Geoff Scudder talked about, and now I have to worry about secret extinctions. But I appreciate Tom's honesty and his reminders that genes are important whether I can see them or not. Biodiversity is sufficiently big and important that we find lots of other contributions, whether it was Larry [Harris] reminding us to think of indigenous species, not exotic, and to be prepared to control and manipulate the exotic, or Jack Ward Thomas and others reminding us of the peripheral species. And then the panel. A lot of things came up. One of the things I don't think had been stressed very much before were some of the moral dimensions to biological diversity.

The second question is, what is the situation in B.C.? At least three of the talks related specifically to B.C. Pojar, Tunnicliffe and McPhail reminded us of the incredible biodiversity in this arbitrarily defined part of the globe and the impressive ignorance associated with that. It reminds me of something Dan Janzen said in 1989. He was talking of tropical forests and he said, "It's as though the nations of the world decided to burn their libraries without bothering to see what's in them." In another session, one on the global values of diversity, Davis, Ward Thomas and Foster reminded us of the incredible range of values biodiversity assumes. In the session on biodiversity at risk, Munro, Chanway, Marshall, Hlady and Harper addressed wider issues but still considered portions of the situation in British Columbia as Scudder did in his talk, placing the situation in B.C. into a wider context.

Then that third question. Given the incredible biodiversity present in B.C. and the values it has or we assign to it, what are we going to do about it? Clearly, there is a range of expectations, as this panel discussion revealed. We had two sessions specifically on the topic, the strategies session and the actions session. Naturally, many points were scattered throughout other sessions in other talks. Some things are under way. Munro revealed the impending Endangered Species legislation; Harper and Hlady talked about the surveys exploiting a huge variety of talents and agencies; Harcombe discussed efforts at organizing existing data, focusing future work and endangered species programs; others talked about building around the utility of preservation that Wareham addressed; Rankin noted legislative approaches. I felt that some really thoughtful approaches were noted, including some that have already been tried elsewhere. Hal Salwasser reminded us that biodiversity is not necessarily a threat at all, in fact it's fundamental to sustainability. I really liked Hal's talk. It supported a lot of my biases. I appreciated Ken Lertzman's insightful overview. But more important was when he addressed the question of what is knowable. Fifteen years ago I wrote a paper called "The Myth of the Omniscient Forester." I guess I could have just as easily entitled it "The Myth of the Omniscient Researcher." We're not going to be able to answer everything. Gunton pointed out improved approaches to planning structures; Larry [Harris] talked about corridors in Florida; Walker listed initiatives and discussed the shifting attitudes and his optimism about the opportunities, although he also presented us with some fairly hard points as a panelist. So that's a review.

I want to try and summarize a little differently instead of just summarizing what we heard. Summarizing thoughtful and heartfelt comments on biodiversity is to some extent akin to summarizing the Bible. Sure it's a bit contradictory in places, but it seems pretty good to me.

I can expand somewhat on that. What is biodiversity? We have a much better idea of its dimensions now and it's a tad frightening. I never saw the Westland television program on biodiversity but I recall the filming. Requested to define biodiversity, Pojar replied, "All things that respire and the processes connecting them." Then Mike Halloran, the host, asked me what I wanted to add. I couldn't think of a thing. But after listening to Suzuki, to Davis, to others, and the panel, I appreciate I was thinking as a scientist about the things I could count. There are a lot of other dimensions to this biodiversity issue. I've said before that we need a definition. We do not need a definition of biodiversity to enjoy it any more than we need a definition of poetry or art or music to enjoy them, but we must have a definition to define legislation, goals and policies. To paraphrase the Cheshire Cat's observation to Alice, if you don't know where you're going, any road will get you there. So we have to decide where we're trying to go. As I listened it reminded me of the two rabbits being chased by a fox, and the fox is gaining. They come to a hole and one rabbit says to the other, "Let's dive down here until we outnumber him." It seems the closer we get, as we gain on the damn rabbits — the closer we get to biodiversity, the faster its dimensions are multiplying. Still, we've had a lot of helpful suggestions, and that's really important — some of the simplifying approaches that Chris Pielou described, some of Larry's [Harris] comments and Hal's [Salwasser] — so I'm pleased because I think we made progress.

The second question, what's the situation in B.C? B.C. is north-temperate, and I think that's important. We heard some things that were downright frightening and could easily be depressing, like McPhail's lakes, but usually they were not in B.C. Yes, we should do what each of us can about the tropics. But we should also get our own house in order and we will probably be more effective here. We certainly have the opportunity, so if you want to feel good, act here, because you can.

What is the situation in B.C.? There are three big parts. First, however you measure it, we have one hell of a lot of biodiversity, among the highest in north-temperate regions. Second, we don't have a really good idea what it is. Third, we still have enormous opportunities compared to other areas. It's true we've hammered some of our really diverse areas, like the South Okanagan, but by and large we've only begun to really stress our resources. That gives us much-needed flexibility within the huge range of tactics available.

Still, I found some points about the B.C. situation troubling to varying degrees. First, some people feel that there are things being left out and it doesn't matter what it is, whether it's cultural diversity, insects or freshwater fish. As Hal [Salwasser] noted, we need unprecedented public participation. We don't want a lot of groups feeling neglected, like Whistler's father. Second, we're responding to this biodiversity issue, and maybe this was what Jim Walker was getting at on the panel, as if we were a bunch of opportunists: we're dedicated to the proposition if the price is right, yet we haven't really figured out what the price is. So I don't know how firm the commitment is. And third, in some ways it seems too good to be true. We have options. So many initiatives are happening right now that you could easily become suspicious. I've said before, if everything's coming your way, watch out, you could be in the wrong lane. Maybe that's what those pleas for humility in the treatment of our land were all about. They were direct from Ward Thomas and Scudder, indirect from Harris and others. Those are relatively small troubles and I'm optimistic about the flexibility, which of course leads us to the third of our original questions: what are we going to do?

It's enormously gratifying to note that we've begun in a variety of important ways. We've begun a variety of processes that I believe will help, from the Endangered Spaces Program to the Conservation Data Centre. The flexibility we have will allow the Endangered Spaces Program and other programs including the Old-growth Task Force to select among the last of the least and the best of the rest. It will allow the Endangered Species Program to retain effectiveness and in most instances it permits us to couple integrated management with more restrictive approaches. And that's a huge political advantage, because it means that at least for some decades yet, we can implement effective tactics to sustain biodiversity while we're working on our attitudes. And we can do it without markedly reducing our standard of living or comfort level, which seems phenomenal.

Of course, to manage the entire array of tactics, we're going to have to co-operate and appreciate the common elements of the goals. And I have some concerns there as well. Some of these I tell my students because these things are generic concerns or generic aphorisms. One is something that Rankin said: the notion is to learn from others' mistakes because you'll never have time to make them all yourself. Harris, Ward Thomas and all those addressing the global picture have exposed mistakes. We can learn and we can revel in the affirmative flames of our very own mistakes (sorry, Jim [Walker]). And I think we will. And I guess a second is that the best way to knock a chip off somebody's shoulder is to pat them on the back when they do well. This issue is far too big for any of us to carry on a lonely crusade. We're going to have to find the common elements. And I believe we will.

The third thing that occurred to me is just how much is going on. I think Jim Walker gave the most complete litany: the Forest Service Wilderness Plan, Parks Plan 90, Old Growth Strategy, Endangered Species, Large Carnivore Conservation Strategies, the Round Table, the Forestry Commission, other task forces. It's all coming our way. Are we in the right lane or not? It's almost too fast, and the young demonstrators that came in here yesterday emphasized that point with their plea. They are to some extent outside of and, I suspect, largely ignorant of some of the larger processes that are going on. So we saw the impatience that we heard about in the panel, the fear, and the lack of trust. They may have a potential realization of the importance of longer-term planning and more comprehensive views but impatience and fear for short-term actions. Everything's happening at once. Some graffiti that I first saw on a wall in a washroom in the Misty Isles said, "Time is nature's way of making sure everything doesn't happen at once." Well it's happening all at once, and I get concerned that we won't find the right blend of urgency and patience. I believe we can and I hope we will.

And here I think one of Piet Hein's grooks sums up eloquently and elegantly our situation with biodiversity. "Problems worthy of attack prove their worth by hitting back." So be it. Maintaining biological diversity is undoubtedly a worthwhile problem. We've learned much at this gathering that will help us towards solutions. I thank you all for your participation.